Ecological Studies, Vol. 202

Analysis and Synthesis

Edited by

M.M. Caldwell, Washington, USA
G. Heldmaier, Marburg, Germany
R.B. Jackson, Durham, USA
O.L. Lange, Würzburg, Germany
H.A. Mooney, Stanford, USA
E.-D. Schulze, Jena, Germany
U. Sommer, Kiel, Germany

Ecological Studies

Further volumes can be found at springer.com

Otto Fränzle • Ludger Kappen
Hans-Peter Blume • Klaus Dierssen
Editors

Ecosystem Organization of a Complex Landscape

Long-Term Research in the Bornhöved Lake District, Germany

 Springer

Prof. Dr. Otto Fränzle
Ökologie-Zentrum
der Christian-Albrechts-Universität
zu Kiel
Olshausenstraβe 40
24098 Kiel, Germany

Prof. Dr. Ludger Kappen
Neue Straβe 141-16
37586 Dassel, Germany

Prof. Dr. h.c. Hans-Peter Blume
Institut für Pflanzenernährung und
Bodenkunde
der Christian-Albrechts-Universität
zu Kiel
Olshausenstraβe 40
24098 Kiel, Germany

Prof. Dr. Klaus Dierssen
Ökologie-Zentrum
der Christian-Albrechts-Universität
zu Kiel
Olshausenstraβe 40
24098 Kiel, Germany

Cover illustration: Land/lake ecotones of Lake Belau with reed belt and adjacent alder carr in the fore- and middle ground discussed in Chapters 3, 4, 5, 8, 9 and 10

ISBN 978-3-540-75810-5   e-ISBN 978-3-540-75811-2

Ecological Studies ISSN 0070-8356

Library of Congress Control Number: 2007941261

© 2008 Springer-Verlag Berlin Heidelberg

Cover design: WMXDesign GmbH, Heidelberg, Germany

Printed on acid-free paper

9 8 7 6 5 4 3 2 1

springer.com

Preface

This book presents the major findings of a 12-year ecological study of the Bornhöved Lake District, situated some 30 km south of Kiel. Historically speaking, the present research scheme, like comparable long-term ecosystem studies at Göttingen, Bayreuth, München, and Berchtesgaden, has been conceived as the core of a comprehensive ecological surveillance system for Germany (Ellenberg et al. 1978). Comprising three interrelated components, namely an ecological monitoring network, comparative ecosystem research, and an environmental specimen bank, this system is intended to promote both ecological science and planning and policy. In this connection the geo- and bioscientifically based ecosystem research aims at understanding the structure and functions of systems, the natural equilibrium and stress tolerance of singular components and the entire system against changes and disturbances from within and from outside, and the relationships between diversity, productivity, and stability. Thus, ecosystem research forms the indispensable basis for the rational analysis of the comprehensive data sets made available by ecological monitoring networks and for the adequate selection of plant, animal, and soil specimens for environmental specimen banking purposes.

The concept of such an ecological surveillance system reflects the basic ideas of UNESCO's Man and the Biosphere Programme as a follow-up action of the International Biological Programme. As an international pilot project of the United Nations Educational, Scientific and Cultural Organization the inter- and transdisciplinary research scheme implemented in the Bornhöved area has made major contributions to the following of the 14 MAB project areas: (3) impact of human activities and land use practices on grazing lands, (5) ecological effects of human activities on the value and resources of lakes, rivers ... and coastal zones, (9) ecological assessment of pest management and fertilizer use on terrestrial and aquatic ecosystems, (13) perception of environmental quality, and (14) research on environmental pollution and its effect on the biosphere.

In accordance with the above MAB core areas of research and in realization of the focal IBP Programme proposition to study landscapes as ecosystems, which had remained unfulfilled at that time, the general aim of the Bornhöved Project has been to study ecosystem organization in a structurally highly diversified landscape which is representative of the whole set of ecotope complexes along the margins of the Weichselian Glaciation. The emphasis was on production and trophic structure, energy flow pathways, biogeochemical cycling, limiting factors, and species diversity. If each of the topics analysed by about 40 working groups, including those of the University of Hamburg, the Max-Planck Society, Fraunhofer Society, the German Meteorological Service, and the Schleswig–Holstein Industrial Inspection Board, was to be justly treated, this book would be too extensive. Therefore, our main purpose was to draw together the findings of the different sub-projects into an integrated view of the structure and function of the particular ecosystems of the Bornhöved Lake District which have been liable to exponentially growing human impact since early Neolithic times.

Thus, the book aims to be a synthesis of research rather than a summary, which means that most of the studies undertaken in the area, and in particular those in and around Lake Belau, get at least a mention. Accordingly, the book is divided into four major sections. In the sense of an introductory methodological reflection, Part I defines the research programme and presents the study area in terms of ecological setting. Part II addresses in nine chapters the structure and function of the different ecosystems at various scales and their interactions. In a transdisciplinary manner Part III finally relates the results of fundamental research to application, emphasizing the practical importance of comparative ecosystem analyses for landscape planning, sustainable land use and rural conservation.

The findings reported in this book result from the work of a great many people, including scientists, administrators, students, and technical staff. As far as possible the discussion is based on published data, and citation is the main form of acknowledgement of the data source. In all cases without such explicit reference the comprehensive relational data bank of the project was directly drawn upon by the authors of the individual chapters. Here the reader may also find additional information on selected topics at http://www.ecology.uni-kiel.de/bornhoeved-report.

Research funding for a project of such an order of magnitude is derived from both the Federal Republic of Germany through the Bundesministerium für Forschung und Technologie (now Bundesministerium für Bildung, Wissenschaft, Forschung und Technologie) and the State Government of Schleswig–Holstein. Extensive preparatory investigations covering the whole of Germany with an end to reproducibly defining a set of representative ecotope complexes for comparative ecosystem research purposes were financed by the Bundesministerium für Umwelt, Naturschutz und Reaktorsicherheit. The editors and all the participants in the research scheme are deeply indebted to the responsible representatives of the ministries mentioned, who exhibited an outstandig amount of interest and commitment, in particular the MinR Dr. Krause and Dr. Schulz as representatives of the Bundesministerium für Forschung und Technologie, and MinR W. Goerke in his double function as a representative of the Bundesministerium für Umwelt, Naturschutz und Reaktorsicherheit on the one hand and Chairman of the German National Committee of the UNESCO MAB Programme on the other.

Sincere thanks are expressed to the Scientific Advisory Board of the Project for its meticulous, very intensive and constructive help and criticism, to the Schleswig-Holstein electric power supply company Schleswag for providing energy to the different sites, and to the burgomasters, farmers, and fishermen of the study area, who so readily and with great personal interest supported the enquiries. We gratefully acknowledge in particular the close and trustful co-operation with the farmers and foresters of the core area around Lake Belau, among whom the late Mr. Banck won special merit by his incessant care for our expensive installations in the field. We are much obliged to the editors of the Ecological Studies Series and to Springer-Verlag for giving us the possibility to publish this book and for their friendly co-operation on a number of details, and we are no less grateful to Dr. Breuer, Ecology Centre of the Kiel University, who provided invaluable services in his capacity as a technical editor.

May this book further the awareness that the ecosystem concept remains the best way of understanding the complex interrelationships and functions of the biosphere. It provides a most valuable structure for the comprehensive analysis of organism–environment interactions and change in order to achieve a science-based compromise between economic needs related to humans' use of the land and an ecology-based care and treatment of landscapes. By developing methods over the long term for managing resources so that all of humankind's needs are provided in a socially acceptable and balanced fashion, ecosystem research also provides a structure whereby complex ideas can be communicated to policy makers and administrators for better stewardship. For both of these reasons, the ecosystem concept, and its application to ecological and environmental problems, has a continuing importance for science and a broad spectrum of society alike.

November 2007   O. Fränzle, L. Kappen, H. -P. Blume, and K. Dierssen

Contents

Part II   Structure and Function of Ecosystems in a Complex Landscape

**3   Ecophysiological Key Processes in Agricultural
    and Forest Ecosystems** 61
Oliver Dilly, Christiane Eschenbach, Werner L. Kutsch,
Ludger Kappen, and Jean Charles Munch

**4   Carbon and Energy Balances of Different Ecosystems
    and Ecosystem Complexes of the Bornhöved Lake District** 83
Werner L. Kutsch, Georg Hörmann, and Ludger Kappen

**13 Ecosystem Research and Sustainable
Land Use Management** 319
Jan Barkmann, Hans-Peter Blume, Ullrich Irmler,
Winfried Kluge, Werner L. Kutsch, Heinrich Reck,
Ernst-Walter Reiche, Michael Trepel, Wilhelm Windhorst,
and Klaus Dierssen

Contributors

Barkmann, Jan
Institut für Agrarökonomie und Rurale Entwicklung, Arbeitsbereich Umwelt-
und Ressourcenökonomik, Platz der Göttinger Sieben 5, 37073 Göttingen,
Germany, jbarkma@gwdg.de

Blume, Hans-Peter
Institut für Pflanzenernährung und Bodenkunde der Christian-Albrechts-
Universität zu Kiel, Olshausenstr. 40, 24098 Kiel, Germany,
hblume@soils.uni-kiel.de

Dierssen, Klaus
Ökologie-Zentrum der Christian-Albrechts-Universität zu Kiel, Olshausenstr. 40,
24098 Kiel, Germany, kdierssen@ecology.uni-kiel.de

Dilly, Oliver
Lehrstuhl für Bodenschutz und Rekultivierung, Brandenburgische
Technische Universität, Postfach 101344, 03013 Cottbus, Germany,
dilly@tu-cottbus.de

Eschenbach, Christiane
Ökologie-Zentrum der Christian-Albrechts-Universität zu Kiel, Olshausenstr. 40,
24098 Kiel, Germany, eschenbach@ecology.uni-kiel.de

Fränzle, Otto
Ökologie-Zentrum der Christian-Albrechts-Universität zu Kiel, Olshausenstr.
40, 24098 Kiel, Germany, ofraenzle@ecology.uni-kiel.de

Heinrich, Uwe
Leibniz-Zentrum für Agrarlandschaftsforschung e.V., Abteilung Landschaftsinfor-
mationssysteme, Eberswalder Str. 84, 15374 Müncheberg, Germany,
uheinrich@zalf.de

Herbst, Mathias, Centre for Ecology and Hydrology, Crowmarsh Gifford,
Wallingford OX10 8BB, UK, mher@ceh.ac.uk

Hölker, Franz
European Commission, DG Joint Research Centre, Institute for the Protection
and Security of the Citizen, AGRIFISH Unit TP 051, 21020 Ispra, Italy,
franz.hoelker@jrc.it

Hörmann, Georg
Ökologie-Zentrum der Christian-Albrechts-Universität zu Kiel, Olshausenstr. 40,
24098 Kiel, Germany, ghoermann@ecology.uni-kiel.de

Irmler, Ulrich
Ökologie-Zentrum der Christian-Albrechts-Universität zu Kiel, Olshausenstr. 40,
24098 Kiel, Germany, uirmler@ecology.uni-kiel.de

Kappen, Ludger
Neue Str. 14-16, 37586 Dassel, Germany, lkappen@t-online.de

Kluge, Winfrid
Königsberger Platz 2, 24214 Gettorf, Germany, winfrid.kluge@web.de

Kutsch, Werner L.
Max-Planck-Institut für Biogeochemie, Postfach 10 01 64, 07701 Jena, Germany,
wkutsch@bgc-jena.mpg.de

Müller, Felix
Ökologie-Zentrum der Christian-Albrechts-Universität zu Kiel, Olshausenstr. 40,
24098 Kiel, Germany, fmueller@ecology.uni-kiel.de

Munch, Jean Charles
GSF-Forschungszentrum für Umwelt - Institut für Bodenökologie, Lehrstuhl
für Bodenökologie, Ingolstädter Landstraße 1, 85764 Neuherberg, Germany,
munch@gsf.de

Nellen, Walter
Dorfstr. 11, 24211 Rosenfeld/Rastorf, Germany, wnellen@uni-hamburg.de

Pfeiffer, Hans-Werner
Umweltbundesamt, Schichauweg 58, 12307 Berlin, Germany,
hans-werner.pfeiffer@uba.de

Reck, Heinrich
Ökologie-Zentrum der Christian-Albrechts-Universität zu Kiel, Olshausenstr. 40,
24098 Kiel, Germany, hreck@ecology.uni-kiel.de

Reiche, Ernst-Walter [†]

Reuter, Hauke
Universität Bremen, UFT Zentrum für Umweltforschung und -technologie,
Abt. 10, Allgemeine und Theoretische Ökologie, Leobener Str., 28359 Bremen,
Germany, hauke.reuter@uni-bremen.de

Schernewski, Gerald
Institut für Ostseeforschung Warnemünde, Biologische Meereskunde, Seestraße
15, 18119 Rostock, Germany, gerald.schernewski@io-warnemuende.de

Schimming, Claus-Georg
Ökologie-Zentrum der Christian-Albrechts-Universität zu Kiel, Olshausenstr. 40,
24098 Kiel, Germany, cschimming@ecology.uni-kiel.de

Schleuß, Uwe
Ministerium für Landwirtschaft, Umwelt und ländliche Räume des Landes
Schleswig-Holstein, Mercatorstraße 3, 24106 Kiel, Germany,
Uwe.Schleuss@mlur.landsh.de

Schrautzer, Joachim
Ökologie-Zentrum der Christian-Albrechts-Universität zu Kiel, Olshausenstr. 40,
24098 Kiel, Germany, Jschrautzer@ecology.uni-kiel.de

Trepel, Michael
Schleswig-Holsteinisches Landesamt für Natur und Umwelt, Abteilung
Gewässer, Hamburger Chaussee 25, 24220 Flintbek, Germany,
mtrepel@lanu.landsh.de

Windhorst, Wilhelm
Ökologie-Zentrum der Christian-Albrechts-Universität zu Kiel, Olshausenstr. 40,
24098 Kiel, Germany, wwindhorst@ecology.uni-kiel.de

Part I
Research Programme and Study Area

Chapter 1
General Concept of the Research Programme and Methodology of Investigations

Otto Fränzle, Ludger Kappen, Hans-Peter Blume, Klaus Dierssen, Ulrich Irmler, Winfrid Kluge, Uwe Schleuß, and Joachim Schrautzer

1.1 Introduction

The integrative biological and geoscientific examination of both the nature of physical and biotic environments which make up ecosystems, and the interactions between these sub-systems, has substantiated the conception that a thorough knowledge of the ecosystem and its unifying position between ecology and environmental science is essential for a coherent understanding of environmental issues and large-scale problems in the biosphere. Recent developments in ecosystem theory have created a challenge to the development of recipes for a sustainable management of ecosystems and regions. It is increasingly considered in the disciplines of conservation biology, in the assessment of ecosystem health, integrity and sustainability, in ecological engineering, and in ecological economics in order to solve practical environmental management problems.

In all of these fields modelling plays an essential role, since models provide for opportunities to develop strong transdisciplinary ties between the ecological community and the agency management personnel. A wide variety of illustrative examples of such approaches is provided by the International Biological Programme (IBP) and, in particular, by UNESCO's Man and the Biosphere Programme (MAB) which, together with the German concept of a comprehensive ecological surveillance system (Ellenberg et al. 1978), contributed to define the concept and structure of the Bornhöved Project. They are summarized in the first part of the present Chapter while, in a concise form, the second part provides information about the practices, procedures and instrumentation used when implementing the ecosystem research scheme in the Bornhöved Lake District.

1.2 History and General Concept of the Research Programme

1.2.1 Conceptual Background and Organizational Framework

In regard to terrestrial productivity in the United States, the focal programme proposition of the IBP was to study 'landscapes as ecosystems', with an emphasis on production and trophic structure, energy flow pathways, limiting factors,

O. Fränzle et al. (eds.), *Ecosystem Organization of a Complex Landscape.*
Ecological Studies 202.
© Springer-Verlag Berlin Heidelberg 2008

biogeochemical cycling and species diversity. Another feature was the proposal to use systems analysis as a mechanism for integrating the results of these studies. In a comparative evaluation of the pertinent biome projects organized in the grass-lands, tundra, deserts, coniferous and deciduous forests, Golley (1993, p. 139) came to the conclusion, however, that only part of these targets were actually reached. The studies "furthered ecological knowledge but failed to essentially contribute to the development of ecosystem theory. The programmes were not designed to sort out competing or contradictory ideas. Rather they were driven, at least initially, by the idea that ecologists could construct a mechanical systems model built on the concepts of trophic levels, the food web, or the food cycle, and then represent the dynamic behaviour of the components by data from organisms or populations that are surrogates of the component. This 'bottom-up' or 'design-up' approach did not prove possible or useful. Further the biome projects did not effectively promote landscape ecology, as Odum had hoped. The biome was the setting for site resarch but was not really addressed as such in an effective manner".

In comparison to the United States biome studies, the German IBP project (located in the Solling Mountains) was organized in a way similar to the Hubbard Brook Project in the White Mountains of New Hampshire (Likens et al. 1977). Both approached ecosystem analysis from the components which could then be linked together systematically in a model-based theory or as a natural object (in Popper's 1959 sense) studied by means of conventional scientific methods. Like the Hubbard Brook investigations, the Solling project took a landscape approach from the beginning, based on an a priori ecological knowledge of the study area. It focused in a comparative way on naturalistic ecosystems, nature-resembling ecosystems, transformed and degraded ecosys-tems. To this end the study sites included acidophilous beech forest and planted spruce forest stands at several different ages, along with permanent grassland and cultivated fields. Research proceeded from a description of climate and soils, the abundance and productivity of vegetation, animals and micro-organisms, to plant physiology, nutrient and energy fluxes (Ellenberg et al. 1986). Unlike the biome projects, the final report of the Solling project did not attempt to force the results into a single synthesis on the basis of an abstract theoretical device or a model. Instead, each part was placed with a site-specific conceptual ecosystem model and developed a theme within its own logic. Owing to this methodology the Solling project can be considered a milestone in the develop-ment of a theory-based ecosystem research.

The general objective of the MAB Programme, as launched by UNESCO in November 1971, and supplemented in 1986 and 1992, has been defined as: "... to develop within the natural and social sciences a basis for the rational use and conservation of the resources of the biosphere and for the improvement of the relationship between man and the environment; to predict the conse-quences of today's actions on tomorrow's world and thereby to increase man's ability to manage efficiently the natural resources of the biosphere" (UNESCO 1988, p. 11). The specific aims of the programme which developed out of the IBP experience are:

- To assess the changes within ecosystems resulting from man's activities and the effects of these changes on man;
- To study and compare the structure, functioning and dynamics of natural and modified ecosystems;
- To study and compare the dynamic inter-relationships between 'natural' ecosystems and socio-economic processes and especially the impact of changes in human populations, settlement patterns and technology on these systems;
- To define scientific criteria as a basis for rational management of natural resources;
- To establish standard methods for acquiring and processing environmental data;
- To promote the development of simulation and other techniques of prediction as tools for environmental management;
- To foster environmental education in its broadest sense and encourage the idea of humans' responsibilty for, and personal fulfilment in, partnership with nature.

Both the conceptual framework of these international programmes and the practical experience gained in interdisciplinary research also exerted a considerable influence on the conception of a comprehensive *Ecological Surveillance System for Germany* (Ellenberg et al. 1978). Composed of three interrelated components, namely an ecological monitoring network, a comparative ecosystem research and an environmental specimen bank, it is intended and largely implemented in Germany to promote both ecological science and, in a transdisciplinary context, planning and policy. In this connection geo- and bioscience-based ecosystem research aims at understanding the structures and functions of systems, the natural equilibrium and stress tolerance of singular components and the entire system against changes and disturbances from within and from outside, and the relationships between diversity, productivity and stability. Operating since 1985, the enironmental specimen bank pursues the systematic and long-term storage of selected environmental materials for deferred analysis and evaluation. It is of prime importance for human toxicology, environmental chemistry and toxicology and for the ecological impact of pollutants.

1.2.2 Selection of Representative Study Areas and Sites

Transdisciplinary research schemes of the above type must be implemented in representative areas in order to permit a large-scale extrapolation of the primarily local results obtained. In this context the term "representative": (1) means reproducing adequately the properties of sets of phenomena in terms of characteristic frequency distributions and (2) relates to specific spatial patterns. The latter aspect merits particular attention since ecosystems, like many other complex phenomena, are not discrete independent and unambiguously identifiable objects or entities; consequently the habitual statistical procedures cannot be applied. The specific problems relating to areal data like mapping units on thematic maps, or satellite imagery and stereo couples, etc. rather "concern (1) the arbitrariness involved in defining a geographical individual, (2) the effects of variation in size and shape of the individual areal units, (3) the nature and measurement of location" (Mather 1972, p. 305).

Difficulties encountered in separating individual areal units from a continuum like soil or vegetation cover or biomes are most frequently, and at least partially, overcome by the selection of grid squares as the basic units, geographical characteristics being averaged out for each grid square. Since grid squares are all of the same shape and size, their use eliminates variability in these properties and thus solves the second problem. The most common solution of the third problem is to make relative location as measured by spatial contiguity the dominant variable in the analysis, allowing individual areas to become members of a region only if they are contiguous to an existing member of that region. It is accomplished by means of geographical diversity analyses (Fränzle 1978) or regionalization procedures based on comprehensive data matrices whose elements were, in the first evaluative step, derived from the digital evaluation of the following ten ecotope-related base maps: soil map 1:1 000 000, land value 1:1 000 000, soil erodibility 1:1000 000, potential natural vegetation 1:2 500 000, length of vegetation period 1:1000 000, orohydrographic map 1:1 000 000, mean annual rainfall 1:2 000 000, mean annual evaporation 1:2 000 000, plant available soil moisture 1:2 000 000, land use and forestry 1:2 000 000.

By means of a specifically developed algorithm (tree analysis), these nominal and metric data yield a ten-dimensional classification of the ecotope assemblages of the German territory which forms the basis for the subsequent assessment of their large-scale spatial differentiation (Fränzle et al. 1987). To this end the (former) Federal Republic of Germany was sub-divided into a 12 706 mesh grid with an average grid square size of $21\,km^2$, each ten-dimensionally characterized by the above ecological indicator variables. In the next evaluative step every square was compared with all others with regard to the variables, equality in a variable being labelled 1, and inequality 0. Averaging the number of comparisons (i.e., about 8.06×10^8), the similarity of two grid squares was then characterized by a figure ranging from 0 (complete inequality) to 1 (equality in terms of indicator variables). Thereafter the vectorial distances of all of the squares were summarized in the form of (virtual) histograms which defined the representativeness of every square by means of the degree of right-skewed asymmetry. The more marked each was, the higher was its the number of high-grade conformities of variables.

Transformation of these histograms into a 12 706-line matrix yields a gradation of (weighted) representativeness indices which, in turn, form the basis of a clustering procedure. It groups the matrix elements into clusters of decreasing representativeness.

In compliance with the second geostatistical requirement the localization of spatially representative grid squares out of the elements of these clusters is based on neighbourhood analysis. The methodology basically consists in determining the individual nearest-neighbourhood relationships of each grid square, i.e., their positive or negative spatial autocorrelation which is a distance-weighted measure for each point in relation to its neighbours. The resultant data matrix permits the definition of average association frequencies of all squares as a basis for a comparison of the individual autocorrelation status with the cluster averages. In terms of spatial structure it ensues that those grid squares, or the ten-dimensionally defined ecotope complexes which they depict, are most representative which differ least in their neighbourhood relationships from the average association pattern of the respective ecotope complex.

The exact locations of the study areas (typically comprising two or more of such complexes) were eventually more precisely determined by applying the same geostatistical methodology to large-scale maps of these areas and their immediate surroundings, the results of which were finally corroborated by visual inspection in the field (Fränzle et al. 1987).

The final step of selection involved the consideration of the environmental stress situation in Germany, defined on the district level into an eight-class gradation by means of multivariate clustering procedures and biplot analyses of ten indicator variables (cf. Fränzle et al. 1986; Fränzle 1988b). The superposition of this stress map and a subsidiary map of the relative acreages of built-up areas onto the locations of the above areas yielded the conclusive differentiation in terms of human impact. The final result of the complex procedure was a list of (potential) representative study areas, each (roughly) 50–100 km² in size. They were situated in the following regions, defined by means of the geographical co-ordinates of the local centre (in degrees longitude E and latitude N): Bornhöved Lake District (central co-ordinates: 10.30/54.15), Bremerhaven Geest (9.05/53.60), Solling (9.45/51.75), Burgwald (8.85/50.90), Eifel (7.10/50.30), Hardt (7.75/49.15), Franconian Basin (10.70/49.90), Bavarian Forest (13.10/49.00), Suabian Alb (9.10/48.35), Bavarian Tertiary Hills (12.25/48.40), Berchtesgaden Alps (12.90/47.55).

For practical reasons, i.e., in view of financial implications and the availability of sufficiently concentrated interdisciplinary manpower, long-term terrestrial ecosystem research has been implemented in four of the above areas in the past two decades, namely the Bornhöved Lake District (Fig. 1.1), the Solling Mountains, the Bavarian Tertiary Hills and the Berchtesgaden Alps. In addition, within the framework of the German Terrestrial Ecosystem Research Network (TERN) long-term studies on forest ecosystems have been carried out in the Fichtelgebirge (cf. Tenhunen et al. 2001) while, for their subject, the Umweltforschungszentrum Leipzig–Halle has inquiries into complex urban–industrial systems and their agrarian environment (Mühle and Eichler 1997).

1.2.3 General and Specific Research Objectives of the Bornhöved Project

Situated some 30 km south of Kiel, the study area has a complicated character of intricately indented ice-marginal deposits which underwent the influence of intense meltwater and dead-ice dynamics. Stratigraphically and geomorphologically speaking three pleniglacial Weichselian advances can be distinguished in the neighbourhood of periglacially remodelled Saalian moraines; they are separated by major stagnation and melting phases during which extensive dead-ice masses were fossilized by outwash sediments. The primary depositional relief of the Weichselian glacial and fluvio-glacial sediments was considerably modified by postglacial soil creep and hillwash processes

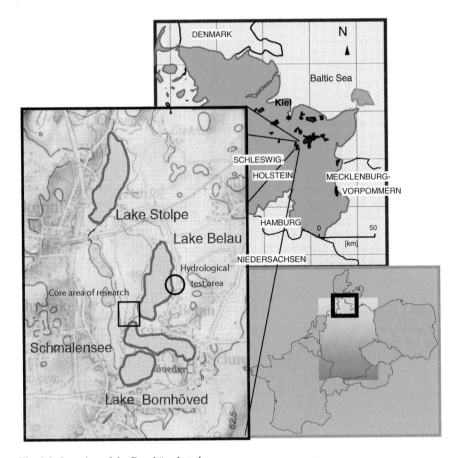

Fig. 1.1 Location of the Bornhöved study area

which were particularly important on the numerous steep hillslopes due to the melting of dead ice during the Allerød and Preboreal periods. Thus, an intricate mosaic of terrestrial and aquatic ecosystems has come into existence during the past 12 000 years; and correspondingly high is the differentiation of land use, forestry and fishery patterns which began in the Early Neolithic and experienced a considerable intensification since the Middle Ages (cf. Fig. 1.2).

In terms of landscape pattern and ecological setting the Bornhöved Lake District reflects, in the light of the above selection procedure, a major proportion of the relevant ecological structures of the German Lowlands. Therefore it appears indicated to distinguish general from specific objectives of the ecosystem research scheme developed. Like in the other research centres of Germany mentioned above (cf. Tenhunen et al. 2001), the general objectives comprise the following:

- Defining and modelling the structure, dynamics and stability conditions of the interrelated terrestrial and aquatic ecosystems with their unusually high number

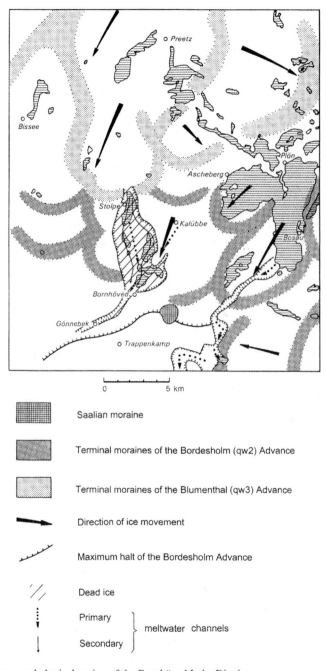

0   5 km

Saalian moraine

Terminal moraines of the Bordesholm (qw2) Advance

Terminal moraines of the Blumenthal (qw3) Advance

Direction of ice movement

Maximum halt of the Bordesholm Advance

Dead ice

Primary
           } meltwater channels
Secondary

Fig. 1.2   Geomorphological setting of the Bornhöved Lake District

of terrestrial and land/inland water ecotones in terms of site characteristics and biocenotic diversity, natural and anthropogenic fluxes of energy and matter, productivity and land use pattern;

- Determining and modelling environmental strains and resilience mechanisms of ecosystem compartments affected by disturbances with a focus on chemical impacts.

The very structure of the research area accounts, in addition to these general aims and ends of ecosystem research, for the following set of specific objectives:

- Modelling of biotic, energetic and material exchange processes between contiguous ecosystems of different land use and fishery patterns (Chapters 3–10);
- Modelling site-dependent relationships between lakes and their drainage basins with a focus on the physical and chemical role of different types of land/water ecotones (Chapters 8–10);
- Developing models of agro- and pasture ecosystems for planning purposes in consideration of national and international production and marketing regulations (Chapter 13);
- Ecotoxicological inquiries into the fate and behaviour of environmental chemicals (Chapters 8–10);
- Inquiries into the efficiency of environmental protection and conservation measures (Chapter 13);
- Testing the validity of spatial extrapolation procedures for simulation models by means of comparative site analyses and geographic information systems (Chapter 13);
- Paleoecological and historical reconstruction of ecosystem evolution during the past 12 000 years (Chapter 2).

1.2.4   Plan of Research

The kind of dynamic balance in ecosystem development which implies a qualitative pluralism and near-continuous change at the same time demands a description of morphogenesis through interactive processes in complementary synoptic approaches. In addition to system theories (cf. Mathes et al. 1996), these approaches amalgamate components of the theory of thermodynamics of irreversible processes (Prigogine 1967, 1985; Nicolis and Prigogine 1977) with elements of catastrophe theory and bifurcation analysis (Thom 1975; Wilson 1981), furthermore with deductions from autopoiesis, information and network theories (Maturana and Varela 1980; Margalef 1995; Ulanowicz 1986, 1995), the theory of games (McMurtrie 1975; Pimm 1982; Ulanowicz 1986), strategy theory (Grime 1979) and hierarchy theory (Allen and Starr 1982; O'Neill et al. 1986; Allen and Hoekstra 1992). Focusing on the comparative assessment of gradients provides for a particularly integrative aspect in evolutionary system analyses, transcending the narrower limits of bifurcation studies (see, e.g., Prigogine 1976; Fränzle 1994; Schneider and Kay 1994a, b; Müller 1998). Thus, for evolving biotic networks primarily isolated phenomena like homogeneity, amplification, synergism (Patten 1992), ascendency (Ulanowicz 1986), power or emergy (Odum 1983) can be appropriately integrated under the generic perspective of gradient formation and related feed-back phenomena (cf. Chapter 11).

A set of conceptual principles ensues from these approaches. They are likely to constitute important elements of a future unified ecosystem theory, but are no less essential for the formulation of the hierarchy of operationalized hypotheses underlying the transdisciplinary Bornhöved research scheme (cf. Leitungsgremium 1991; Fränzle 1998a):

- A state of sufficient non-equilibrium is maintained within living systems and in their relations with the environment.
- There exist several (and possibly multiple) stable regions or dynamic regimes for living systems and, in switching between them, systems have the ability to undergo structural transformation (Fränzle and Markert 2000). Near the maximum sustainable non-equilibrium fluctuations are numerous and stability is low (Holling 1976); inversely, high stability normally implies low resilience – the corresponding system is geared to short-term efficiency and productivity.
- The 'thrust of evolution' seems to further flexibility of the individual system at all levels; this implies that the development of a capability to respond to the unexpected is favoured over short-term efficiency (Jantsch and Waddington 1976; Lichtenthaler 1998).
- The comparative analysis of hierarchically nested physical, chemical and biotic gradients opens up commendable ways to a unified vision and understanding of ecosystem structure and development.
- Rapid and unpredicted system responses can result from minor changes in forcing conditions or from gradual and continuous environmental change.
- Such non-linear systems behaviour introduces an important element of uncertainty into ecological economics and, more specifically, into resource management, which is reflected in integrative modelling approaches with an emphasis on the valuation of ecological functions and processes or strategic decision-making (Duchin and Lange 1994; Costanza and Daly 1992; Jansson et al. 1994; Dierssen 2000).
- Ecosystem and landscape management are particular facets of human systems management, and as such they have to be adequately analysed as processes of catalytic reinforcement of a dynamic self-organization and bonding of human components (Zeleny and Pierre 1976; Krohn and Küppers 1990).

1.2.5 Organizational Structure of the Project

The implementation of the research scheme started in 1988 and united about 150 scientists and technicians of the following institutions: Kiel and Hamburg Universities, Max-Planck-Gesellschaft (Institute of Limnology, Plön), Fraunhofer-Gesellschaft (Institute of Atmospheric Environmental Research, Garmisch-Partenkirchen),

Deutscher Wetterdienst (Agrometeorological Research Unit, Quickborn), Gewerbeaufsichtsamt (Industrial Inspection Board, Itzehoe). They represented 25 disciplines: agricultural management, bacteriology, chemistry, climatology, computer science, ecophysiology, ecotoxicology, geobotany, geomorphology, human geography, hydrogeology, ichthyology, limnology, mathematics, meteorology, paleobotany, paleozoology, physical geography, physics, planktology, plant nutrition, soil microbiology, soil science, toxicology, zoology.

The organizational structure of the project is summarized in Fig. 1.3. It illustrates the expedient combination of two principles, namely a vertical structure grouping closely related sub-projects into six divisions of research, and a horizontal one comprising a series of interdisciplinary working groups with integral functions transcending the realm of individual ecosystems, for example the analysis and modelling of energy and material fluxes (e.g., N and C cyles) through ecosystem complexes or ecotones.

Owing to its innovative concept, transdisciplinarity and organizational structure, the Bornhöved Project was declared an international pilot project of the MAB Programme in 1991. As such it has contributed to the following MAB project areas (cf. Section 1.2.1):

(2) Ecological effects of different land uses and management practices on temperate forests;
(3) Impact of human activities and land use practices on grazing lands;

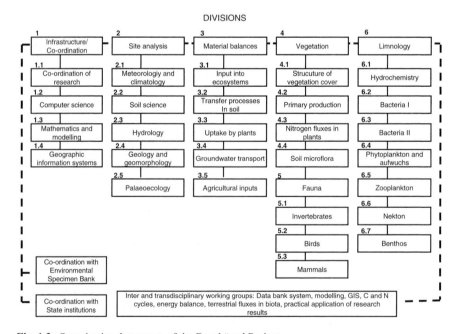

Fig. 1.3 Organizational structure of the Bornhöved Project

(5) Ecological effects of human activities on the value and resources of lakes and rivers;

(9) Ecological assessment of pest management and fertilizer use on terrestrial and aquatic ecosystems;

(13) Perception of environmental quality;

(14) Research on environmental pollution and its effect on the biosphere.

In view of the fundamental importance of heterogeneity and the related problems of representative observation, measurement and sampling at different spatial scales, basically the same methodology as described in Section 1.2.2 was applied to define representative ecotope complexes for comparative systems analyses within the catchment of the Bornhöved Lakes (Venebrügge 1988). Within such sub-units, the networks were organized in compliance with geostatistical requirements (viz. variogram analysis and related techniques) in order to derive ecologically meaningful spatial averages from primarily punctiform measurement data. Figure 1.4 illustrates the methodology by means of a large-scale map of the measuring fields and catenary sets of installations in a representative ecosystem complex west of Lake Belau which covers the major part of the core area of research.

In order to cope with scaling problems typically involved in heterogeneity studies, pedological and phytosociological investigations in the beech forest (sites W1, W2) were carried out on the basis of three nested reference grid systems of 50 m, 10 m and 1 m mesh width as shown in Fig. 1.5.

It ensues from the importance of the above heterogeneity-inducing mechanisms that geological, geomorphological, pedological and hydrological investigations as a basis for, and indispensable addition to, biological inquiries attained an extent unprecedented in the past, but in consistent compliance with one of the most essential, although at that time factually unattainable, objectives of the IBP which had claimed to study landscapes as ecosystems (cf. Golley 1993).

Information about practices, procedure and instrumentation of the research scheme is provided in the following Section.

1.3 Methodology of Investigations

Ecosystem analysis was organized under hierarchical aspects exhibiting several operational levels and a set of temporal and spatial scales, ranging from the micro-scales of the site or rapid meteorological processes to the meso-scale of the landscape or the whole drainage basin of the Bornhöved lake system. The following Sections provide a brief account of the different methodological approaches applied in the enquiries. The review focuses on principal aspects rather than on technical details, starting with the methodology of meteorological and hydrological investigations, then dealing with the techniques of large-scale soil surveys and analyses, descriptions of the geobotanical and geozoological assessment techniques used, ecophysiological analyses, the measurement of material fluxes through the ecosystems, and ending with a summary of the different modelling approaches applied.

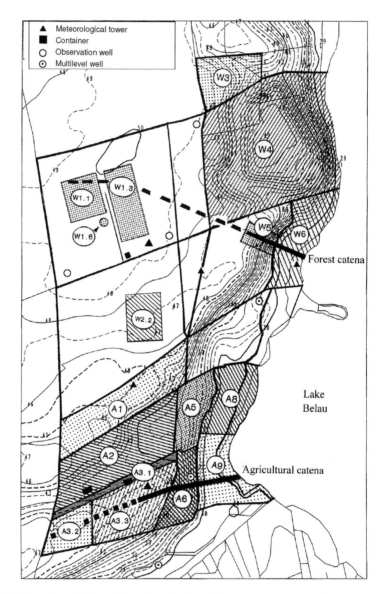

Fig. 1.4 Measuring fields and related installations in the western core area of research. W1, W2 (sub-divisions W1.1, etc.): beech forest; W3, W4: spruce forest; W5: mixed conifer forest; W6: alder carr; A1: northern field (extensive land use: maize); A2: middle field (intensive land use: maize); A3: southern field (crop rotation); A5, A6 dry grassland: pastures; A8, A9: wetland: pastures

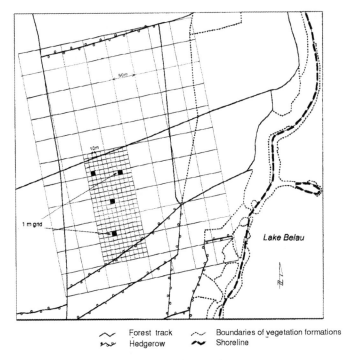

	Forest track		Boundaries of vegetation formations
	Hedgerow		Shoreline

Fig. 1.5 Position of the nested 10 and 50 m grids overlaid on the beech forest in the framework of heterogeneity-related pedological and phytosociological studies

1.3.1 Instruments and Methods of Meteorological and Hydrological Observations

Topo- and micrometeorological data are based on recordings of a 35-m tower in the beech forest (W1.3) and a 16-m telescope mast in the farmland under tillage (A3.3 in Fig. 1.4) and the long-term Ruhwinkel station of the German Meteorological Service (DWD) situated in the immediate vicinity of the core area (i.e., 54°05′ N, 10°13′ E; 36 m a.s.l.). The corresponding instruments, parameters selected and levels of measurement in the crop field (A3) and the beech forest (W1) are listed in Table 1.1. In addition, micrometeorological stations with analogous equipment were installed at sites W6, A2, A5 and in the reed belt near W6, which recorded wind speed and direction, air temperature and humidity, net radiation, photon flux density (PPFD), soil temperature and heat flux. The resultant comprehensive data sets and additional data from the neighbouring DWD stations Kiel, Eutin, Neumünster and Plön served to characterize the ecosystem types of the core area.

Hydrometeorology is concerned with the study of the atmospheric and land phases of the hydrological cycle, with an emphasis on the interrelationships covering the wide spectrum from leaf and plant water relations over canopy water fluxes,

Table 1.1 Micrometeorological parameters for comparative site characterization

Parameter	Unit	Levels above ground (m)			
		Grassland	Maize field	Reed belt	Alder forest
Wind speed	m s^{-1}	–	4	5	–
Wind direction	°	–	4	5	–
Air temperature	°C	0, 0.05, 2	0.05, 0.5, 1, 2, 4	0.05, 1, 2, 5	17.5, 15.2, 14.4, 13.2, 11.6, 9.6, 7.5
Air humidity	%	0.05, 2	0.5, 1, 2, 4	0.05, 1, 2, 5	17.5, 15.2, 14.4, 13.2, 11.6, 9.6, 7.5
Net radiation	J cm^{-2}	2	0.1, 4	5	–
PPFD	µmol m^{-2} s^{-1}	0.05, 2	0.5, 1, 2, 4	0.05, 1, 2, 5	17.5, 15.2, 14.4, 13.2, 11.6, 9.6, 7.5
Soil temperature (*)	°C	−0.1, −0.3, −0.5	−0.1, −0.3, −0.5	−0.05, −0.1, −0.3	−0.02, −0.1, −0.3
Soil heat flux	J cm^{-2}		−0.1	–	–

soil and groundwater balance to comprehensive budget approaches on the drainage basin level. In accordance with this definition, descriptions are provided of the instruments and methods for determining the primary elements of the hydrological cycle, i.e., precipitation, interception, stemflow, evaporation and evapotranspiration, soil moisture and groundwater, water-levels of Lake Belau and streams.

Precipitation was determined by means of standard gauges with a 200 cm^2 opening of the collector, recording gauges of the weighing type in all of the ecosystems studied. In addition, for purposes of interception and throughfall measurements in forest stands and maize fields (cf., e.g., Herbst and Tamm 1994), smaller gauges (50 cm^2 opening) and several cross-shaped four-gutter devices leading the rain through a conventional collecting funnel into a container with a tilting bucket type recorder, were used. On the site level the location of the gauges has been defined in compliance with geostatistical criteria of representativeness as described in Section 1.2.5. Within the framework of water balance estimates on the ecosystem and larger scales (cf. Chapter 9) the structure of the vegetation cover requires corrections of the primary gauge data by the DVWK (1996) approach. The higher the proportion of forests in the landscape, the more the gauge data appear biased in the sense of a reduction of the amount of water collected, which may attain a value of no less than 12% as is the case in the forested areas west of Lake Belau.

Plant transpiration and leaf conductance as related to microclimate were measured in situ (in the trees from a 20-m metal stage) by means of a stationary gas exchange equipment, a mobile CO$_2$/H$_2$O diffusion porometer (Walz Mess- und Regeltechnik, Germany) and a H$_2$O steady-state porometer (Li-1600; Lincoln, Neb., USA). Parallel to these measurements the diurnal course of leaf water potentials was determined (Scholander pressure chamber; Roth Gerätebau, Baiersdorf, Germany). Based on these measurements (Eschenbach 1995; Vanselow 1997), the transpiration of the canopy layer or a stand could be deduced by means of model-based up-scaling procedures (Herbst 1994; Section 5.3). Calculations of pertinent evapotranspiration rates were then based on the combination of bottom-up modelling procedures with lysimeter and Bowen ratio measurements. A combination of the above approaches allows an estimation of the partitioning of evaporation rates between plants and soil (Herbst and Vanselow 1997). Soil surface conductance for bare and vegetation-covered soils was calculated by the Penman–Monteith equation from soil evaporation data, using net radiation estimates at 0.1 m and saturation deficit measurements of the air at 0.5 m above ground. In addition to this formula the Haude approach which couples potential evaporation with phenology was used to determine the water turnover of agrarian systems (DVWK 1996). For estimating the actual evaporation rates of the beech and alder stands a two-layer model of the Shuttleworth type (Shuttleworth and Wallace 1985) was developed and applied by Herbst et al. (1999) which fits much better to the plant performance than the above calculations. Input paramaters to the model which also provide estimates of the conductance values of both the canopy and the soil surface are standard weather data.

Soil moisture was determined at 15, 60, 150 and 180 cm depths below ground by means of tensiometers (pressure transducers) whose suction potential had been calibrated by gravimetric moisture determination. Soil water content then ensued from a soil water tension/water content relationship based on laboratory and field data (Bornhöft 1993). Soil water movement in both vertical and horizontal directions as related to the soil water potential under unsaturated or saturated conditions, respectively, has been formulated in a variety of models (cf. Section 5.3).

Water flow and associated chemical fluxes between the lakes and their catchment areas are largely controlled by riparian wetlands which have, in terms of hydrological functions, the character of relatively small interfaces or ecotones with correspondingly increased hydrophysical and hydrochemical or biocoenotic gradients (Fränzle and Kluge 1997). Therefore, the research strategy along the banks of Lake Belau focused on analysing the hydraulically relevant alternation of mineral and organic soils (Schleuß et al. 2001) and the resultant complex interplay of different water pathways, such as surface runoff, interflow from slopes, flooding from the lake, seepage, capillary rise of groundwater and its discharge in different depths below ground (Piotrowski and Kluge 1994, Dall'O et al. 2001).

Groundwater levels were recorded automatically with transmitters at ten different locations in the alder stands and the wet grassland every ten minutes and, in addition, manually at 30 locations once a week in order to calibrate the automatic sensors. Groundwater exchange, the major hydrological interaction between Lake Belau and the contiguous ecosystems (Chapter 9), was subject to measurements on the basis of a network comprising 50 piezometer tubes, seven full wells and 12 multilevel sampling tubes (Piotrowski 1991; Rumohr 1996). For comparison purposes the sub-surface dispersive flow and the pertinent hydrogeochemical transformation processes were specifically analysed in a test field immediately west of the village of Belau from 1993 to 1995, using five multilevel tubes and 14 piezometers (Rumohr et al. 1996; Niedermeier-Lange 2000). The in situ flow velocities, water–mineral interactions and structure of the chemocline in the groundwater column were determined by tracer experiments and analytically. The evaluative integration of the resultant data pools and the stepwise up-scaling procedures from the site to the catchment scale were accomplished by a set of models (Section 1.3.6). For instance, the water balance of the wetland sites and the lateral water exchange processes at different depths between the slopes around Lake Belau and the limnetic water body were calculated by means of the novel riparian wetland model FEUWAnet.

1.3.2   Soil Surveys and Analyses

The soils of the study area were analysed and examined in the laboratory with regard to their morphology, genesis, classification and ecological site specifications and in respect of the availability of nutrients and water. Morphological, physical and chemical properties and classification criteria were deduced from detailed descriptions of soil columns of at least 1 m length, in accordance with standard field methods (cf. Schlichting et al. 1995). Soil classification during the primary survey period, i.e. in 1987 and 1988, was based on the pertinent pedological mapping manual of the German Geological Surveys (AG Bodenkunde 1982; Schleuß 1992). Later the comprehensive data set was updated according to the AG Bodenkunde (1994) and harmonized with the world reference base for soil resources (FAO 1998).

Soils were mapped in relation to the scales of both the entire drainage basin and the core area contiguous to Lake Belau. The maps depict soil associations and describe pertinent soil units and corresponding parent materials. Reference profiles were studied in detail in 1–2 m soil pits along two transects, namely the forest catena (500 m) and the crop field catena (250 m; see Fig. 1.4).

In order to more precisely define the spatial representativeness of the profile descriptions for extrapolation purposes and for assessing the effects of different land use patterns, sites with identical geopedological setting under agricultural and forestry management were compared. The physical and chemical properties of representative samples from different soil horizons were analysed following the International Organisation for Standardisation (ISO) methodology described by Schlichting et al. (1995) unless stated otherwise. After eliminating organic carbon, carbonates and aggregating oxides and dispersing the fraction <2 mm with $Na_4P_2O_7$, the particle size distribution was estimated by means of the sieve-pipet method (ISO 11277). The exchangeable (or available) cations Ca^{2+}, Mg^{2+}, K^+, Na^+, H^+ and Al^{3+} were extracted with $BaCl_2$ at pH 8.2 and measured by atomic adsorption spectrometry (ISO 11260), while the cation exchange capacity (CEC) was determined by desorption of adsorbed Ba^{2+} with Mg^{2+}. Carbonates were assessed by volumetric CO_2 measurement (ISO 10693). Total organic carbon (TOC) was measured coulometrically after dry combustion at 1200 °C and subtraction of carbonate-C. Total nitrogen was determined by means of the Kjeldal approach (ISO 11261). Pedogenic oxides were extracted with oxalate (Fe_o, Al_o) and dithionite-citrate (Fe_d). For mineral characterization purposes X-ray diffractometry (cf. Wetzel 1998) and for the specification of soluble element fractions the Zeien and Brümmer approach (1989) were used. Bulk density (ISO 11272) and water content at pF 0.6, 1.8 and 4.2 were determined in 100 cm³ core samples after saturation with water, providing for drainage by low or high pressure (pF 4.2) and heating at 105 °C (ISO 11276).

In order to characterize the organic matter transfer potentials a distinction was made between litter and humic compounds of the soil organic matter. Fats, waxes, polysaccarides, hemicellulose, cellulose, lignin and proteins were analysed as litter components; fulvic acid, humic acid and humin together with black carbon fractions were extracted in accordance with the classic NaOH procedure and summarized under the label humic compounds (for details the reader is referred to Beyer et al. 1993). The results were corroborated by CPMAS ^{13}C-NMR spectroscopy and by pyrolysis–field ionization mass spectrometry (cf. Wachendorf 1996).

1.3.3   Biocoenotic Investigations

A detailed account of the microbiological methods applied in the determination of soil microorganisms is provided by Bloem et al. (2006). The components of the microbial communities were counted under the microscope after soil dilution and DAPI staining without separating living and dead bacteria and microfungi. Active oligotrophic and heterotrophic bacteria were isolated on PVGV agar, microfungi on Czapek–Dox agar, acinomycetes on glycerol–arginine agar, cellulose decomposing fungi on cellulose agar, and chitin decomposers on chitin agar. The most probable number of living micro-organisms was estimated by statistical methods and served to identify denitrifying bacteria, ammonium oxidisers and nitrite oxidisers. The cultivation techniques are selective but useful to separate physiological groups (Dilly et al. 2001). In addition, fumigation-extraction, substrate-induced respiration and ATP measurements were applied as indirect biomass estimates and correlated to total DAPI-stained and culturable organisms. Since each biomass estimate refers to specific microbial communities, they were combined to achieve information about the functional composition of the biomass (Dilly and Munch 1998). Complementary to the biomass estimates, activity measurements were made, in particular with reference to basal respiration. The functional interpretation was additionally based on enzymatic measurements such as protease, urease, phosphatase and β-glucosidase activity (Dilly and Nannipieri 2001). The activity and biomass data were combined to characterize the ecophysiology of the microbial biomass.

The frequency of *Frankia* in the alder carr was measured by counting the nodules in situ down to 30 cm soil depth within a reference area of 0.25 m². Nitrogenase activity was determined in situ using the acetylene-reduction method with 12 to 13 nodules in PVC chambers. Acetylene was injected and air samples were analysed after 15, 30, and 60 min of incubation using gas chromatography. Nodules were incubated with cold $^{15}N_2$ and the incorporated $^{15}N_2$ was quantified by mass spectrometry (Dilly et al. 1999).

The terrestrial vegetation was classified and mapped following the Zürich–Montpellier (Braun–Blanquet) approach (e.g., Dierssen 1990; Kent and Coker 1992;

Dierschke 1996). The classifications follow the survey for Schleswig–Holstein by Dierssen et al. (1988). The mapping results were presented by means of GIS (Arc/Info, ver. 5.0) on the basis of topographic maps at 1:5000. GIS makes it possible to quantify units of identical or similar vegetation types.

Surveying representative areas repeatedly every second or third week allowed a description of the phenological performance of the vegetation according to Dierschke (1996). The phenological key was adjusted to describe the development of leaves and generative organs of the woody elements of the hedgerow.

Development and structure of the reed vegetation was quantified by measuring the basal shoot diameter, shoot length and stand density (Vogel 1980). For the grassland the plant cover was analysed in plots ($0.625\,m^2$) discerning between grasses, herbs and litter. Grasses were differentiated according to leaf, axis, sheaths, flowers and fruits. Each fraction was separately counted, dried and weighed. Every fourth week the properties of 40 randomly selected plants were determined (length, number of leaves, oven-dry weight). similarly in the maize field samples were taken five times in the growing season and fractions such as leaf, sheath, axis, flowers and cob were discerned.

As leaf properties, particularly for deciduous woody species, specific leaf weight (g dry weight m^{-2} leaf area), leaf thickness, stomata density, chlorophyll content (per g dry weight or leaf surface area), and chlorophyll a/b ratio were measured. Chlorophyll was analysed from 80% acetone extracts containing tris-hydoxymethyl-aminomethane buffer in a spectral photometer U-2000 (Colora, Hitachi) at 647, 664 and 750 nm and the chlorophyll content was calculated according to Ziegler and Egle (1965). The leaf area index (LAI) was determined by means of a LAI 2000 probe (Licor; Lincoln, Neb., USA) in 1992 and 1993. Also the number of leaves per unit volume (leaf area density; LAD) was estimated by counting procedures in representative canopy sectors (Eschenbach and Kappen 1996).

Development and extension of roots in the soil was analysed by determining the root length density (Böhm 1979). Samples were taken by means of a Soil Column Cylinder (Eijkelkamp Agrisach Equipment, The Netherlands). Five equal segments of 1 m soil were investigated. Root material was separated by washing in a hydropneumatic elutriation system (Gillison V.F., Benzonia, USA). After the material was repeatedly floated and decanted, different fractions of roots could be counted under the binocular, using the intersect method of Tennant (1975).

Soil fauna in the forests of the main research area was determined by hand sampling (macro-fauna), heat extraction (meso-fauna) and wet funnel technique (Enchytraeidae, Nematoda). Biomass was determined directly as fresh mass or indirectly by special coefficients for different soil faunal groups (Behre 1983; Irmler 1995).

Among the fauna particular functional groups have been isolated, identified and quantified. Density of Testacea (Protozoa) was measured for the upper 20 cm soil stratum by direct counting using the technique of Laminger (1980). Enchytraeid and nematode densities were measured from $2.5\,dm^2$ samples of the same soil stratum by means of the wet extraction method (Didden et al. 1995). Meso-faunal density was determined in the $2.5\,dm^2$ litter and top soil samples by heat extraction in the Macfadyen apparatus (Macfadyen 1962). The density of macro-fauna in the litter was determined from $0.1\,m^2$ samples by hand sorting and additionally by heat extraction. Samples were harvested monthly (1988–1990, 1992) or every second month (1991, 1993–1995). Spiders and ground beetles were additionally recorded using pitfall traps which were replaced every second week.

The vertical stratification of Empidoidea and Coleoptera in the canopy of the beech and the alder forests at 1.5, 9.0,18.0, 27.0 m, and at 1.5, 5.0, 17.0 m, respectively, was determined by means of intercept traps between 26 April and 14 October, 1992. Emergence was recorded by means of emergence traps constructed as metal boxes (with an aeration mechanism) covering 0.25 m^2 or 1.0 m^2. Sampling was carried out every two weeks in the grassland sites and every four weeks in the forests, or generally in winter times. Flight intercept traps of 1 m^2 size at 1.5 m above ground and of 0.5 m^2 size at higher levels in the air were used to record flying arthropods during the growing season. The running floor fauna was collected by means of pitfall traps of 17.6 cm circumference filled with formaldehyde and a liquid detergent. A plastic shelter was installed to protect these traps against rainfall. Additionally, section traps were posted in the marginal regions of ecosystems, using a v-shaped metal frame with a pitfall trap in the inner angle. The ground beetle density was determined in 4 m^2 exclosures surrounded by 50-cm metal frames with nine pitfall traps in each frame. Every four weeks exclosures were moved to another part of the ecosystem.

Arthropods living in the ground vegetation of the grassland and forests were collected in a suction trap of 0.25 cm^2 size at two-week intervals during the growing season. Each sampling lasted 10 min with an efficiency of 80% of the individuals per area. The shrub and tree fauna was collected from branches that were enveloped in plastic bags and then cut off. Twig length and the number of leaves were recorded and the density of the arthropods was correlated to the leaf-area index, with 40–60 randomly collected branches forming one sample. The samples were taken every two weeks over the period April to July and every four weeks in August to October. Arthropods of dead wood were also recorded from the emergence traps. A total of 1.7 m^3 of dead wood was investigated.

For the purpose of measuring the interactions of faunal elements between ecosystems, observations were made between 1988 and 1990 by means of several methods. In each habitat three replicate pitfall traps were installed. Along five borderlines between habitats section traps were exposed, catching animals only moving from the open side into the pitfall trap (Fig. 1.4).

Breeding birds were counted by direct observation. More detailed investigations dealt with robin and lark. For these species territory size, reproduction parameters and the composition of food for nestlings and adults were recorded by observations at the nest. The seasonal composition of the food for the adult robin was identified in the faeces (Davies 1977; Bryant 1978). The food for nestlings was identified from photographs of the adults carrying material to the nest (Grajetzky 1992).

Small mammals were recorded by mark and recapture techniques. Trapping was performed every second week during the growing season and at two month intervals in winter time.

Fish density in the lake was determined by means of a sonic depth finder (Lowrence eagle, Type Mach 1) six times in the period March–October during 1989–1991. In addition, the moving net technique in the profundal zone and electric fishing in the littoral zone were applied on these days (Pfeiffer 2000). Fish echoes were counted over the lake profile in consecutive 2-m steps.

1.3.4   Element Fluxes in Air, Water and the Soil–Vegetation Complex

In order to estimate bulk deposition rates funnel-shaped samplers ($320\,cm^2$ opening) were installed in the forest and field catenas and were emptied once a week. Sampling by this common technique is normally, however, rendered difficult by contamination or adsorption of compounds in the samplers. These problems were tackled by the use of identical samplers, placed closely together and operated in the same way. Thus, the standard deviation calculated from the results of eight samplers stayed constant or went down at increasing concentrations. Comparative determinations showed (cf. Slanina 1983; Fränzle 1993) that two samplers are sufficient to obtain mean values within 5% of the value found for eight samplers for parameters such as rainfall, sulfate, nitrate, fluoride and lead. The relative standard deviation is 5% or less. The results for H^+, NH_4^+, Ca^{2+}, Mg^{2+}, K^+, Na^+ and Cl^- indicate that a minimum of three samplers is necessary to guarantee that the mean value is within 5%, while heavy metals present greater problems. For instance, at least four or five samplers are needed to reach the 10% accuracy level for Zn and Cd. In the light of these technical difficulties sampling always involved at least ten identical samplers to obtain mean values in the above range of validity (Spranger 1992; Branding 1996).

Stemflow was collected in polyurethane gutters wound spirally round the trunks of four neighbouring beech trees which led the water together into containers equipped with the same tilting bucket type recorders as used for throughfall measurements. Sub-samples of 15 ml, stored in polyethylene bottles, were used for chemical analyses.

Total deposition in the forest stands was estimated as the sum of bulk and dry deposition, The amount of dry deposition was calculated by means of a canopy interaction model (Spranger 1992) which combined both the Ulrich (1983) and van der Maas et al. (1990) approaches. In the study area the assumptions underlying Ulrich's model proved to hold for K, Ca and Mg but not for Na, ammonia, or sulfur and nitrogen oxides. For validation purposes, therefore, a systematic comparison of the primary results with those of two resistance models (Wesely 1989; Erisman et al. 1993) was successfully carried out during a four-week campaign (Branding 1996).

Litterfall as an important component of material flux rates was assessed by sampling of the material in 24 pyramid-shaped collectors ($0.25\,m^2$ opening). Samples taken every two weeks, and the totals accumulated for every four weeks were fractionated into leaves, wood, buds, flowers and fruits, then dried at 40 °C and pulverized for chemical analysis (Lenfers 1994).

The analysis of seepage water was based on under-pressure sampling in triplicate parallels from five soil layers by means of ceramic suction cup devices (KPM,

Berlin, Germany). The positioning in the Ah and Ap horizons, i.e. at 5 cm and 12 cm depth in the forest and at 10 cm in the arable land, accounts for the levels of both the maximum mineralization of organic matter and the maximum activity of the root system due to tillage and fertilization effects. In the B horizon the sampling levels at 50 cm and 150 cm were designed for elucidating the influence of the root system on the chemical character of the soil solution on the one hand, and for detecting vertical gradients of soil acidification on the other. Below 150 cm the influence of roots decreases rapidly, and thus the chemistry of the seepage water at a depth of 400 cm (which corresponds to the average decalcification limit under forest) yields additional information on matrix effects in the aeration zone above the aquifer.

In addition to hydrochemical analyses of seepage water in the mineral soil a triplet of humus layer mini-lysimeters with 250 m² opening was installed which contained undisturbed soil monoliths. Later, the set was supplemented by ceramic plate samplers in order to obtain interstitial water from the humus layer. Sampling took place every two weeks over a period of 10 days (Aue 1993). To this end a moderate under-pressure was provided by vacuum pumps which were electronically triggered by tensiometer data. Owing to a programmed discontinuation of the sampling procedure for 4–5 days, the resilience of the soil system in the vicinity of the ceramic cup could be warranted.

The pH value and electric conductivity of the soil solutions were measured in the laboratory. Membrane-filtered and acidified samples served to determine metal element concentrations by atomic absorption spectrometry. The analysis of anionic compounds and ammonia was performed by means of spectro-photometric and auto-analyser techniques. If necessary, the samples were stored deep-frozen before analysis.

Element flux calculations are based on element concentrations and water flux rates. The latter were modelled by means of the VAMOS model (cf. Section 1.3.5) using the soil water tensions monitored with tensiometers in the field. Thus, cumulative element flux rates in the five sampling depths mentioned above result from multiplying element concentrations by water flux rates.

Element transfer due to harvesting was calculated from the element concentrations in the dry biomass. Element concentrations of the different types of manure allocated to the fields were estimated according to the reference data provided by the agricultural consulting services of Schleswig–Holstein; for fertilizer composition the standard formulations of the producers were drawn upon.

Carbon uptake by plants was estimated from long-term continuous in situ leaf CO_2 gas exchange measurements on attached leaves in two cuvettes of a stationary gas exchange equipment (Walz Mess- und Regeltechnik, Germany) and by means of a Walz CQP 130 leaf diffusion porometer (cf. Section 1.3.1). The canopy net photosynthesis was calculated by multiplying leaf gas exchange rates with LAI in three crown layers of the beech forest and two layers of the alder forest and the maize field, respecting the microclimatic parameters. Stem and branch respiration of alder was measured in branches of three diameter (age) classes (>10 cm, 3–10 cm, <3 cm) with specifically constructed branch boxes attached to the Walz CO_2 gas exchange equipment (Steinborn et al. 1998). Branch respiration in the alder canopy was calculated on a volume basis for the different branch size classes with their different specific respiration rates as a function of air temperature, in the beech canopy the data of Kakubari (1988) and Gansert 1998) were used. Above-ground respiration of maize was calculated according to Penning de Fries et al. (1987).

Soil respiration was measured in situ by means of an open-top dynamic system (Kutsch 1996; Kutsch and Kappen 1997; Kutsch et al. 2001c). The continuous measurements were carried out in the beech forest (W1: 1997–2000), the alder forest (W6: 1992), and in the crop fields A2 and A3 (1990–1992) Rhizomicrobial respiration was estimated either by calculating the difference between parallel measurements of total soil respiration in and between the planted rows in the maize field (A2; Dilly et al. 1999) and in the alder forest (W6) directly on the fine roots in specially designed chambers (Kutsch et al. 2001c).

Net ecosystem CO_2 exchange was calculated combining the modelled results for canopy gas exchange, stem and branch respiration and soil respiration. The calculated annual net ecosystem production was compared with field measurements of net primary production (NPP) in the ecosystems. NPP resulted from quantifications of litterfall, wood increment calculated by trunk biometry (Schrautzer and Wellbrock, unpublished data) and tree-ring analysis (Werner 1994). Below-ground biomass dynamics were estimated in W6 by sequential coring and in situ with mesh bags in which fine roots could invade (Middelhof 2000). Repeated phytomass harvesting served to calculate the above-ground NPP in maize field (A2) and wet grassland (A8; Wachendorf 1995; Weisheit 1995; Kutsch 1996; Sach 1997). Total standing phytomass of *Corylus avellana* and alder was estimated from harvesting the whole plants with roots (Eschenbach 1996; Weisheit and von Stamm 1996). NPP of the maize field included the farmer's impact such as harvesting and fertilizer application (Kutsch et al. 2001c).

Energy fluxes were calculated at the field, farm and regional scales. Data at the farm level were collected from 60 farms in the research area during 1990 and 1998. In 1990 only yield data were collected, while in 1998 additionally energy consumption at farm level (electric energy, fuel) was evaluated by interviewing the farmers. The data at the regional scale were extrapolated from regular energy reports for Schleswig–Holstein (Ministry of Finances and Energy S-H 1999) and annual statistical reports. Energy contents (J per reference unit) were taken from GEMIS data base (Rausch et al. 1998) and from literature (Heinloth 1996; Kaltschmitt and Reinhard 1997).

1.3.5 Ecological Modelling

The way followed to unravel, to a certain extent, the complex network of relationships and feedback loops in the ecosystems analysed is model building. It has the advantage of admitting a variety of broadly different approaches, ranging from empirical models for practical purposes, to rather abstract ones aiming at qualitative general insights. At one end of the spectrum there is a detailed and pragmatic description of specific systems such as single adsorbents in interaction with chemicals in aqueous solutions or the primary productivity of individual leaves. At the opposite end are relatively general models which have to sacrifice numerical precision for the sake of general principles. They need not correspond in detail to any single 'real-world' process, but aim to provide a framework for the discussion of phenomena or simply of contentious issues. Rationally handled, these different approaches mutually reinforce each other, thus providing reciprocally new and deeper insights (Fränzle 1993; Fränzle and Jensen 1999).

In the framework of the Bornhöved research scheme specifically adapted or newly developed models have played a particular role in the core areas for defining intra- and intersystemic material and energetic fluxes, structure and functioning of biocenoses, and land use. They are summarized in Fig. 1.6.

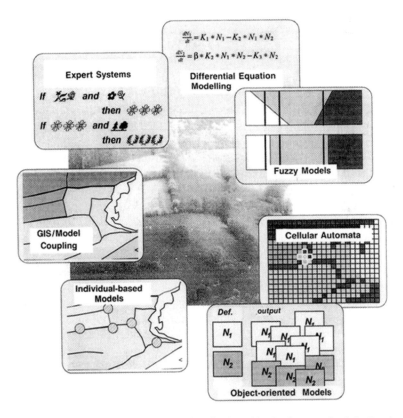

Fig. 1.6 Synopsis of ecological models used or developed in the framework of the Bornhöved ecosystem research (Fränzle 2001 after Breckling, unpublished). In the text the affiliation of individual models to these model classes is mostly indicated by the following acronyms: DE (differential equation modelling), FM (fuzzy models), CA (cellular automata), OO (object-oriented models), IB (individual-based models), GM (GIS/model coupling), ES (expert systems)

With regard to groundwater flow and water balance estimates of the entire lake system or its lower-order components the following set of models, belonging to the DE class unless otherwise stated, were used: VAMOS and SIMPEL (one-dimensional water movement in small-scale soil units; Bornhöft 1994; Hörmann 1997), MURKEL (OO: soil moisture balance of forests; Clemen and Hörmann 1996), SVAT and CoupModel (soil–vegetation–atmosphere transfer; Geyer and Jarvis 1991; Kniess 2001) and SWAT (DE/GM: soil and water assessment tool; Srinivasan et al. 1995), FLOWNET, FEUWA, FEUTRANS (water balance of lowlands and exchange processes in lotic and lentic ecotones; Piotrowski and Kluge 1994; Kluge and Theesen 1996; Kluge et al. 1997; Dall'O et al. 2001; Kluge et al. 2003), TWODAN (analytical flow model for the determination of groundwater sub-basins; Kluge and Jelinek 1999), MODFLOW/MT3D (two- and three-dimensional groundwater flow and solute exchange models; McDonald and Harbaugh 1988; Rumohr 1996; Rumohr and Scheytt 1996; Rumohr et al. 1996), MULAT (multi-layered aquifer transport model; Verruijt 1991), and WASMOD (DE, GM: two-dimensional simulation of water movement at different scales; Reiche 1996). Based on comprehensive ecophysiological data sets specific transpiration models were conceived for alder, beech and reed (Herbst 1994; Eschenbach et al. 1996; Herbst and Hörmann 1998; Herbst and Kappen 1999).

For energy flow and balance modelling, in addition to the above SVAT models and selected DILAMO components (Reiche et al. 1999), several ECOPATH procedures (energy fluxes through food webs; Opitz et al. 1997) were useful; furthermore wind action (Naujokat 1991; Schernewski 1992) and energy balance models (DE, GM: Davies et al. 1988; Venebrügge 1996) were applied at both regional and local scales. A high-resolution simulation of the wind-induced circulation patterns of Lake Belau was achieved by means of the hydrodynamic model MAST 2D (Huttula 1992; Schernewski 1999).

Numerous models served to describe and simulate element fluxes and balances. For simulating atmospheric deposition a revised resistance model (DE; Wesely 1989; van der Maas et al. 1991; Erisman and Draaijers 1995; Branding 1996) proved well suited, while LEACH-P models appropriately simulated water-borne element fluxes in soil. The geochemical simulation model PHREEQE (DE; Parkhurst et al. 1980) helped to illuminate the complex groundwater/aquifer reactions in terms of ion saturation indices and the related speciation of inorganic aluminium (Niedermeier-Lange 2000). A combination of a calcite and phosphate precipitation model (Rossknecht 1977, 1980) with the MIT reservoir model (Hurley Octavio et al. 1977) and Jørgensen's (1983) eutrophication model led to a unified simulation of the thermal stratification and essential mineral precipitation and co-precipitation phenomena in Lake Belau (DE; Schernewski et al. 1994). STOMOD in combination with WASMOD and GIS proved particularly useful for deriving material balances for a wide range of terrestrial ecosystems and spatial scales (Reiche 1996; Meyer 2000). Lagrangian fluid mechanics provided the basis for developing a comprehensive path approach, describing water and material fluxes in and between ecosystems or drainage basins and related groundwater and surface water systems, respectively. Thus, it has been possible to model both transport and transformation processes of sets of ecologically relevant substances on their way from source to more or less distant sink areas with a particular emphasis on forcing functions and time delay effects in reactions or processes (Trepel and Kluge 2002a, b; Kluge et al. 2003). On this basis WETTRANS was formulated as a decision support system for the assessment of water and related nitrogen exchange processes in riparian peatlands; it is a matrix model combining flowpaths and nitrogen transformation processes in the framework of a quasi-stationary mass balance approach (Trepel and Kluge 2004).

In order to get estimates of the plant-carbon input into the ecosystems micrometeorological plant gas exchange data (Section 1.3.1) from the canopy at various height levels were integrated into a bottom-up modelling approach of carbon fixation and transpiration of whole canopies (Kutsch and Kappen 1991; von Stamm 1994; Eschenbach 1996). The same methodology had already been applied in previous ecosystem studies by Tenhunen et al. (1976), Janecek et al. (1989) and Stickan et al. (1994).

The influence of temperature increase on soil respiration and organic carbon content of arable soils was modelled by means of a hierarchical respiration model which was calibrated by laboratory measurements of the basal respiration of litter and humic substances of the forest soils and both the readily biodegradable and recalcitrant organic matter of the crop fields (DE; Kutsch and Kappen 1997). Coupling of nitrogen and base balance approaches yielded proton balances of soil solutions, while material balances for lotic and lentic ecotones were established by means of a wetland nitrogen model (Schleuß et al. 2001). Hazelnut bushes, alder trees and agrarian ecosystems were subject to carbon modelling. CERES, FAGUS and SWACRO models were applied to simulate and predict the productivity of cereals, beech and a series of other useful plants (DE; Hoffmann 1996). Eutrophication and carbon balance models were specifically developed for Lake Belau (Schernewski 1998). Considering long-term soil erosion phenomena the ABAG/USLE approach (Schwertmann et al. 1987; Meyer et al. 1998) and the DILAMO model system were used (DE, GM: Meyer 2000); for short-term erosion events EROSION 2D (DE: Schmidt et al. 1996; Jelinek 2000) yielded conclusive results.

Within the framework of biocoenotic modelling state-of-the-art applications of fuzzy logic (Salski et al. 1996) permitted the simulation of population dynamics of the yellow-necked mouse (*Apodemus flavicollis*) in a beech forest. Furthermore they led to the development of an individual-based model of the reproductive success of the European robin (*Erithacus rubecula*) and skylark (*Alauda arvensis*) in complex environments (FM, IB: Daunicht et al. 1996; Reuter 1996). As a novel model of the IB type FAust (from German *Faunenaustausch*) was developed to analyse the dispersal of ground-dwelling arthropods (Reuter 2001). Finally fuzzy logic formed the essential component of comprehensive decision support systems for wet grassland management (ES: Asshoff 1999) and general landscape planning purposes (ES: Herzog 2002).

Object-oriented approaches were on the one hand devoted to the simulation of fish populations and the related types of shoal formation (Reuter and Breckling 1994; Hölker and Breckling 1998; Hölker 2000; Hölker and Breckling 2005) or to the modelling of both movement and migration patterns of ground-dwelling invertebrates (Jopp et al. 1998); and on the other they permitted detailed descriptions of the manifold plant–environment interrelationships on the individual basis of morphological structure and ecophysiology (Breckling 1996, 1998). Productivity models were developed for leaves, plants and vegetation stands of hazel and black alder (cf. Stamm 1992; Eschenbach 1996, 1998) and the simulation of competing root systems of black alder trees (Middelhoff 1998).

ECOPATH procedures (Polorina and Ow 1983) provided coherent descriptions and quantitative assessments of the complex littoral foodweb structure of Lake Belau in comparison with other lakes of the Bornhöved District (Opitz et al. 1997).

Socio-economic modelling had two major objectives. The first one was distinctly planning-related and aimed at simulating the influence of short and long-term economic changes on structure and functioning of agrarian ecosystems and land-scapes (ES: Dibbern 2000). To this end linear optimization was used for the selection of a crop mix and for the determination of cultivation intensity of the different agro-ecosystems of the study area. The second approach followed generalized thermo-dynamic orientations. Thus, it provided a novel tool for the comparative estimate of the entropy balance of agro-ecosystems as an integrative effect indicator of land use practices (DE: Steinborn 2000).

Chapter 2
Ecological Setting of the Study Area

Hans-Peter Blume, Otto Fränzle, Georg Hörmann, Ulrich Irmler,
Winfrid Kluge, Uwe Schleuß, and Joachim Schrautzer

2.1 Introduction

Due to Weichselian meltwater and late-glacial dead ice dynamics the Bornhöved Lake District comprises six lakes, in two broadly parallel alignments, namely the Bornhöveder See, Schmalensee and the Belauer See in the southeast and the Fuhlensee, Schierensee and Stolper See in the northwest. Hydrographically speaking, they form part of the Schwentine system which includes a major proportion of ditches in the southern part of the drainage basin. Lake Bornhöved has two outlets, the western one feeding a stream flowing past the Fuhlensee and through the Schierensee to finally emptying into lake Stolpe, while the eastern one forms the Schwentine River proper which connects the adjacent Schmalensee with Lakes Belau and Stolpe (Fig. 1.1).

As substantiated in Section 1.2.2, the study area reflects the relevant ecological structures of an essential part of the German Lowlands. It is the purpose of the present Chapter, therefore, to provide background information on the large-scale setting of the study area in terms of sediments, soils, climate, drainage systems, vegetation cover, and the development of the socio-economic structure of the cultural landscape. On this basis, the following Sections present the major findings of the research scheme, starting with the results of nano- and microscale enquiries into eco-physiological key processes of ecosystem functioning, then turning to energy and material balances of ecosystems and the related biocoenotic dynamics at the small and medium scale in order to finally consider structures and functions of the terrestrial and aquatic ecosystem complexes of the study area on the macroscale.

2.2 Geological Setting

2.2.1 Rocks and Relief Features

As a result of geological history and human activity a complicated pattern of glacigenic, fluvioglacial, limnetic, organic, and anthropogenic deposits characterizes the study area which was subject to detailed geological and geomorphological mapping

O. Fränzle et al. (eds.), *Ecosystem Organization of a Complex Landscape.*
Ecological Studies 202.
© Springer-Verlag Berlin Heidelberg 2008

in preparation and execution of the ecosystem research programme (Stephan and Menke 1977; Fränzle 1981, 1988a; Hiebner 1985; Garniel 1991; Piotrowski 1991).

The sedimentological characteristics of these deposits are essential for soil formation (cf. Section 2.2.3). The predominant till facies of the northern part of the study area (Fig. 1.2) is lodgement till of the second Weichselian advance whose textural differentiation, in particular the degree of consolidation, varies according to the continuity or discontinuity of the lodgement process, the amount of water in the subglacial environment, and syndepositional processes; also the primary carbonate content is quite variable. Ablation till with a high proportion of flow till is associated with the extensive occurence of kame moraines in the area south of Wankendorf. On either side of Lake Belau the lodgement till gradually thins out and the underlying Kalübbe Sandur forms the land surface in the southern part of the lake system.

Limnetic sediments comprise sand, silt, and detritus as a result of littoral wave action and deposition of suspended matter coming from the inflowing streams, and calcareous gyttja. The latter comprises different minerogenic (i.e. sandy, silty, clayey) and minero-organogenic facies (mixtures of calcite crystals with planktonic and non-planktonic diatoms; Zeiler 1996; Håkansson et al. 1998). Peat is widespread due to Holocene formation of low fens in the numerous depressions of the area, in particular between Lake Stolpe, Schierensee and Lake Bornhöved. Soil colluvia whose thickness may locally exceed 2.5 m, but generally is less than 1.0 m, cover the basal parts of steeper slopes (Schleuß 1992). They were formed since early Neolithic times due to ploughing and related soil creep, and to a lesser extent by splash and rill-wash erosion. Technogenic substrates like rubbish heaps or rubble occur in the neighbourhood of settlements, forming the parent material of specific soils unless they are covered by natural soil material which also forms the predominant substrate of hedgerows.

2.2.2   Climate

In terms of an explanatory-descriptive system, e.g. Flohn (1957), the humid meso-thermal climate of the study area is characterized by the prevalence of moist maritime air masses and frequent cyclonic storms. According to the Lauer and Rafiqpoor (2002) classification the study area is covered by a $Cmh\beta1$ type of climate, i.e. mid-latitude sub-maritime, mesothermal, humid. Table 2.1 summarizes the basic climatic data of the long-term reference station Ruhwinkel of the Deutscher Wetterdienst in terms of annual means of the 1989–1998 period and the 30 years average. Owing to its topographic situation and the vicinity to the lake system the station can be considered representative of the Bornhöved Lake District as a whole, since the total acreage of forests is such that corrections of the precipitation and evaporation figures by means of the DVWK (1996) approach do not appear necessary. The situation is different, however, for the core area where the mean precipitation value of the above reference period based on rain gauge records has to be increased by approximately

Table 2.1 Annual means of the 1989–1998 period and a 30-year mean of climatic parameters of the Ruhwinkel station of the Deutscher Wetterdienst

Year	Air temperature °C Mean	Max.	Min.	Relative humidity %	Precipitation mm	Evaporation mm	Windspeed m s^{-1}	Sunshine H	Net radiation J$\times$cm^2 day^{-1}
1989	9.5	13.4	5.5	79	801	509	1.9	1642	
1990	9.8	13.4	6.0	82	977	442	2.6	1574	1005
1991	8.7	12.4	5.0	82	761	409	2.3	1653	1007
1992	9.6	13.3	5.6	80	756	536	2.4	1666	980
1993	8.0	12.2	3.9	85	838	432	2.3	1155	892
1994	8.6	12.8	4.4	81	972	478	2.2	1436	1029
1995	8.0	12.7	3.5	81	687	494	2.2	1507	1044
1996	6.3	10.9	2.1	81	490	435	2.1	1365	858
1997	8.1	12.7	3.9	81	662	419	2.2	1534	1045
1998	8.2	12.3	4.5	83	957	343	2.1	1251	1003
Mean	8.5	12.6	4.4	82	790	450	2.23	1478	985
30-year mean	8.1	11.6	4.7	83	697	–	–	1646	–

12%, which yields an effective (i.e. balance-true) area precipitation of 890 mm year^{-1} (cf. Chapter 9).

In the first years of observation until June 1993 the mean air temperatures were warmer than average, especially from December to late spring. The year 1992 was referred to as "The Northern Summer" because the summer was exceptionally hot and dry. The following years were intermediate in character; 1993 and 1998 had a warm spring, but very cool summers. Long and cold periods occurred in early winter 1993, during the whole winter 1995, and to a lesser extent in 1996.

Monthly precipitation varied between a few millimetres and 150 mm month^{-1} but no clear seasonal pattern existed. Three longer periods of low precipitation stand out: from autumn 1995 to December 1997, from winter 1990 to spring 1991, and in the first half of 1993. The most marked anomaly began in 1995, as groundwater recharge became effectively zero and the groundwater level decreased for two years (cf. Chapter 9, Fig. 9.5). Despite an average annual precipitation (801 mm) the summer 1989 has to be considered dry since a quarter of the total rainfall was due to two heavy showers in July and August.

Evaporation at the landscape scale, calculated as potential evaporation for unstressed grassland (Haude 1958), yielded slightly more than 500 mm year^{-1}. Exceptional were the warm years 1992 and 1998 or the cold and wet years 1991 and 1993, respectivly. Higher-resolution estimates of evaporation and evapotranspiration are based on the combination of bottom-up modelling with lysimeter and Bowen ratio measurements. For the different meteorological approaches as also used in Chapter 5 the reader is referred to Section 1.3.

2.2.3   Soil Pattern

The soil associations of the study area comprise terrestrial, semiterrestrial, and anthropogeomorphic soil units which are briefly characterized in terms of the WRB approach (IUSS Working Group WRB 2006) and the German classification scheme (Wittmann 1997) on the soil map in Fig. 2.2. The soils were mainly formed from glacial till and glacial or fluvioglacial sands in the study area (cf. Fig. 2.1). During 13 000 years of soil formation a deep decalcification of the terrestrial soils has occurred: the loamy soils from glacial till, with <20% carbonates have been decalcified down to 2 m depth, the sandy (originally with <5%) down to 3–5 m depth.

Soils in a landscape influence each other by manifold lateral energy, water, mass, and solute exchange processes, and thus are interrelated (Blume 1984). Four different types of pedogenic coupling exist as related to slope, valley or plateau positions; in addition artificial (man-made) coupling patterns occur. The slope structure (ss) characterizes unidirectionally overlapping soil processes like erosion/sedimentation as well as solute fluxes on slopes. Valley soils (vs) influence each other on the one hand by solutions flowing to the receiving waters during periods of low water level, on the other by flooding and sedimentation. On flat or gently undulating substrates in plateau position (ps) reciprocal influences between neighbouring soils are of little importance and limited to minor distances. The artificial coupling type (as) characterizes villages and towns with >30% sealed areas and is associated with higher soil temperatures, marked influence of dust and other emissions; it includes many Anthropic Regosols, which are partly poisoned.

Figure 2.2 shows a soil association map of the Bornhöved Lake area. Each unit is characterized by the principal soils associated, the main substrate(s) and the coupling structure.

The units 5, 6, 13, 21–24, 26 exhibit plateau coupling; lateral translocations are low under these conditions. Under tillage loamy Luvisols and sandy Arenosols are levelled by ploughing. Under forest, soils around tree stems are more acid and sometimes podzolized due to a higher input of "acid rain" by stemflow. Soils of clearings are wetter and nutrients like Fe, Mn, and P move laterally along a redox gradient into soils under trees (Blume 1984). Forest paths, reinforced with calcaric gravel, changed the pH of soils along the path verges (Reiche and Dibbern 1996).

The units 1 and 14–17 exhibit slope coupling: Soils of unit 1 which occupy an upper-slope position are liable to erosion, whereas the soils of the other units are formed from colluvia. But under tillage below each hedgerow erosion and above each hedgerow accumulation takes place.

The units 2–4, 8–10, 12, 18–20, 25, 29, 32 represent the valley coupling type. Those of river and lake banks are eutrophic to calcitrophic due to flooding by nutrient-enriched waters on the one hand, but acidified in footslope position contiguous to forest soils due to the influence of acid groundwater flow on the other.

The units 11, 27, 28, 30, 33 exhibit artificial coupling due to strong human influence: 30–60% of their soils are sealed by buildings or roads, which influence the water and element status of the soils. Some of them are enriched with rubbles, others with organic waste, which may result in microbial activity may result under reductive conditions. Soils from waste material (like bricks, mortar, sludge, waste or slag) are classified as Technosols. Long-term gardening resulted in the formation of Hortic Anthrosols (unit 13) with a strong enrichment of organic matter (especially black carbon) and nutrients.

Among the class of Terrestrial Soils which are are not liable to groundwater influence in the upper metre of the pedon different subunits of loamy Luvisols (Parabraunerde) are widespread. In the north-east of the study area they predominate in the form of Stagnic Luvisols and Stagnosols (Pseudogley), together with Dystric Luvisols (Dystrophe P.) under forest, and Eutric Luvisols (Eutrophe P.) under arable land.

For illustration purposes Tables 2.2, 2.3 summarize the results of soil analyses of an Arenic Umbrisol from sandy till under beech forest.

For comparison purposes, tables of analytical details can be found at http://www.ecology-uni-kiel.de/bornhoeved-report for an Eutric Arenosol under tillage (Tables 2.10, 2.11), Dystri-rheic and Eutri-fibric Histosols (under alder carr or meadow, respectively; Tables 2.12–2.15), Dystri-colluvic Anthrosols (Table 2.16) and an Anthropic Arenosol.

2.2.4 Drainage Systems

2.2.4.1 Lakes and Water Courses

Hydrographically speaking, the lakes of the study area form part of the fluvial system of the River Schwentine whose tributaries include a considerable number of ditches in the southern part of the catchment. In terms of Strahler's (1957) stream-ordering notation they form a third-order system discharging $0.11\,m^3\,s^{-1}$ into Lake Bornhöved (Nowok 1994). Its outlet forms the Schwentine River connecting the adjacent Schmalensee with the downstream Lakes Belau and Stolpe. The Fuhlenau stream drains the western part of the catchment, flowing past the Fuhlensee and through the Schierensee to finally flow into Lake Stolpe. All the lakes belong to the class of hardwater lakes and are characterized in Table 2.4 by their morphological, physical and chemical properties.

2.2.4.2 Groundwater

The topographic catchment areas of the Bornhöved Lake District as shown on hydrological maps are only roughly suitable for water balance purposes since they normally differ from the subterranean catchments considerably as a consequence of the complicated hydrogeological situation. With the groundwater flow model MODFLOW (Guiguer and Franz 1997) and TWODAN (Fitts 1995), the subterraneous catchment areas as shown in Fig. 9.3 can be defined (Ruhmohr 1996; Kluge and Jelinek 1999). The mean subterranean catchment area of the lake district totals

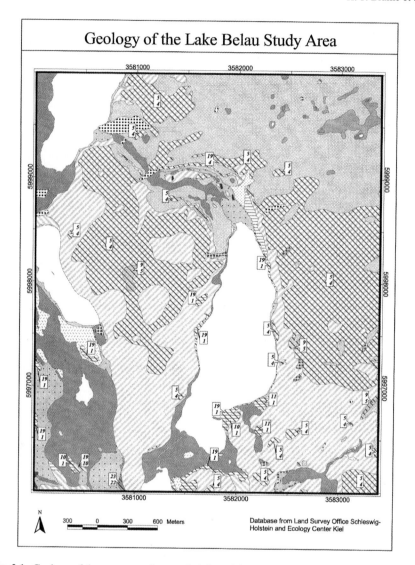

Fig. 2.1 Geology of the core area of research (adapted from Piotrowski 1991)

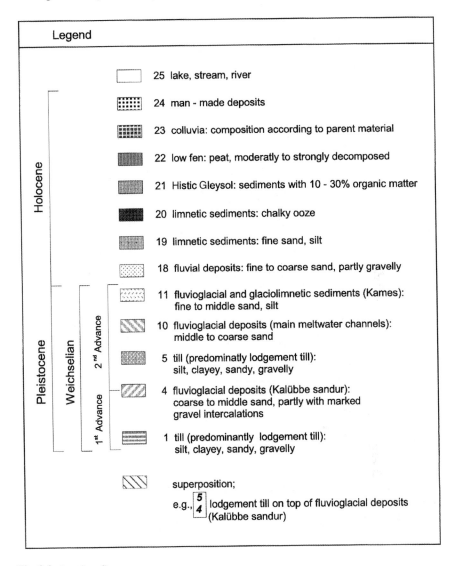

Fig. 2.1 (continued)

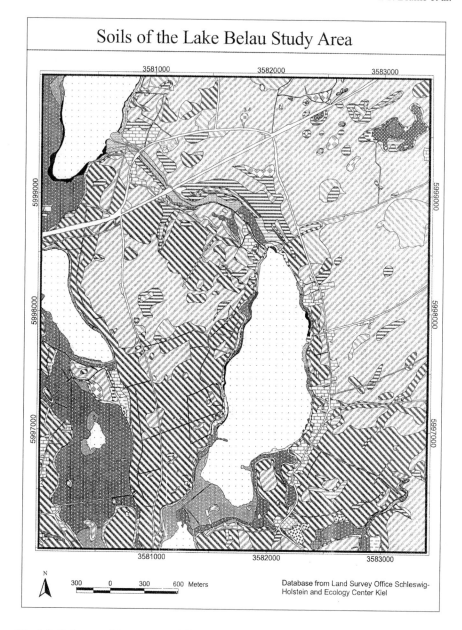

Fig. 2.2 Soil map of the core area (after Schleuß 1992). Soil units after IUSS Working Group WRB (2006)

Soil of the Lake Belau Study Area

Soils / substrates / structures of coupling
(w with; su sub units; a artificial; p plateau; s slope; v valley; s structure)

1 Histic + Humic Gleysols w calcari-, eutri- su / till / vs

2 Areni-gleyic + Histic Fluvisols / river sand / vs

3 Arenic Gleysols + Gleyic Arenosols w calcari-, dystri-, eutri- su / sand / vs

4 Dystri-ombric Histosols / peat / vs

5 Cambisols (agric.) + Umbrisols (forest) w stagnic, haplic su / till / ps

6 Cambic Arenosols (agr.) + Arenic Umbrisols (forest) sand / ps

7 Lakes and Rivers w Limnic Fluvisols (> 1 m water depths)

8 Gleyic Luvisols + Cambisols w calcari-, dystri-, eutri- su /loam / ss

9 Technic Gleysols + Gleyic-Technosols w calcaric su / anthropog. sand or loam /vs

10 Arenic Gleysols + Gleyic Fluvisols w dystri-, eutri-, humi-, histi- su /sand /vs

11 Gleyic Technosols + Technic Gleysols w calcari-, eutri- su/ technogenic rubbles / as

12 Gyttji-limnic Fluvisols + Ombric Histosols / mud + peat / vs (< 1 m below lake surface)

13 Hortic Anthrosols + Humic Regosols / sand / ps

14 Gleyi-cumulic Anthrosols / loamy colluvia above peat / ss - vs

15 Gleyi-cumulic Anthrosols / sandy colluvia deposite above peat / ss-vs

16 Hapii-cumulic Anthrosols / loamy colluvia / ss

17 Hapii-cumulic Anthrosols / sandy colluvia / ss

18 Gleysols w calcaric, mollic, dystric, eutric su / till /vs

19 Areni-histic + Humi-arenic Gleysols w calcari-, eutri- su / sand / vs

20 Ombric Histosols + Histic Fluvisols / peat+sand / vs

21 Luvisols w stagnic, dystric, haplic su / clay loam / ps

22 Luvisols w arenic, dystric stagnic su / sandy loam / ps

23 Calcaric Arenosols w cambic, gleyic su / sand / ps

24 Podzols + Arenic Umbrisols w gleyic su / sand / ps

25 Ochri-limnic Fluvisols / lake sediments / vs(< 1 m below lake surface)

26 Stagnosols w luvic, mollic umbric su / clay loam / ps

27 Anthropic Regosols + Arenosols w gleyic, humic, haplic su / excavat. sand or loam / as

28 Urbic Technosols + Technic Arenosols w calcaric, skeletic su / sand + technogenic deposits / as

29 Thioni-limnic Fluvisols / mud / vs (< 1 m below lake surface)

30 Village w sealed soils + Technosols + Hortisols / sandy + loamy excavates / as

31 Other sites

32 Eutri-ombric Histosols + Histic Gleysols / peat + sand / vs

33 Traffic areas w sealed soils + Technosoils + Technic Regosols / mixture of rubbles + excavates / as

Fig. 2.2   (continued)

Table 2.2 Arenic Umbrisol (Dystrophe Normbraunerde), spodic features, from sandy till above fluvioglacial sands with fine humus rich mor (Moder: 3.5 cm organic layer) in a level to lightly sloping moraine plateau (R: 2581.310; H: 5997.160), 49 m a.s.l.; no groundwater influence (W1). Structure[a]: AB angular block-like, GR granular, PL platy, SG single-grained, SB sub-angular block-like, PR prismatic, CO coherent. C_{org} Organic carbon, N_t total nitrogen, Fe_o oxalate extraction iron, Fe_d dithionite extraction iron, Al_o oxalate extraction aluminium, CEC cation exchange capacity (pot potential, eff effective; pH 7.3), b.s. base saturation

Horizon symbol[a]

	Depth	Colour	Bulk density	Total pore volume	Vol% water at pF				Particle size distribution[b]							
	cm	Munsell	g cm⁻³	vol%	0.6	1.8	2.5	4.2	X	cS	mS	fS	cSi	mSi	fSi	Cl
									%							
Ah	0–7	10YR2/2	1.10	57	55.4	30.3	15.2	6.7	10	11	35	32	9	5	1	7
RAp	–38	10YR3/3	1.39	47	44.2	20.0	8.8	3.5	14	13	40	20	11	7	0	9
Bw1	–68	10YR4/6	1.51	43	36.5	13.4	5.3	2.6	58	28	38	18	8	5	0	3
Bw2	–91	10YR4/6	1.50	43	36.8	24.8	10.5	4.6	30	25	38	18	10	5	2	2
RBg	–112	7.5YR4/6	1.47	44	37.9	21.9	15.5	3.3	45	28	51	14	2	2	0	3
Bw3	–125	10YR5/4	1.47	44	37.9	21.9	15.5	3.3	48	33	52	7	2	2	1	5
BdtC	–148	10YR5/6	1.40	47	34.6	10.3	5.6	0.6	8	31	58	7	0	0	0	3
BwC	–390	10YR6/3	1.40	47	34.6	10.3	5.6	0.6	2	7	79	10	0	1	1	3

Horizon symbol[a]

	pH	Lime	C_{org}	N_t	C/N	Fe_o	Fe_d	Fe_o/d	Al_o	Exchangeable cations (cmol_c kg⁻¹)					CEC		b.s.
	Structure CaCl₂	%	%	‰		‰	‰			Ca	Mg	K	Na	H+Al	pot	eff	%
Ah	GR 3.0	0	3.4	2.0	17	1.5	6.3	0.24	0.9	11.4	2.4	0.9	1.4	197	213	98	8
RAp	GR 4.0	0	0.9	1.0	9	1.4	5.3	0.26	1.5	12.7	0.8	1.0	0.8	92.7	108	48	14
Bw1	SB 4.4	0	0.4	0.4	10	1.2	5.1	0.24	1.7	16.1	0.2	0.6	0.4	57.2	75	35	23
Bw2	SB 4.5	0	0.4	0.2	20	0.9	3.7	0.24	1.9	10.6	0.2	0.7	0.4	37.6	50	24	24
RBg	SG 4.5	0	0.1	0.1	10	0.7	5.2	0.13	0.8	9.1	0.2	0.1	0.4	29.1	39	27	25
Bw3	SG 4.4	0	<0.1	<0.1	13	0.5	2.4	0.21	0.7	11.8	0.7	0.5	0.1	34.8	48	34	27
BdtC	SG 4.6	0	<0.1	<0.1	8	0.6	5.3	0.11	0.6	17.8	0.9	0.5	0.3	38.5	58	38	34
BwC	SG 4.4	0	<0.1	<0.1	7	0.4	2.0	0.20	0.3	10.2	0.4	0.3	0.1	26.8	38	29	29

[a] Horizon symbols (WRB 1998, with subsequent amendments): L fresh litter, O organic layer, H peat layer, A upper mineral horizon, RA Relictic A, cA/cB colluvial material, B subsoil, C soft parent material (highly decomposed organic matter), b buried genetic horizon, c concretions, d densic layer, dt thin clay bands, e moderately decomposed organic matter, g strongly mottled, h organic matter accumulation, i slightly decomposed organic matter, k pedogenic carbonates, p plough layer, r strongly reduced, s illuvial sesquioxides, t illuvial clay, w moderately weathered. Soil description after FAO (2006).
[b] X Gravel + stones, S sand, Si silt, Cl clay (c coarse, m medium, f fine)

Table 2.3 Humus body of an Arenic Umbrisol with fine humus–poor mor (after Wachendorf 1996)

	Depth cm	C_{org}	C/N %	pH	Lipids	Sugar/ starch	Hemi-cellu-lose	Cellu-lose	Pro-tein	Lig-nin	Fulmic acid	Humic acid	Humin
								% of C_{org}					
Litter		51.6	127		3.7	0.5	18.0	30.0	1.8	42.0	(1.8)	(1.5)	(0.8)
Leaves		50.9	23		16[a]	5.0	17.0	18.0	8.1	17.0	(9.3)	(3.9)	(4.9)
Horizon L	4–3	53.2	48	4.4	12.0	1.2	15.0	18.0	3.4	21.0	10.0	15.0	4.6
Oi1	−2	51.3	31	3.8	5.7	0.8	12.0	12.0	6.6	22.0	13.0	16.0	11.0
Oi2	−0.5	49.0	26	3.6	5.1	0.7	9.4	12.0	7.1	16.0	16.0	23.0	11.0
Oe	−0	39.8	23	3.4	5.3	0.8	8.9	12.0	7.8	15.0	14.0	19.0	16.0
Ah	0–3	5.4	16	3.1	6.2	0.5	12.0	2.1	8.7	8.9	25.0	22.0	14.0
RAp	−31	1.6	18	3.6	7.4	2.1	13.0	3.8	9.3	58	28.0	16.0	15.0

[a] Includes chlorophyll.

Table 2.4 Morphological, physical, and chemical characteristics of the Bornhöved lakes

Parameter	Unit	Lake Bornöved	Schma-lensee	Lake Belau	Schie-rensee	Lake Stolpe
Water area[a]	km²	0.73	0.88	1.13	0.27	1.4
Water volume[a]	10^6 m³	3.38	3.60	10.18	0.97	9.59
Average depth[a]	m	4.6	4.1	9	3.6	6.9
Stratification type		Dimictic to mono-mictic	Pleomictic	Dimictic to mono-mictic	Pleo-mictic	Dimictic to mono-mictic
Directly surrounding drainage basin	km²	5.6	9.4	4.4	7.6	12.1
Drainage basin of inflow water systems	km²	11.7	18	28.4	12.2	59.9
Water exchange time	Year	0.45	0.31	0.72	0.12	0.31
Calcium[b]	mg l⁻¹	64	55	53	78	64
pH[b]	–	8.7	8.7	8.4	8.1	8.4
Total nitrogen[b]	µg l⁻¹	3000	2200	1700	4200	2300
Total phosphorus[b]	µg l⁻¹	70	60	100	110	100
Trophic status		Hyper-trophic[c]	Eutrophic[c] to hyper-trophic	Eutro-phic[c]	Hyper-trophic[d]	Eutrophic[d] to hypertro-phic

[a] Data from Müller (1981).
[b] Mean values of the spring circulation of the 1994–2000 period, data from a monitoring programme of the Environmental Agency of the Federal State of Schleswig–Holstein.
[c] Classification corresponding to OECD (1982) and Schernewski (1999).
[d] Evaluation with data from Hofmann (1997).

73.4 km², including 4.6 km² (6%) of surface waters. The average height of the ground surface is 44 m above sea level and varies between 83.3 m in the south and 27.5 m at the outlet of Lake Stolpe in the north. The mean subterranean discharge within the catchment area amounts to almost 400 mm year⁻¹, i.e. 12.71 km⁻² s⁻¹ or 27.5 million m³ year⁻¹ while the direct overland runoff totals less than 5 mm year⁻¹, i.e. 0.161 km⁻² s⁻¹ or 0.4 million m³year⁻¹. The influx from ditches and to a lesser extent from tile drains, supplies the streams and lakes with a total volume of about 4 million m³ year⁻¹ (cf. Chapter 9).

2.2.5   Vegetation

Vegetation mapping (arable land and settlements excluded) covers an area of 479 ha (Fig. 2.3). Communities of the *Alnetea glutinosae* occur mainly as narrow belts along the banks of the lakes. According to environmental conditions these forests are adapted to at least temporarily flooded Histosols with varying base saturation and different degrees of decomposition (cf. Chapter 12). Alder carrs (*Carici elongatae–Alnetum*, typically with *Carex elongata* and *Solanunm dulamara*) and birch carrs (*Betula pubescens*-community) in which a more or less fragmentary shrub layer (e.g. *Ribes nigrum*) and species-rich herbaceous and moss layers has developed, occur as narrow belts along the banks of the lakes (Fig. 2.3). Alder–elm forests (*Alno–Ulmion*) are characterized by high proportions of *Prunus padus* and *Fraxinus excelsior*. Most of the sites are assigned to the *Alnus glutinosa*-community occurring adjacent to the alder carrs from which they have developed due to drainage and eutrophication. The beech forests (*Fagion sylvaticae*) of the study area are assigned to the association *Galio–Fagetum*. The herb layer consists mainly of species of the *Querco–Fagetea* (e.g. *Polygonatum multiflorum, Stellaria holostea, Galium odoratum*) and is often dominated by the grass species *Milium effusum*. Three subassociations are distinguished representing different soil conditions (see Chapter 12). Forest stands with coniferous and deciduous trees which could not be assigned to a defined community of the *Querco–Fagetea* are summarized under "Other Forests". This forest type covers more than double the area of the beech forests, which illustrates the high anthropogenic influence prevailing in the whole Lake District (Fig. 2.3).

Reed swamps (*Phragmition australis*) dominate the riparian zones of the lakes. Tall sedge reeds (*Magnocaricion elatae*) have only a minor coverage in the study area. Primary stands are seldom and occur at lake margins, while most tall sedge reeds colonize drained fens with a different degree of humification (Schrautzer 1988).

Weakly drained wet grassland (*Calthion*) is rare in the Bornhöved Lake District. Most sites occur on Histosols and are grazed or abandoned. Characteristic species are *Lotus uliginosus, Cirsium palustre*, and *Angelica sylvestris*. Stands of moderately and intensely drained wet grasslands (*Lolio–Potentillion*) cover the largest area of the wet grassland. The only association of this type is the *Ranunculo–Alopecuretum geniculati*. Most sites are located in peatlands with ground subsidence due to drainage. Grasslands without groundwater contact (*Cynosurion*) is the most frequent vegetation

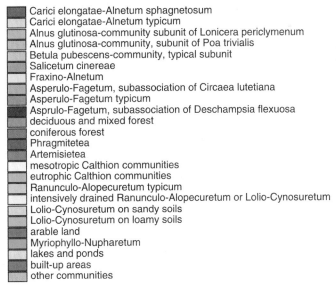

Carici elongatae-Alnetum sphagnetosum
Carici elongatae-Alnetum typicum
Alnus glutinosa-community subunit of Lonicera periclymenum
Alnus glutinosa-community, subunit of Poa trivialis
Betula pubescens-community, typical subunit
Salicetum cinereae
Fraxino-Alnetum
Asperulo-Fagetum, subassociation of Circaea lutetiana
Asperulo-Fagetum typicum
Asprulo-Fagetum, subassociation of Deschampsia flexuosa
deciduous and mixed forest
coniferous forest
Phragmitetea
Artemisietea
mesotropic Calthion communities
eutrophic Calthion communities
Ranunculo-Alopecuretum typicum
intensively drained Ranunculo-Alopecuretum or Lolio-Cynosuretum
Lolio-Cynosuretum on sandy soils
Lolio-Cynosuretum on loamy soils
arable land
Myriophyllo-Nupharetum
lakes and ponds
built-up areas
other communities

Fig. 2.3 Vegetation pattern of the core area of research

type of the pastures comprising only one association, the *Lolio–Cynosuretum* with predominating *Lolium perenne*.

Most of the study area is covered by arable land (Fig. 2.3) where intensive land use has led to a largely fragmentary development of vegetation types. Due to shading of the crop plants and the use of herbicides the wild flora has only a low coverage. Some widespread communities in the study area are *Digitarietum ischaemi* in maize fields, *Veronico–Fumarietum* in fields with root vegetables, and the *Stellarietea*-basic community.

2.2.6 Development and Socio-Economic Structure of the Cultural Landscape

Based on palynological evidence and radiocarbon dating, examples of forest clearance have been identified in the study area since early Neolithic times (Wiethold 1998). The first phase of significant human impact by clearing extensively primeval alder–oak–linden forest and with agricultural activity is associated with the Funnel Beaker Culture (±3400–2750 BC). It is indicated by marked soil erosion signals in the lake sediments and several megalithic tombs west and south of Lake Belau (Hingst 1985). The second period with remarkably increased fluxes of the trace elements K, Rb, and Pb exhibits progressive deforestation of the alder–oak–birch forest, beginning 2110 BC and lasting until 1490 BC. It is paralleled by decreasing Mn and Fe fluxes as a consequence of a new eutrophication cycle of Lake Belau. During the Late Neolithic a regeneration of the woodland and a corresponding shrinking of settlements can be observed; erosion and eutrophication tracers indicate, however, that a certain degree of agricultural activity subsisted.

The Bronze Age (1490–784 BC) brought a re-intensification of the agricultural activity which came to an end close to the Sub-Boreal/Sub-Atlantic boundary. The Iron Age (520 BC–205 AD), when the indigenous alder–oak forest around the settlements was gradually replaced by alder–beech forest, is characterized by a newly intensified settlement activity. The resulting strong erosion in the catchment of the lake system reached its maximum around the turn of the millenium. After 200 AD all sedimentary tracers reflect a distinctly reduced human activity in the study area which reached its post-migration minimum in the Early Middle Ages (520–656 AD), coupled with a widespread expansion of the natural beech–hornbeam–birch forest. The following Slavonic Period with its simple agricultural techniques and the low-productivity field-grass rotation brought about only minor changes in the agrarian landscape of the lake district. It marks, however, the spread of village organization east of the Limes Saxoniae which seems to have started in the Bornhöved area (Schwerin v. Krosigk 1976). Since 1143 AD things changed with the immigration of German settlers who introduced new agricultural practices, in particular more advanced crop rotation systems with fallow, which eventually paved the way to modern high-yield production systems. These agrotechnical innovations formed the basis for a growth of existing settlements and the foundation of new ones, among which Bornhöved achieved a particular political importance. The Late Middle Ages (1350–1520 AD) were, however, characterized by plague, war and famine which caused the well documented series of agricultural crises of the time, the desertion of arable land and villages and a concomitant expansion of the beech–oak woodland (Wiethold 1998).

The present-day agricultural landscape is the result of political reforms and economic innovations introduced during the following centuries, among which the Danish agrarian reforms since 1766 with the introduction of enclosures (hedgerows) and the abolition of bondage were most important. Since the nineteenth century the arable land was limed and Stagnosols were systematically drained with pipes; since the 1920s an intensive N, P, and K fertilization has been practised. The last considerable changes in possessory rights were due to post-war legislative acts which were aimed at breaking down the extensive landed property in favour of the rural refugees from the former German territories east of the river Oder. Justifiable in terms of social policy, the economic consequences of these land reform acts fell short of expectations (cf. Rumohr 1973). With regard to the study area the result of this development is, roughly speaking, a subdivision into a distinctly less diversified northern and northeastern part with a predominance of extensive estates, and a central and southern part with a comparatively dense network of hedgerows separating the medium-sized fields of the peasant-owned farmsteads which are grouped together into relatively compact rural villages. The different percentages of the essential structural elements of the landscape are summarized in Fig. 2.4, showing some interannual fluctuation in acreages as a result of rotation practices and changes in livestock.

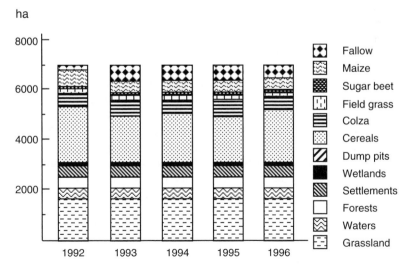

Fig. 2.4 Time-series diagram of the acreages of the essential landscape elements of the study area in the representative 1992–1996 period (after Meyer 1996; Fränzle 1998)

The present landscape is characterized by a predominance of arable land interspersed with pastures and meadows on the sandur and till surfaces with their Brunic Arenosol, Luvisol, and Stagnosol associations (cf. Section 2.2.3), while pastures and wetlands with Gleysols and Histosols are concentrated in the depressions, in particular along and between the lakes. Due to an accelerated removal of a considerable part of the nineteenth century hedgerows during the past century, the arable land now forms largely an open-field system with non-contiguous plots scattered in various directions and distances around the villages where the farmers dwell. Exceptional in this respect is the northern part of the study area which belongs to two estates whose plots are much larger, forming contiguous blocks of arable land and forests.

A questionnaire-based inventory which involved 63 out of a total of 100 farms provided a detailed insight into the socio-economic structure of the area in the reference years 1987/1988 and 1996/1998 (Reiche 1991; Dibbern 2000). The total acreage of these farms amounts to 3590 ha, comprising 2290 ha of arable land, 914 ha of permanent grassland, 178 ha of forests, 50 ha of wasteland, and 158 ha of surface waters and settlements; thus the average farm size is 57.8 ha.

The significant feature of the regional type of farming is that it has usually more than one source of cash income but with one of them commonly predominant. Outstanding is the high degree of differentiation in crop rotation, which involves the following successions:

Colza – cereals – cereals – (cereals); $n = 16$. Among the cereals winter wheat and winter barely are common; in addition rye and oats are grown.

Colza – cereals – legumes, vegetables – cereals; $n = 13$. Common legumes and vegetables are beans, peas, fodder and sugar beet.

Vegetables/legumes – cereals – cereals – legumes, vegetables – cereals; $n = 7$.

Field-grass (3 years) – cereals; $n = 7$. Predominant cereals are rye and oats.

Field-grass – (field-grass) – maize – cereals – (cereals); $n = 12$. The number of years with grass or cereal production is variable.

Maize – cereals or field-grass; $n = 9$. Single-crop farming for maize is practised in four cases.

Full-time family farms predominate, only in 12 cases does a considerable part of the family income result from activities outside farming. Apart from food production agriculture provides tasks which are growing in importance in the modern industrial society, i.e. preservation of nature, securing and cultivating a varied landscape for people to live, relax and recuperate in, and supplying agricultural raw materials for industrial purposes. The family farm appears especially suited to adapt to overall economic changes and to secure these different functions.

2.3   Site and Community Characteristics in Catenary Analysis

The core area of research around Lake Belau comprises representative ecosystems of the Bornhöved Lake District which were selected on the basis of large-scale mapping campaigns and additional field inspection in compliance with geostatistical requirements on the one hand and the catenary principle on the other (cf. Section 1.2). In the present Section these ecosystems are described in catenary order (cf. Fig. 1.5) with emphasis on soils, vegetation and fauna to provide background information for the analysis of biotic and abiotic processes in Chapters 3–7.

2.3.1 Beech Forest

At the western end of the forest catena grows a 110-year-old beech forest (W1, W2) which has replaced former farmland. During the past 30 years the stand was cleared at intervals of 5–10 years. In phytosociological terms the beech forest constitutes an *Asperulo–Fagetum*. Major parts of the area (ca. 14 ha) exhibit the typical subassociation on moderately dry sites whereas only a few fresh sites are covered by the subassociation of *Circaea lutetiana* (cf. Section 2.5). The subassocciation of *Deschampsia flexuosa* indicates local soil acidification.

The shrub layer is poorly developed. Measurements of the tree diameters revealed a standing crop of the tree layer amounting to about 2000 g m^{-2}. *Milium effusum* is the dominant species of the herb layer. The net above-ground production of this species amounted to 13.5 g m^{-2} year^{-1} in 1990. Additional characteristic species are *Stellaria holostea* and *Oxalis acetosella* (Table 2.5). Indicators of nutrient-rich conditions, i.e. *Rubus idaeus* and *Urtica dioica*, frequently occur resulting from the high internal nutrient availability of the system and the high external nutrient input from the adjacent agricultural areas (cf. Chapter 8).

With regard to the fauna, the ground beetle assemblage of the southern part of the forest (W2) indicates the fresh type of beech forest, whereas the northern area (W1) is predominantly of the dry type. Biomass and species richness are similar to W1 plots, where the microfauna attains the highest biomass of the investigated sites in the core area, based on the largest species number of Testacea (Tables 2.6, 2.7). The micro-arthropods exhibit a lower species richness and a lower density than in the forests with more humid soils. Abundant species of springtails (Collembola) and oribatid mites are *Hypogastrura denticulata, Isotomiella minor* or *Rhysotritia duplicata, Nothrus silvestris*, respectively (Irmler 1995). A few groups of the macrofauna, e.g. millipedes (Diplopoda), woodlice (Isopoda), and earthworms (Lumbricidae) exhibit very small densities, which is typical for forests on acid soils (Schaefer and Schauermann 1990; David et al. 1993). The macrofaunal Diptera and click beetles (Elateridae) occur with relatively high abundances and are predominated by a few species. *Bradysia confinis* contributes to more than 40% of the dipteran family Sciaridae (Heller 1996). The click beetles are almost exclusively represented by *Athous subfuscus* (Irmler 1995). The total biomass of the ground fauna amounts to 2.46 g dm m^{-2} which is in the lower range of beech forests on acid Umbrisols in Europe (Peterson and Luxton 1982; Weidemann and Schauermann 1986).

Density and species richness of the phytophagous fauna of the herb layer are extremely low (Tables 2.6, 2.7). Sioli (1996) found a maximum density between 53 individuals m^{-2} in *Dryopteris* stocks and 8 individuals m^{-2} in *Rubus* stocks. Also the vertebrate assemblage is represented by only a very few species. Yellow-necked

Table 2.5 Percent coverage of frequent vascular plants of permanent plots of the forest catena in 1995

	Ecosystem type		
	Beech forest	Alder carr (wet, eutrophic)	Alder carr (drained, mesotrophic)
		Catena location	
	W1, W2	W6	W6
		Vegetation type[a]	
Species	GF	CAT	AG
Tree layer			
Fagus sylvatica	63		
Alnus glutinosa		63	63
Shrub layer			
Lonicera periclymenum	18		
Padus avium			4
Sorbus aucuparia			8
Herb layer			
Milium effusum	18		
Galium odoratum	2		
Stellaria holostea	2		
Oxalis acetosella	2		
Carex acutiformis		88	
Solanum dulcamara		8	
Carex paniculata		2	
Phragmites australis		2	
Glyceria maxima		2	
Juncus effusus		2	
Sium erectum		2	
Mentha aquatica		2	
Dryopteris dilatata			38
Rubus fruticosus agg.			8

[a] Vegetation types: *GF Galio–Fagetum, CAT Carici elongatae–Alnetum, AG Alnus glutinosa* community

mouse (*Apodemus flavicollis*) and bank vole (*Clethrionomys glareolus*) contribute to 60% and 33% of the mammals in the forest. Among the birds, ubiquitous species dominate, e.g. chaffinch and blackbird. The species richness of the beech forest proves to be largely due to the saproxylic fauna and predators that contribute about 18% to the total species richness (Irmler et al. 1998).

Among the soils of the beech forest Arenic Umbrisols of different base saturation (5–50%) have developed in association with Hypoluvic Umbrisols and Dystric Luvisols. Humus accumulation, decalcification (at present by 350–400 cm below soil surface), acidification, dealcalization, brownification (see pH, base saturation, colour in Table 2.2) and clay formation indicate incipient podzolization [bleached sand grains, Al_o minimum in the A(e)h]. The actual Ah is 5–10 cm thick and partly covered by a litter layer of 1–5 cm; it has developed in a relic 25 cm Ap-horizon

Table 2.6 Mean density (n m^{-2}) and biomass (mg dry weight m^{-2}) of the most important soil animals in the investigated sites in the main research area. Insect data include larvae; data refer to the litter and a soil depth of 5 cm of mineral layer or 30 cm for earthworms. Averages are based on four (W11, A33, A91) or two (W54, W61, A61) replicate samples.

Site													
W11		W54		W61		W61		A33		A61		A91	
Location													
Hilltop		Hillside		Downslope		Lake margin		Hilltop		Hillside		Lake margin	
n	mg	n	mg	n	mg	n	mg	n	mg	n	mg	n	mg
Microfauna													
Testacea[a]													
60	257	89	251	108	193	50	81	97	39	62	46	163	198
Nematoda[b]													
5	230	4	176	4	188	3	134	3	167	5	267	4	221
Mesofauna													
Collembola[c]													
5600	70	9300	110	14 100	130	5850	90	8361	92	5216	100	6236	126
Collembola[c]													
7800	130	9000	270	15 500	420	11 750	240	450	6	2297	45	504	11
Collembola[c]													
2900	70	4260	104	5300	130	4020	110	961	18	1326	29	807	24
Enchytraeidae[d]													
39 300	309	18 000	161	8200	92	2900	71	2200	20	8100	72	3300	30
Macrofauna													
Gastropoda[e]													
8	220	35	240	110	550	76	300	–	–	6	13	76	277
Gastropoda[e]													
0.1	0.3	39	210	140	110	63	150	–	–	1	14	–	–
Isopoda[e]													
6	4	110	60	580	110	700	140	–	–	–	–	–	–
Lepidoptera[e]													
9	56	7	4	5	21	4	4	1	1	2	4	7	55
Trichoptera[e]													
–	–	7	4	8	4	6	2	–	–	–	–	–	–
Lumbricidae[e]													
11	320	14	160	200	670	320	1090	17	47	47	785	220	1809
Diptera (sapr.)[c,e]													
1250	220	620	317	6720	600	3740	1120	680	152	728	304	753	564
Elateridae[e]													
117	150	76	70	90	115	20	20	–	–	11	62	–	–
Carabidae[e]													
20	70	1	2	25	18	8	23	9	8	18	29	9	30
Staphylinidae[c,e]													
160	95	170	80	500	100	270	80	93	13	140	42	159	87
Cantharidae[e]													
170	15	51	13	106	12	11	7	–	–	2	85	1	2

(continued)

Table 2.6 (continued)

Site													
W11		W54		W61		W61		A33		A61		A91	
Location													
Hilltop		Hillside		Downslope		Lake margin		Hilltop		Hillside		Lake margin	
n	mg	n	mg	n	mg	n	mg	n	mg	n	mg	n	mg
Dermaptera[e]													
13	120	3	23	4	42	4	20	–	–	–	–	–	–
Opilionida[e]													
11	8	30	20	75	23	30	20	–	–	–		2	3
Araneida[e]													
127	50	120	30	156	23	45	20	37	6	153	40	197	69
Chilopoda[e]													
20	20	130	64	40	20	50	20	–	–	–		12	5
Diptera spp[c,e]													
139	50	220	116	410	175	320	130	34	6	14	4	78	80

[a] Laminger method (1980), see Matthiesen (1995).
[b] Determined by means of Baermann funnel.
[c] Heat extraction in Macfadyen apparatus (Irmler 1995.
[d] Wet extraction according to Didden et al. (1995).
[e] Handsorting from 0.1 m^2 reference plot (Irmler 1995).

which, due to agricultural use, was till to the end of the nineteenth century. Partly the Ah/RAp horizons have a thickness of up to 40 cm where shallow depressions have been filled with translocated soil materials. The rusty coloration of the RGBv or RGo is relictic and probably due to water stagnation above a frozen subsoil during the late glacial period. Ecologically speaking, there are (very) deep, dry sites with low to medium nutrient reserves which have low pH values and low contents of available nutrients. The contents of available nitrogen are higher due to an annual deposition of 20–40 kg ha^{-1}.

The humus body of the soils (cf. Table 2.3) originated from vegetation- and animal-borne substances, i.e. proteins, celluloses, and lignins (for methodological reasons including living microfauna and -flora) as well as fulvates and humates. The humus of the A horizons had its origin in dead roots and was mixed by bioturbation and ploughing in the past. Humate contents of living vegetation reflect the limited selectivity of the methods applied.

2.3.2 Mixed Forest

East of the beech forest a 30- to 40-year-old mixed forest is following downslope (W5, Fig. 1.5) Dominant trees of the stand are *Pseudotsuga menziesii* (douglas fir), *Picea abies* (fir), *Larix kaempferi* (tamarack), *Quercus robur* (oak), and *Quercus*

Table 2.7 Species richness of four ecosystems in the main research area

Group of organisms	Beech forest n	%	Alder wood n	%	Maize field n	%	Wet grassland n	%	Total n	%
Plants[a]	43	2.2	89	4.5	15	0.8	42	2.1	157	8
Testacea[c]	18	0.9	31	1.6	9	0.5	19	1	39	2
Lumbricidae[b]	6	0.3	9	0.5	4	0.2	9	0.5	10	0.5
Gastropoda[b]	14	24	24	1.2	–	–	15	0.8	27	1.4
Araneida[d,e]	115	5.8	146	7.4	169	8.6	139	7.1	237	12
Oribatida (partim)[h]	44	2.2	54	2.7	10	0.5	22	1.1	59	3
Isopoda[b]	4	0.2	7	0.4	–	–	2	0.1	7	0.4
Myriapoda[b]	7	0.4	10	0.5	–	–	3	0.2	13	0.7
Collembola[h]	53	2.7	62	3.2	40	2	49	2.5	77	3.9
Auchenorrhyncha[e,g]	22	1.1	24	1.2	9	0.5	39	2	59	3
Heteroptera[e,g]	21	1.1	27	1.4	15	0.8	31	1.6	58	2.9
Coleoptera (ex. Carabidae)[d,e,g,h]	318	16	400	20	193	9.8	218	11	657	33
Carabidae[d]	35	1.8	52	2.6	81	4.1	47	2.4	104	5.3
Diptera (partim)[d,e,f,h]	164	8.3	261	13	49	2.5	144	7.3	397	20
Vertebrata[a]	43	2.2	33	1.7	20	1	22	1.1	67	3.4
Sum	907	69	1229	62	614	31	801	41	1968	100

[a] Direct sampling or registration during 1988–1995 (Irmler 1995).
[b] Direct sampling by handsorting during 1988 to 1992/1995 (Lumbricidae up to 30 cm depth).
[c] Sampling in 1989, for method see Matthiesen (1995).
[d] Pitfall traps during 1989 to 1997/1998.
[e] Emergence traps during 1989 to 1997/1998 (Itmler et al. 1996, 1997).
[f] Window traps from the years 1989, 1990, 1992 (Irmler et al. 1996; Rief 1996; Irmler 1998).
[g] Suction traps 1989–1994 (Sioli 1996).
[h] Different extraction methods from 1988 to 1995 (Heller 1995; Irmler 1995).

rubra (red oak). The most frequent species of the herb layer is *Rubus fruticosus* which profits from the high light availability of the young plantation.

The soil fauna in the mixed forest is similar to that of the beech forest, whereas the composition of the aboveground fauna is distinctly different and reflects an intermediate position between the hilltop and the alder carr at the foot of the hillside (Tables 2.6, 2.7). Several macrofaunal groups, e.g. Gastropoda, Diplopoda, and Isopoda, occur with characteristically higher abundances than in the beech forest. Most frequent representatives of snails and millipedes are *Discus rotundatus*, *Aegopinella nitidula*, and *Allajulus punctatus* which are absent in the beech forest and attain their highest densities on the footslope. Altogether, the fauna comprises more species with preference for higher soil moisture and lime content than in the beech forest (Irmler 1995).

There are different soils in slope position which were under tillage till to the end of the past century. On the moderately inclined moraine plateau sandy acidic Umbrisols from glacial sands and Dystric Luvisols from glacial till have developed. In upper slope position mainly eroded Umbrisols from fluvioglacial sands occur. Near the lake in middle and lower slope positions transitions to sandy Colluvic

Anthrosols predominate which have developed under tillage (Table 2.16 at http://www.ecology.uni-kiel.de/bornhoeved-report). The soils have a high penetrability for roots, a low available field capacity and low contents of available nutrients. On the footslope the Colluvic Anthrosols are partly groundwater-influenced and have a higher plant-available water supply by capillary rise. Almost all sites of the mixed forest exhibit incipient podzolization.

2.3.3   Alder Carrs

Adjacent to the mixed forest two facies of alder carr have developed due to the lowering of the groundwater table in the 1930s. A wet eutrophic alder carr occurs as a narrow strip behind the reed swamps subsequently followed by a drained alder carr (W6, Fig. 2.3). The shrub layer of the drained site consists mainly of *Padus avium* and *Sorbus aucuparia*, whereas *Dryopteris dilatata* is the dominant species of the herb layer. The herb layer of the wet eutrophic alder carrs (*Carici elongatae–Alnetum typicum*) in the riparian zone of Lake Belau is dominated by the stoloniferous tall sedge *Carex acutiformis* (Table 2.5). In addition, the typical species composition of this vegetation type is characterized by indicators of wetness and flooding, i.e. *Glyceria maxima*, *Sium erectum* and *Phragmites australis*. Mesotrophic alder carrs (*Carici elongatae–Alnetum sphagnetosum*) are characterized by species like *Viola palustris* and *Lysimachia thyrsiflora*. This vegetation type, however, occurs only on sites which are neither flooded by lake water nor influenced by nutrient-laden groundwater (Wiebe 1998).

Microarthropods, snails, millipedes, and woodlice exhibit maximum density and biomass in alder carrs (Table 2.6). In addition, earthworms occur in high densities at locations with high lime content in sandy soils, e.g. *Aporrectodea caliginosa* with 14 individuals m^{-2} on average. In wet depressions species preferring humid soils are found, e.g. *Carychium minimum* (Gastropoda) and *Eiseniella tetraedra* (Lumbricidae). In the wet alder carr, hygrophilous species contribute to 90% or 70% of the ground beetles and rove beetles (Nötzold 1996) with a high dominance of *Agonum fuliginosum* (19%) and *Elaphrus cupreus* (15%). The biomass of the ground-fauna is comparable to that of the drained site consisting mainly of earthworms and large tipulid larvae, e.g. *Molophilus bihamatus* and *Nephrotoma analis* (Rief 1996). Concerning the microarthropods, this alder carr exhibits the typical composition of the *Gustavia fusifer* assemblage of oribatid mites (Strenzke 1952). The endogeic and epigeic woodlouse species *Haplophthalmus danicus* and *Ligidium hypnorum* are abundant with about 350 individuals m^{-2} and 13 individuals m^{-2}, respectively.

The fauna of the tree layer can be divided into the populations of the marginal zone adjacent to the lake and those of the canopy layer in the centre of the stand (Ambsdorf 1996). Plant-sucking species attain a frequency of 41 individuals m^{-2} in the marginal zone and 25 individuals m^{-2} in the canopy layer. Both leaf-mining and leaf-feeding species occur in both zones with, respectively, 16 individuals m^{-2} and

19 individuals m^{-2} and only 2 individuals m^{-2}. The gall mite *Eriophyes laevis* (100–300 galls m^{-2}) is most frequent.

Adjacent to the lake, peats have developed in the alder carrs above fluvioglacial sands. Rheic (= groundwater influenced) Histosols with medium (= hemic) to distinctly (= sapric) humidified peats include sand layers caused by strong wave movements of the lake. They are associated with Histic, Humic and Arenic Gleysols. The peat of the Dystri-fibric Histosol exhibits low to medium humification and acidification due to drainage since the 1930s. The contents of available nutrients are medium to low (Table 2.12 at http://www.ecology.uni-kiel.de/bornhoeved-report).

The humus body of the Histosol (Table 2.13 at http://www.ecology.uni-kiel.de/bornhoeved-report), compared with that of the Umbrisol and the Colluvic Anthrosol, has a litter richer in protein due to the occurrence of alder as opposed to beech and spruce. The relative contents of easily decomposable sugars decrease with depth, while the relative contents of cellulose and lignin decrease less because the decomposition is delayed under anaerobic conditions. Thus, the formation of carboxylic and phenolic hydroxyl groups is impeded but carbonication is supported.

2.3.4 Agroecosystems

The major part of the study area consists of agroecosystems. On field A3 (Fig. 1.5) fertilization generally followed the recommendations of the "Landwirtschaftskammer Schleswig-Holstein" and was adapted to the expected nutrient removal by plants; additionally, manure was applied. The use of agricultural pesticides was in conformity with the recommendations for integrative cultivation of the local advisory board. Thus, a moderately intensive land use was practised.

In contrast, field A2 was characterized by permanent cultivation of silage-maize in monoculture. The management was always manuring with slurry at the end of April, with subsequent ploughing and the preparation of the seed bed. After seeding of maize mineral nitrogen and phosphorus was given depending on weather conditions. In 1988 and 1989 atrazine was used as a weed control and then substituted by other pesticides in the following years. The crop was harvested normally in October. Only in 1990, was rye sown as an intercrop.

Field A1 was managed in nearly the same way as A2 until 1993. Maize was cultivated as a monoculture on the field for a period of 20 years with different fertilization regimes. After the 1993 harvest, a more extensive management was initiated, which involved reduced fertilization.

The spontaneous component of the field vegetation changed according to the kind of management. A typical plant community co-occurring with maize was the *Digitarietum ischaemi* with species more or less resistant to herbicides such as. *Digitaria ischaemum* and *Setaria viridis* (cf. Table 2.8). Additional plant communities with frequent species, e.g. *Chenopodium album*, *Viola arvensis*, *Stellaria media*, and *Fallopia convolvulus* belong to the Chenopodietalia, Aperetalia, and Stellarietea (Table 2.8).

Table 2.8 Percent cover of the most frequent vascular plant species of different vegetation types on permanent plots in the agricultural catena I in 1995. *d x* Differential species

Species	Ecosystem type[a]					
	I	I	II	II	III	IV
	Vegetation type[b]					
	LC	LC	RA	RA	CB	SP
	Land use					
	Grazed	Abandoned	Grazed	Mown/grazed	Abandoned	
d 1						
Lolium perenne	18					
Geranium molle	38					
d 1–2						
Achillea millefolium	4	38				
Elymus repens	18	38				
Poa pratensis	8	18				
Dactylis glomerata	2	4				
d 3–4						
Glyceria fluitans			63	38		
Agrostis stolonifera			38	38		
Alopecurus geniculatus				8		
d 5						
Carex acutiformis					38	2
Equisetum fluviatile					2	
Anemone nemorosa					2	
d 6						
Phragmites australis						88
Peucedanum palustre						18
Lemna minor						8
Mentha aquatica						8
Epilobium palustre						8
Stellaria palustris						8
Cicuta virosa						2
Lycopus europaeus						2
Non-differential species						
Taraxacum officinale	18	2	4			
Cerastium holosteoides	4	2	2			
Festuca rubra		18	4	8		
Cardamine pratensis			2	2	2	2
Ranunculus repens			18	38	18	
Poa trivialis			18	4	2	
Rumex acetosa			8	2	18	
Holcus lanatus			38	18		
Alopecurus pratensis			18		38	

[a] Ecosystem type: *I* grassland without groundwater contact, *II* moderately drained wet grasslands, *III* weakly drained wet grasslands, *IV* reed swamps.
[b] Vegetation type: *LC Lolio–Cynosuretum*, *RA Ranunculo–Alopecuretum geniculati*, *CB Calthion* basic community, *SP Schoenoplecto–Phragmitetum*, sub-association of *Cicuta virosa*.

With respect to the fauna, the small-sized fields on loamy or sandy-loamy soils were dominated by the ground beetles *Platynus dorsalis* (13%) and *Pterostichus melanarius* (20%). The soil fauna was comparatively poor. A few groups, e.g. Gastropoda, Diplopoda, and Isopoda, were totally absent (Tables 2.6, 2.7). Concerning the micro-arthropods, the collembolan species *Isotoma tigrina* and *Cryptopygus thermophilus* were relatively numerous, whereas the density of oribatid mites was very low. The field exhibited a high dominance of a few species, such as the terricolous nematoceran species *Scatopsciara atomaria* which attained 75% abundance (Heller 1996). In addition, the field community was remarkable because of a comparatively low constant occurrence of species. While in the beech forest 24% of the ground beetle species and 11% of the spider species occurred permanently during nine and eight years respectively, only 6% and 9% were continuously present on the field between 1989 and 1997/1998 (Irmler et al. 2000).

On the agricultural sites studied, the initial substrates of soil formation are sandy to loamy till and fluvioglacial sands as well as colluvial material. Rubic Arenosols, Eutric Luvisols and Colluvic Anthrosols have developed as soil units. The agricultural site A3 has an Eutri-rubic Arenosol, characterized by humus accumulation, decarbonization down to a depth of 3 m, acidification, and brownification. In this soil unit, the penetrability for roots is high, the available water capacity medium and the contents of available nutrients are medium to low (Table 2.10 at http://www.ecology.uni-kiel.de/bornhoeved-report).

The humus body of the Eutri-colluvic Arenosol is comparatively rich in proteins, but poor in other litter substances (Table 2.11 at http://www.ecology.uni-kiel.de/bornhoeved-report). More than 60% of the humus body consists of humified substances, whereas the average of Schleswig–Holstein amounts to 45% on arable land. The high albuminous content is supposed to be caused by a high content of living biomass (Dilly 1994).

2.3.5   Grassland without Groundwater Contact

In the core area, this ecosystem type is situated on the steep slopes adjacent to the arable land on sandy soils (A6). During the study period this site was used as a pasture except a 30 m strip abandoned since 1988. Management changed from standing to rotation grazing. The number of grazing periods increased from two to four per year between 1988 and 1993, coupled with a temporal reduction of the single grazing periods.

The vegetation is primarily a Lolio-Cynosuretum typicum. Some areas could not be fertilized due to steep slope and thus developed a vegetation subunit with *Hypochoeris radicata*. On the fertilized pasture *Lolium perenne*, *Geranium molle*, *Elymus repens*, and *Taraxacum officinale* dominated (Table 2.8). An experimental abandonment using enclosures resulted in a change in the species composition. Grazing-tolerant species like *Lolium perenne* disappeared and species with below-ground stolons (*Elymus repens*, *Achillea millefolium*) extended their coverage.

The soil faunal biomass of this grassland type was very low (Table 2.6). The dominant ground beetle species were *Nebria brevicollis* (21%) and *Calathus fuscipes* (16%). With respect to micro-arthropods, the Collembola were frequent with more than 5000 individuals m^{-2}, in particular herb dwellers, e.g. *Lepidocyrtus cyaneus* (1300 individuals m^{-2}) and *Isotoma tigrina* (1000 individuals m^{-2}), whereas the mainly endogeic oribatid mites occurred only at low densities. The macro-fauna was mainly represented by Diptera, with 700 individuals m^{-2} on average, and during a few years with an abundance of more than 1500 individuals m^{-2}. The sciarid midge *Bradysia rufescens* was the dominant species with 36% (Heller 1996). Another important representative was the bibionid midge *Dilophus febrilis*, that attained 300 adult individuals m^{-2} and an average larval density of 1600–1800 individuals m^{-2} on the average. The phytophagous groups, i.e. leaf hoppers (Auchenorrhyncha), bugs (Heteroptera), and grasshoppers (Orthopteroidea) were represented with mean densities of 15.0, 0.7, and 0.8 individuals m^{-2}, respectively.

The prevailing soil unit of the pastures without groundwater contact is the Eutri-colluvic Anthrosol (Table 2.17 at http://www.ecology.uni-kiel.de/bornhoeved-report). Below the Ah, colluvial material accumulated due to former tillage of the slopes. Furthermore, this ecosystem type is characterized by a deep penetrability for roots, a moderately high available water capacity and a high content of available nutrients.

The humus body of the Eutri-colluvic Anthrosol reveals an extremely high content of humified substances (Table 2.18 at http://www.ecology.uni-kiel.de/bornhoeved-report); in particular, the valuable humates are dominant. In the subsoil lignins were mainly decomposed, whereas cellulose was even decomposed in the topsoil. The relatively high contents of proteins, lipids, and hemi-celluloses in the subsoil were a consequence of the high biological activity at this site (Dilly 1994).

2.3.6 Grassland with Groundwater Contact

This ecosystem type is located between the reed swamps and the dry grasslands on the eastern slopes of the Weichselian meltwater channel at the village of Belau. Two subtypes of wet grassland can be distinguished, due to their hydrology: The subtype "moderately drained" borders directly the slopes (A9) and the subtype "weakly drained" follows lakewards (Fig. 1.5). The largest area of the weakly drained subtype is characterized by eutrophic conditions, whereas mesotrophic sites are rare.

Management practices differed in the wet grassland. Site A9 was previously mowed and not fertilized. The change in management (fertilization, regular grazing, temporary rolling) drastically changed the species composition (see below). The grazing period was short, the number of cattle was low and changed from year to year. A small area within A8 (A8.3) represents the weakly drained, eutrophic subtype which was used for haymaking in the past. To observe the consequences of the secondary progressive succession on the species composition, the meadow was abandoned since 1988. A neighbouring site (A8.1) was intensively used as a pasture until 1986. In connection with an extensification contract in 1987 land use changed from heavy grazing to mowing without fertilization.

The vegetation of site A9 is a *Ranunculo–Alopecuretum* with typical species including *Glyceria fluitans*, *Alopecurus geniculatus*, and *Agrostis stolonifera* (Table 2.8). In 1988 the vegetation of site A8 was assigned to a basic *Calthion*-community. Due to abandonment, species composition changed during the study period to a tall sedge reed. The dominant species of the stand is the highly competitive sedge *Carex acutiformis* which has displaced numerous other species. Initially, also the vegetation of site A8 could be assigned to the *Ranunculo–Alopecuretum*. The change in management in 1987 from grazing to cutting without fertilization resulted in a spreading of higher-growing grasses including *Holcus lanatus* and *Alopecurus pratensis*.

The fauna of site A9 represents the moderately wet grassland type with high dominance of the ground beetles *Carabus granulatus* (25%) and *Pterostichus diligens* (19%). The hygrophilous *Oodes helopioides* (1–2%) and *Agonum afrum* are but rarely present or absent. The soil fauna attaines an essentially higher biomass with $3.6\,g\,dm\,m^{-2}$ on the average than in the two other agrarian ecosystems (Table 2.6). Mainly earthworms and Diptera contribute to the high biomass, with the earthworm species *Lumbricus rubellus* (27 individuals m^{-2}) and *Lumbricus castaneus* (11 individuals m^{-2}) and sciarid midges with 445 individuals m^{-2} (most frequent species *Bradysia nitidicollis*) and limoniid midges with 50–80 individuals m^{-2} (most frequent species *Ericonopa trivialis*). Collembola are the most frequent group of microarthropods, with hygrophilous epigeic species dominating, e.g. *Isotomurus palustris* (1500 individuals m^{-2}) and *Lepidocyrtus lanuginosus* (1200 individuals m^{-2}).

Phytophagous arthropods exhibit low densities of 32 individuals m^{-2} (Auchenorrhyncha, dominant species *Arthaldeus pascuellus*), 1.2 individuals m^{-2} (Heteroptera, dominant species *Stenodema laevigatum*), and 0.6 individuals m^{-2} (Orthopteroidea, dominant species *Chorthippus monatnus*). The intensity of land use in the wet grassland caused different reactions of the faunal groups. Gastropoda are significantly more frequent and have the highest species richness in the fallow (Neumann 1998), while for Diptera no significant difference is found (Heller 1996; Rief 1996).

The fen basis of site A9 is formed by undulating pleistocene sands. With the beginning of the limnetic sedimentation minerogenic late-glacial muds were deposited, above which post-glacial calcareous muds follow which in turn are covered by alder peat. A top layer of a 80–100 cm thick brown moss peat then marks the end of the sedimentary sequence. Due to drainage this site is strongly humified in the topsoil. The humus body of the Eutri-fibric Histosol is relatively rich in protein. The contents in fulvates and especially humates are significantly higher than those of the Histosols of the alder forest while high humin contents occur in the deeper subsoil (Tables 2.14, 2.15 at http://www.ecolog.uni-kiel.de/bornhoeved-report).

2.3.7 Hedgerows

Fields and grasslands of the core area are frequently bordered by hedgerows, originating from the end of the eighteenth and the beginning of the nineteenth centuries. One hedgerow was intensely studied throughout the first phase of the project (cf.

Chapters 3, 6); its dominant shrub is *Corylus avellana* (cf. Stamm et al. 1996). Frequent species of the herb layer are *Rubus idaeus* and particularly on the northern flank *Lonicera periclymenum* and *Dryopteris filix-mas*.

Arthropods of the beech forest are scarcely represented in this hedgerow. The spider and ground beetle assemblage comprised mainly species of the adjacent fields, e.g. the spider *Erigone atra* and the ground beetle *Bembidion lampros*. The vertebrates also reflect the ecological situation between agrarian and forest systems, which prevents mammals and birds from developing stable populations in the hedgerow (Irmler et al. 1996). The robin as a typical bird of the hedgerow suffers from high predator pressure, and mice populations exhibit high seasonal fluctuations. In the shrubs, caterpillars and weevils are most abundant, using approximately 4% of the leaf mass. Dominant species on hazelnut bushes are the weevils *Phyllobius argentatus* and *Polydrusus cervinus* and the leaf wasp *Rhogogaster punctulatus*.

Most of the hedgerows were set up on sandy deposits far above the groundwater level but occasionally also in groundwater-influenced positions. Increasing humus contents down to greater depths as well as a strong acidification (in spite of the regular liming of neighbouring agricultural soils) appear typical of the soils of such hedgerows. Analytical data for a sandy Anthropic Arenosol overlying a former Dystric Arenosol are summarized in Table 2.19 at http://www.ecology.uni-kiel. de/bornhoeved-report.

2.3.8 Reed Swamps

This ecosystem type is transitional between the alder carrs or the wet grasslands and Lake Belau (cf. Chapter 9). The vegetation is a *Schoenoplecto–Phragmitetum* and can be subdivided into two subassociations. The typical subassociation borders the littoral down to a mean water depth of roughly 1 m, whereas the subassociation of *Cicuta virosa* (Table 2.8) occurs at sites with a mean water depth from 0 cm to 10 cm, frequently in contact with alder carrs. The second subtype is richer in species than the typical subassociation. The species composition of drained reed swamps is characterized by the nitrophyte *Urtica dioica* and the loss of typical reed species of the Phragmitetea.

The phytophagous invertebrate fauna was sudied in the typical subassociation of the aquatic reed zone and in the the *Cicuta virosa* subassociation of the terrestrial reed belt (Grabo 1991). In front of both the wet grassland and the alder carr, diptera contributed with 33 species and 22 species, respectively, to the total species richness of 44 species. The most frequent were the aphid *Hyalopterus pruni*, leaf hoppers of the genus *Chloronia* and the gall midge *Giraudiella inclusa*. Density, biomass, and species composition of the phytophagous fauna correlated with the land-water gradient. The aphid *Hylopterus pruni* had maximum values in the typical subassociation in front of the alder carr with 2.7 g dm m^{-2}. The gall midge *Giraudiella inclusa* attained its maximum density in the aquatic reed zone with 100 individuals m^{-2}. The population dynamics of *Giraudiella inclusa* was largely con-

Table 2.9 Eutri-rheic Histosol covered by a Gyttji-limnic Fluvisol of silty calcaric lake sediments

Horizon	Depth cm	Soil colour Musell	Particle size distribution[a] X	S	Si	Cl	Humus	Lime %	C_{org} %	N_t ‰	C/N	pH	CEC mmol$_c$ kg^{-1}	b.s. %
Fo	0–20	5Y6/2.5	0	5	95	0		52.8	0.8	2.2	4.0	7.8	3.43	100
nH	–80	7YR3/4	0				H3	28.1	31	22	14.0	7.2	3.06	100

[a] Particles: *X* gravel + stones, *S* sand, *Si* silt, *Cl* clay.

trolled by the blade diameters of *Phragmites*. While the maximum gall density occurred on thin blades, hatching success was highest on thick blades, which may be due to higher mortality caused by predators, parasitoids and winter impact on thin blades. *Archanara* species also showed a relationship between density of larvae and thickness and density of blades. The maximum density of 18 individuals m^{-2} was found on blades with a diameter >6mm and a density >90 blades m^{-2}. Thus, *Archanara* species consumed between 9.0±1.8% (alder carr) and 7.3±1.9% (wet grassland) of the reed blades in the terrestrial zone and between 2.0±1.0% (alder carr) and 2.3±1.3% (wet grassland) in the aquatic zone.

In the littoral zone of Lake Belau Limnic Fluvisols have developed. Table 2.9 characterizes a typical soil of this areas with a layer of limnetic sediments above peat. The site exhibits high pH values, a high potential CEC and high carbonate contents.

Part II
Structure and Function of Ecosystems in a Complex Landscape

Chapter 3
Ecophysiological Key Processes in Agricultural and Forest Ecosystems

Oliver Dilly, Christiane Eschenbach, Werner L. Kutsch,
Ludger Kappen, and Jean Charles Munch

3.1 Introduction

Elemental cycling in ecosystems depends on the structure of the biocoenoses and the ecophysiological key processes regulating the input, allocation and mineralization of compounds, elements and energy (Stocker et al. 1999). The four key processes driving elemental and energy fluxes in the vegetation are: (1) the reduction of atmospheric CO_2 during photosynthesis which provides the energy for growth and maintenance of plant, animal and microbial communities, (2) allocation of assimilates to plant organs and to the environment via exudates, leachates and litter and (3) requirements for maintenance, growth and storage. These three processes are regulated by endogenous and exogenous factors and adjust to environmental conditions to stabilize the vegetative and generative development of the biocoenoses. Plant exudates and residues are deposited onto the soil and: (4) decomposition then completes the element cycling. Decomposition provides energy and nutrients to the heterotrophic macro- and microbiota and, in addition, ensures the nutrient availability of auto- and heterotrophic organisms. The mineralization liberates CO_2 from soil derived from microbial and faunal activity. In addition, respiration of below-ground components of plants considerably contributes to the in situ soil respiratory activity in ecosystems.

The ecophysiology of the vegetation and soil microbiota is largely controlled by external factors like microclimate and the composition of organic matter, such as the lignin/N ratio (Richards 1987). However, the biotic response to external factors suggests that adjustment and interactions of plants and soil microbiota can reduce the role of external constraints. For instance, the strategy of the microbial consortium may promote the decomposition rate rather than temperature when passing thresholds (Schulze 2000).

Anthropogenic ecosystem activities and related moderate and strong human impact with respect to, for instance, N input is particularly considerable as it modifies both primary production and decomposition processes and also the interactions between ecosystem components. Therefore, this Chapter addresses the adjustment of key processes during C assimilation, partitioning and decomposition in agricultural (arable land, grassland) and forest ecosystems in the Bornhöved Lake District.

O. Fränzle et al. (eds.), *Ecosystem Organization of a Complex Landscape.*   61
Ecological Studies 202.
© Springer-Verlag Berlin Heidelberg 2008

The structural and physiological characteristics of plants and soil microbiota are compared in different ecosystems in order to more precisely evaluate the role of biological adjustment to variable environmental conditions (cf. Section 2.3). In particular, the degree of human activity on key processes is evaluated in regard of regulating sources and sinks of CO_2.

Several plant species representing various ecophysiological types were examined for their gas-exchange characteristics and biomass dynamics. *Alnus glutinosa*, *Fagus sylvatica* and *Corylus avellana* were considered as the typical woody plants of the study area, *Lamiastrum galeobdolon* and *Galium aparine* as representative herbaceous species in this forest, *Elymus repens* and *Dactylis glomerata* as dry grassland and *Alopecurus pratensis* and *Holcus lanatus* as wet grassland species. *Zea mays* and *Secale cereale* represented cash crops. For the soil microbiota C availability, microbial biomass, metabolic-responsive biomass and biomass-specific activity levels were estimated in the different ecosystem types, and soil and root respiration and also organic matter decomposition were analysed by in situ techniques.

3.2 Control of Carbon Input

3.2.1 Leaf Gas Exchange

Plant CO_2 uptake is the major pathway of C input into most ecosystems. One bottleneck is the stomatal aperture of the leaves. Therefore, leaf H_2O and CO_2 exchange and ambient microclimatic conditions [irradiance (PPFD), temperature and water vapour pressure deficiency (ΔW)] were measured for selected plant species using both stationary equipment with two independent gas exchange chambers and a steady state porometer. Diurnal leaf CO_2 and H_2O gas exchange of *Alnus glutinosa* and four grass species with their ambient microclimatic conditions are shown in Figs. 3.1 and 3.2. The net photosynthesis of *A. glutinosa* was limited by low leaf conductance during a summer day at maximal PPFD of approximately 800 μmol m^{-2} s^{-1} and 1600 μmol m^{-2} s^{-1} in the outer and inner part of the crown, respectively. The temperature was in the optimal range, increasing from approximately 14 °C in the morning to 26 °C in the afternoon. Leaf temperature differed less than 0.5 K from the ambient air temperature.

In the periphery of the crown, leaf conductance decreased from 300 mmol H_2O m^{-2} s^{-1} in the morning to 150 mmol H_2O m^{-2} s^{-1} at noon due to high differences in leaf–air vapour pressure. Accordingly, assimilation reached values of 17 μmol CO_2 m^{-2} s^{-1} in the morning and 13 μmol CO_2 m^{-2} s^{-1} during the whole day. The assimilation was approximately 4 μmol CO_2 m^{-2} s^{-1} for inner, shaded leaves and increased to 10 μmol CO_2 m^{-2} s^{-1} in the light. The CO_2 fixed daily was 0.79 mol CO_2 m^{-2} for the peripheral leaves and 0.16 mol CO_2 m^{-2} for the inner shaded leaves (data not shown). The transpiration rates of peripheral leaves were 4 mmol H_2O m^{-2} s^{-1} and those of the inner leaves 2.5 mmol H_2O m^{-2} s^{-1} but increasing to 4 mmol H_2O m^{-2} s^{-1}.

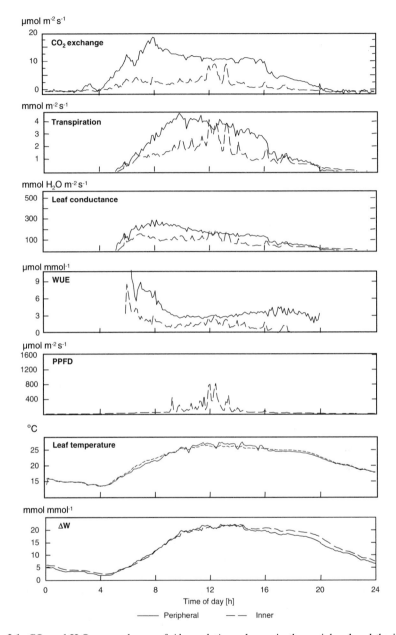

Fig. 3.1 CO_2 and H_2O gas exchange of *Alnus glutinosa* leaves in the peripheral and the inner crown and the ambient microclimatic conditions on June 27, 1992 in W6 (Eschenbach 1995). *WUE* Water use efficiency, *PPFD* photosynthetic photon flux density, *ΔW* leaf/air water vapour difference

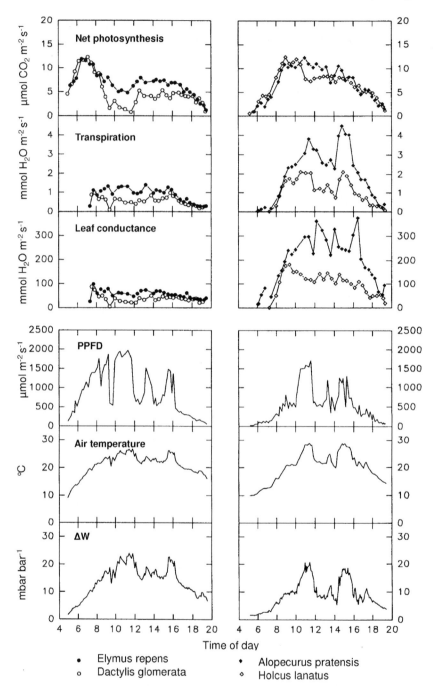

Fig. 3.2 Leaf gas exchange of four grass species in dry (A5) and wet grassland (A8) ecosystems (Weisheit 1995). *Elymus repens* and *Dactylis glomerata* were measured in the dry grassland on 28 May 1991, *Alopecurus pratensis* and *Holcus lanatus* in the wet grassland on 6 June 1991. *PPFD* Photosynthetic photon flux density

Water use efficiency (WUE) was 3.5 μmol CO_2 mmol^{-1} H_2O for the peripheral leaves and 1 μmol CO_2 mmol^{-1} H_2O for the leaves in the inner crown during most of the day. At the periphery, WUE was highest in the morning and lowest at noon and increased again during afternoon. WUE was lower for inner-crown leaves but showing the reverse pattern.

Elymus repens and *Dactylis glomerata*, typical for dry grassland (A 5), and *Alopecurus pratensis* and *Holcus lanatus*, typical for wet grassland (A8), showed similar maximum net photosynthetic rates ranging over 12.5–13.0 μmol CO_2 m^{-2} s^{-1} (Fig. 3.2). Low stomatal conductance (<100 mmol H_2O m^{-2} s^{-1}) reduced net photosynthesis throughout the day in the dry grassland. The reduction was larger for *D. glomerata* than for *E. repens*. The maximum transpiration rates were 1.0 mmol H_2O m^{-2} s^{-1} for *Dactylis* leaves and 1.4 mmol H_2O m^{-2} s^{-1} for *Elymus* leaves. In contrast, stomata of *A. pratensis* and *H. lanatus* of the wet grassland responded less sensitive to ΔW showing transpiration rates of approximately 4.0 mmol H_2O m^{-2} s^{-1} and 2.1 mmol H_2O m^{-2} s^{-1}, respectively. Net photosynthesis of leaves of the two species was not reduced by stomatal closure and was regulated mainly by PPFD. WUE was highest in the morning, reached its minimum of 3 μmol CO_2 mmol^{-1} H_2O at noon and increased again in the afternoon.

The leaf CO_2 and H_2O exchange were controlled by climatic conditions, plant acclimation and phenology. Characteristics of leaf gas exchange, and spatial and temporal variations were determined by fitting the data obtained at different times during the growing season and at different locations within the canopy (von Stamm 1994; Eschenbach 1996a). Dark respiration (R), light compensation points (LCP), initial slope (= quantum yield, *k*), saturating irradiance (*Is*) and maximal net photosynthesis (A_{max}) were calculated from light response curves of net photosynthesis calculated for each time interval. The seasonal photosynthetic capacity of *Alnus glutinosa* and *Fagus sylvatica* (Fig. 3.3) showed that the photosynthetic capacity at maximum PPFD and ambient CO_2 concentration (A_{max}) of these deciduous tree species increased during spring, reached a maximum in summer and decreased with

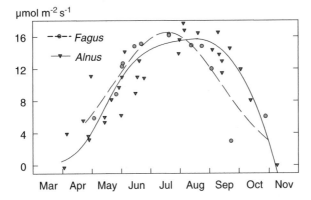

Fig. 3.3 Seasonal variation of the photosynthetic capacity (A_{max}) of beech (W1) and alder (W6)

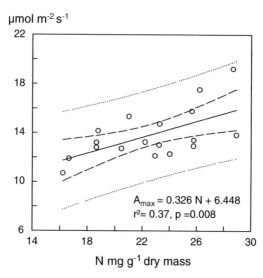

Fig. 3.4 Linear correlation between maximal assimilation rate at maximum PPFD (A_{max}) and leaf N content for 18 leaves of *Fagus sylvatica* (after Kutsch et al. 2001b, modified). *Dashed and dotted lines* represent 95% and 99% confidence limits, respectively

senescence in autumn. Maximum photosynthetic capacity in summer was similar for the two species with 16 µmol CO_2 m^{-2} s^{-1}. However, *A. glutinosa* sustained high assimilatory rates until leaf fall, whereas the A_{max} value of *F. sylvatica* stayed constant only from July until the beginning of August and decreased afterwards. This difference between the species is attributed to the poor N re-translocation from *A. glutinosa* leaves indicated by the small increase of the C/N ratio from 19 in fresh leaves to 27 in litter. In contrast, the litter C/N ratio of *F. sylvatica* increased to 59 in autumn (Kutsch et al. 1998). No significant difference in A_{max} between peripheral and inner leaves was observed for alder (Eschenbach 1995), whereas beech had a strong light acclimation (Schulze 1970; Kutsch et al. 2001b).

Figure 3.4 shows that the A_{max} values in beech corresponded to leaf N content. The regression slope from the Bornhöved data was lower and the net photosynthesis level higher in comparison with leaves investigated by Peterson et al. (1999). Our high A_{max} values may be due to high N deposition. The initial slope of the light response curve of net photosynthesis (k) which is proportional to the maximum quantum yield of the photosynthesis, changed during the growing season (data not shown), but varied spatially which is attributed to acclimation response. Sun leaves of *F. sylvatica* had a lower maximum of quantum yield ($k = 0.038$) than shade leaves ($k = 0.055$). Leaves of *A. glutinosa* approached 0.056 for both peripheral and inner leaves, equivalent to 18 mol photons mol^{-1} fixed CO_2. Stickan and Zhang (1992) calculated $k = 0.07$ for shade leaves of beech in the Solling Mts. Thus, alder leaves showed a high C use efficiency at high irradiance, being equivalent to that of shaded beech leaves.

The stomatal conductance showed the close relationship between net photosynthesis and transpiration which was controlled by microclimatic conditions. Stomatal conductance followed a light saturation function, a hyperbolic function for ΔW and optimal temperature (von Stamm 1994; Herbst 1995; Eschenbach 1996a; Vanselow 1997; Eschenbach and Kappen 1999; Kutsch et al. 2001b). Stomatal conductance of alder, beech and hazel was light-saturated at approximately 300 µmol photons m^{-2} s^{-1}.

The phenological development of the stomatal function was accounted for by empirical coefficients of the hyperbolic function used in the model by von Stamm (1994) and Eschenbach (1996a) that describe the minimum conductance at high ΔW and the sensitivity of the stomata to air humidity. Minimum stomatal conductance was nearly constant throughout the year with approximately 70 mmol H$_2$O m^{-2} s^{-1} for beech and 80 mmol H$_2$O m^{-2} s^{-1} for hazel. It was higher for alder (125 mmol H$_2$O m^{-2} s^{-1}) and increased during the growing season indicating the adaptation of *Alnus glutinosa* to wet conditions (Eschenbach and Kappen 1999). The stomatal sensitivity increased during spring, reached a summer plateau and declined during senescence in autumn. Some phenological deviations were observed for both hazel and beech. In 1989 and 1993, which were years with high temperature and irradiance during May, the values were lower in June and July and exceeded the usual range after the cold and cloudy spring 1992. In 1990, the seasonal pattern of hazel was delayed. The seasonal dynamics represented the acclimation to the climatic conditions of the respective years (Kutsch et al. 2001b).

For alder, beech and also the grassland species, the dependence of net photosynthesis on stomatal conductance could be modelled by a saturation-type curve with maximum rates of net photosynthesis at values higher than 200 mmol H$_2$O m^{-2} s^{-1}. This correlation was constant throughout the growing season and significant variations between the species could not be detected.

Characteristics of leaf gas exchange for the plant species in the Bornhöved Lake District are given in Tables 3.1 and 3.2. Leaves of *A. glutinosa* were not able to acclimate to the light conditions within the canopy. All properties of the inner and the peripheral leaves except dark respiration were uniform. In contrast, *F. sylvatica* and also *Lamiastrum galeobdolon* and *Galium aparine* exhibited a high plasticity with reference to the light factor. In contrast, the grassland and the crop species developed leaves adjusted to the sunlight and, therefore, photosynthetic capacities of crop and grassland species were higher than those of the forest plants.

Tables 3.1 and 3.2 indicate that photosynthetic plant characteristics adapted to environmental conditions on the individual level, while the community level reflected the properties of species and life forms. The photosynthetic capacity varied between 1.3 µmol CO$_2$ m^{-2} s^{-1} and 54.0 µmol CO$_2$ m^{-2} s^{-1}, being lowest for the shade-adapted *G. aparine* plants and highest for *Zea mays*. The light compensation point was lowest for shaded *Corylus avellana* leaves and highest for *Alopecurus pratensis* indicating the adjustment to the light resource. Correspondingly, the dark respiration was lowest for shaded *C. avellana* leaves and highest for the grass species. Sites without light limitation favour plants with high A_{max} values since plants have a low capability to adjust to low light intensity. *A. glutinosa* uses its high photosynthetic capacity in wet biotopes with low competition, while crop species such as

Table 3.1 Gas exchange characteristics of forest plants from W1 and W6 in the Bornhöved Lake District. *n.d.* No data

| | Fagus sylvatica | | Corylus avellana | | | |
Alnus glutinosa	Sun leaf	Shade leaf	Sun leaf	Shade leaf	Galium aparine	Lamiastrum galeobdolon
Photosynthetic capacity (μmol CO_2 m^{-2} s^{-1})						
15.5	16.0	6.2t	13.0	4.1	1.3–11.4	3.4–12.0
Maximal optimum temperature (°C)						
23.2	n.d.	n.d.	23.2	19.9	30.0	22.0
Light compensation point (μmol photons m^{-2} s^{-1})						
16	11	11	10–15	3	5–20	10–26
Light saturation (μmol photons m^{-2} s^{-1})						
420	590	200	460	150	100–300	250
Quantum yield (mol CO_2 mol^{-1} photons)						
0.056	0.043	0.055	0.038	0.06	0.038	n.d.
Dark respiration (μmol CO_2 m^{-2} s^{-1})						
Periphery: 0.98						
Centre: 0.35	0.46	n.d.	0.46	0.24	0.25–1.26	1.2–3.1
Period of physiological optimum						
1–15 Aug	Jul	Aug	Aug	Aug	May	May

Table 3.2 Gas exchange characteristics of plant species from dry (A5) and wet (A8) grassland and arable land in the Bornhöved Lake District

Elymus repens	Dactylis glomerata	Alopecurus pratensis	Holcus lanatus	Secale cereale	Zea mays
Photosynthetic capacity (μmol CO_2 m^{-2} s^{-1})					
20.8	19.8	21.5	15.5	28.0	54.0
Maximal optimum temperature (°C)					
20	20	20	20	n.d.	27
Light saturation (μmol photons m^{-2} s^{-1})					
17	14	23	16	21	n.d.
Quantum yield (mol CO_2 mol^{-1} photons)					
1200	1100	1150	1000	1500	2000
Dark respiration (μmol CO_2 m^{-2} s^{-1})					
0.033	0.041	0.036	0.049	0.031	n.d.
Dark respiration (μmol CO_2 m^{-2} s^{-1})					
0.85	0.6	0.55	0.4	0.655	n.d.
Period of physiological optimum					
Apr–Aug	Apr–Aug	Apr–Aug	Apr–Aug	May	July

Z. mays and *Secale cereale* benefit from agricultural management practices. In contrast, beech was capable to acclimate to low light and nutrient availability since the canopy protects the floor from sunlight, and nutrients are bound in the biota and the organic matter (Kutsch and Dilly 1999; cf. Chapter 4).

3.2.2 Upscaling to Canopy Level

For upscaling of C fixation and water loss from leaf gas exchange to the stand level, a bottom-up modelling approach was developed by and Kutsch et al. (2001c), von Stamm (1994), Eschenbach (1996a), Eschenbach et al. (1997), Kutsch and Kappen (1997) and Vanselow (1997). Similar approaches had been applied to other species (e.g. Tenhunen et al. 1976, 1987; Caldwell et al. 1986; Meister et al. 1987; Janecek et al. 1989; Webb 1991; Stickan et al. 1994). The following three hierarchically linked units were considered: (a) a leaf model referring to the short-term response of net photosynthesis and stomatal conductance to microclimate, (b) a main model varying the coefficients of the leaf model during the growing season and (c) a canopy model. The leaf model consists of two sub-units unifying the stomatal conductance related to irradiance and ΔW and the CO_2 exchange depending on irradiance and temperature. The two subunits are connected by the interaction between net photosynthesis and stomatal conductance. The main model integrates the variations in stomatal responses, leaf H_2O and CO_2 exchanges during the growing season as time-dependent changes of empirical coefficients. The canopy model links seasonal and spatial variation in photosynthesis and transpiration rates with the seasonally varying leaf area index (LAI). The shifting proportion of sun-lit and shaded foliage, obtained from the interrelation of the cumulative LAI and the light extinction in the canopy was also taken into account. Annual simulations of CO_2 exchange and water loss were modelled using microclimatic data measured at 10-min intervals at several levels in the canopies. For years without such data, the simulations were based on pertinent data of the main station in the crop field. Correlations of microclimatic data calculated between the arable land and the other sites were used for interpolation purposes (Table 3.3). Between 1991 and 1998, the beech canopy (W1) fixed 1352 g C m^{-2} year^{-1} and the leaves respired 247 g C m^{-2} year^{-1} during the night. The canopy in the alder forest fixed 2450 g C m^{-2} year^{-1} in 1992 and 1993 and losses by leaf dark respiration were 300 g C m^{-2} year^{-1}. This high assimilation was due to the specific location because the alder carr grows along the shore of Lake Belau and therefore receives a large amount of light. The maize canopy (A2) assimilated 1807 g C m^{-2} year^{-1} between 1991 and 1995 and leaf respiration induced losses of 315 g C m^{-2} year^{-1}. For the wet grassland (A9) an average gross primary production of 1500 g C m^{-2} year^{-1} was estimated for 1992 and 1993.

3.3 Partitioning of Carbon

Partitioning of accumulated C towards plant organs and other compartments was estimated by growth and analysing tree-rings (Werner 1994) and is given for alder in 1992 in Fig. 3.5. The trunk and branch biomass was 387 g C m^{-2} year^{-1} according to the development of trunk-rings. The plants invested 104 g C m^{-2} year^{-1} into branches, 174 g C m^{-2} year^{-1} into leaves, 151 g C m^{-2} year^{-1} and 67 g C m^{-2} year^{-1} into cones and catkins and 25 g C m^{-2} year^{-1} into buds. Total aboveground production

Table 3.3 Annual C flux rates (g C m^{-2}) in the ecosystems W6, W1 and A2 calculated from CO$_2$ exchange measurements. Fluxes into the system have a negative sign, respiration is labelled positive

	Alder forest	Beech forest	Maize field
	1992–1993	1991–1998	1990–1994
Gross primary production (a)	−2450	−1352	−1807
Plant respiration (incl. exudates) (b)	1607	640	675
Net primary production (a) − (b)	−843	−712	−1132
Yield (c)	0	0	891
Manure (d)	0	0	289
Net plant biomass (e)	−292	−233	0
Total soil respiration (f)	1688	634	718
Rhizomicrobial respiration (g)	1181	243	266
Soil C$_{org}$ mineralization (f) − (g)	487	391	452

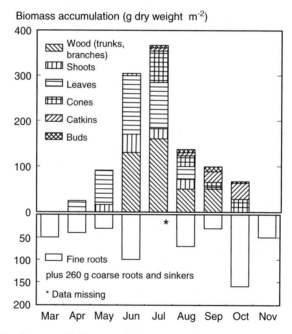

Fig. 3.5 Monthly above- and belowground biomass increment of different organs and parts of *Alnus glutinosa* in 1992 in W6 (after Eschenbach et al. 1997)

was 543 g C m^{-2} year^{-1}, and total belowground biomass of fine roots, coarse root and sinkers was 307 g C m^{-2} year^{-1} (Middelhoff 2000). The total net primary production was approximately 850 g C m^{-2} year^{-1} for the vegetation period. These biometric values fitted to the modelled net CO$_2$ exchange.

The portion of the gross primary production transferred to heterotrophic organisms and soil were ecosystem-specific. The high gross primary production of alder was attributed to rhizo-microbial respiration (Dilly et al. 2001; Kutsch et al. 2001a) and a high C allocation into the rhizo-microbial system in order to overcome nutrient limitations and to support N_2 fixation by the *Frankia* symbiosis.

Maize transferred the highest fraction of gross assimilated C into a net primary production of 1132 g C m^{-2} year^{-1}, out of which 891 g C m^{-2} year^{-1} were harvested. Thus, only 241 g C m^{-2} year^{-1} remained at the site as a substrate for heterotrophic soil respiration.

3.4   Respiration as Carbon Loss from the System

The respiration of the heterotrophic plant organs uses a high proportion of the photosynthetic yield linking growth and nutrient uptake. The respiration of stems and branches were estimated only for *Alnus glutinosa*.

Temperature was the major factor controlling respiration. Branches with a diameter below 2.5 cm had a significantly higher respiration rate per volume than bigger branches and the trunk (Steinborn et al. 1997). Based on these measurements, the respiration of stems and branches was calculated to be 127 g C m^{-2} year^{-1} which was similar to the value of 160 g m^{-2} year^{-1} calculated by Kutsch et al. (2001c) for beech, using data from Gansert (1994) and Möller et al. (1954).

Rhizo-microbial respiration refers to the activity of roots, mycorrhizae and other microorganisms in the rhizosphere and was controlled by temperature and root structure as studied in the alder carr (W6: Kutsch et al. 2001a). Between August and November 1995, the specific respiration rate was related to the ratio between main-to-lateral root biomass represented by the weight classes >10 and <7 (g g^{-1}). Considerable respiration rates occurred already at temperatures below 0 °C and increased exponentially with temperature (Fig. 3.6). The rhizo-microbial respiration rate was approximately 20 nmol CO_2 g^{-1} root s^{-1} for the class '>10' and 10 nmol CO_2 g^{-1} root s^{-1} for the class '<7' at 15 °C. The modelled rhizo-microbial respiration revealed high soil respiration rates during spring and summer but main growth of lateral roots during spring. Root biomass decreased significantly in late autumn (Middelhoff 2000). The total rhizo-microbial respiration of 1181 g C m^{-2} for 1992 and 1993 (see Table 3.3) accounted for approximately 50% of the canopy photosynthesis (Dilly et al. 2000; Kutsch et al. 2001a).

The mean rhizo-microbial respiration of *F. sylvatica* in the 1991–1998 period was 243 g C m^{-2} year^{-1} and thus approximately 17% of the canopy photosynthesis. The value for *Zea mays* was 266 g m^{-2} year^{-1} for 1990–1994 period, which corresponds to 15% of the canopy photosynthesis (Table 3.3).

Soil respiration rates could be explained by a temperature effect except during spring when root growth and activity were interfering as was evident for the alder carr (Fig. 3.7). A fit of the values for the alder forest (W6) to an exponential curve would result in a Q_{10} value of 14.6, which is unlikely since root and microbial activity

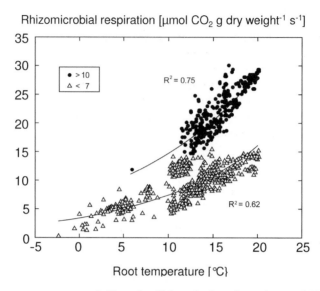

Fig. 3.6 Temperature response of rhizo-microbial respiration of two classes of *Alnus glutinosa* roots after Kutsch et al. (2001a). The fine-to-long-root weight ratios (g g^{-1}) are >10 and <7, respectively

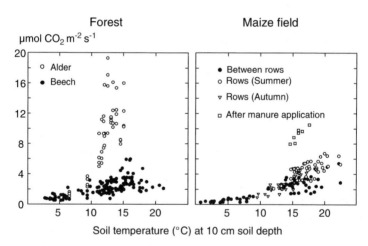

Fig. 3.7 Soil respiration related to soil temperature in the beech (W1) and the alder (W6) stands and in a maize monoculture (A2). Values above 10 °C in the alder soil are those of spring and early summer

usually show Q_{10} values between 1.5 and 3.0 (Kirschbaum 1995). Thus, the respiration rates seem stimulated by factors such as root growth and high C allocation to fine roots (Middelhoff 2000). In addition, the decreasing groundwater table in spring may have favoured the microbial activity in deeper soil horizons. It was evident that physiological factors modified the temperature control (Högberg et al. 2001). The correlation with temperature was higher for beech forest (W1) and the

Q_{10} value was 2.7. Soil respiration did not exceed $6\,\mu$mol CO_2 m^{-2} s^{-1} and was limited by low water availability at temperatures higher than 17 °C. This happened particularly in the L and Of horizons.

The maize field showed high values in spring 1991 when farmyard manure was added to the regular slurry application. In spring, a gradient of root density was observed based on the soil respiration rates under the plants and the area between the rows in the maize field (A2). This spatial pattern ceased in fall due to equal distribution of the roots in the soil (Tardieu 1988; Kutsch 1996).

3.5 Decomposition and Mineralization of Organic C

Enquiries into litter decomposition and soil organic matter mineralization were carried out in the alder carr (W6), the beech forest (W1) and the crop rotation field (A3), as shown in Fig. 3.8. In nylon bags of 5 mm mesh size, litter decomposition proceeded rapidly in the alder carr and the agricultural soils (Dilly and Munch 1996; Irmler 1996; Dilly et al. 1999). In the alder carr, particularly close to the lake, more than 50% of leaf litter decayed during the first 12 months (Dilly and Munch 1996).

The litter quality was defined in terms of the carbon availability index (CAI, after Cheng et al. 1996), an approach that combines two microbiological assays, namely the microbial basal respiration and respiration in the presence of readily available glucose. The CAI was highest at the beginning and lowest at the end of a period of 305 days, in correspondence with decreasing litter quality.

The ratio between basal respiration rate at 22 °C and organic carbon content, calculated to estimate the current C mineralization capacity with reference to the totally available organic C, varied between $10\,\mu$g CO_2-C g^{-1} C_{org} h^{-1} and $250\,\mu$g CO_2-C g^{-1} C_{org} h^{-1} (Fig. 3.9). Although the basal respiration and organic C content were correlated when considering a broad set of soils ($r^2 = 0.83$, $P < 0.001$; Dilly 1994), this ratio varied significantly between soils with respect to the recalcitrance of the organic matter and the C use efficiency of the microbial communities. The ratio indicates that soils with large organic C stocks and containing a large amount of fresh litter in the L and Of horizons in the beech forest have a high current mineralization capacity. High CO_2/C_{org} ratios and, thus, accelerated degradation of the organic matter were also found at the agricultural sites as compared with the respective forest topsoils. This result may be important for modelling C cycling (see Chapter 4) in soil and for proving the sensitivity to variations of the decomposition factor 'k' in soil C models (Dilly 2001; Kutsch et al. 2001c).

3.6 Structure and Activity of Soil Microbiota

The intensity of microbial colonization, as indicated by the ratio of microbial biomass content to C_{org} content, varied from 6 mg C_{mic} g^{-1} C_{org} to 26 mg C_{mic} g^{-1} C_{org} in the Bornhöved topsoils (Fig. 3.10). The C_{mic}/C_{org} ratio decreased in the order dry

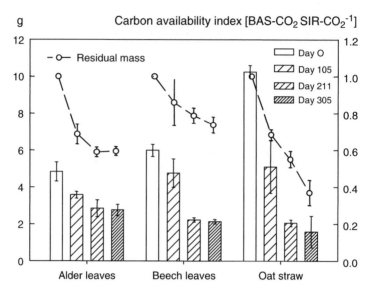

Fig. 3.8 Dry mass (*g; lines*) and carbon availability index, CAI (*bars*), calculated after Cheng et al. (1996) for leaf litter of alder (W6), beech (W1) and oat straw (A3) during decomposition in litter bags. Error bars exceed the 95% confidence limits

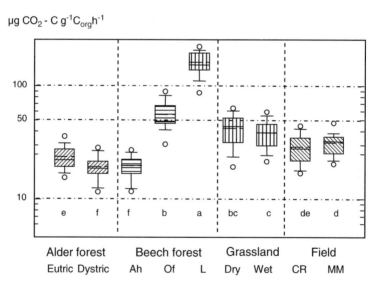

Fig. 3.9 Basal respiration rate related to organic C content in topsoil horizons of the ecosystems W6, W1, A5, A8, A3 and A2. Different letters indicate significant differences if all pairwise multiple comparison procedures (Dunn's method) at $P < 0.05$ are applied. *CR* Crop rotation (A3). *MM* Maize monoculture (A2)

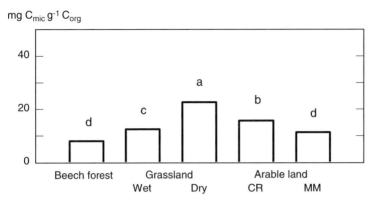

Fig. 3.10 Mean microbial biomass versus organic C content (C_{mic}/C_{org}) in the A horizon of beech forest (W1), grassland (A5, A8) and arable land (A2, A3) soils (different letters indicate significant differences (Student–Newman–Keuls method, $P < 0.05$). *CR* Crop rotation, *MM* maize monoculture (after Dilly et al. 2001)

grassland (A5) > crop rotation field (A3)> maize monoculture (A2) > wet grassland (A8) > beech forest (W1). The microbial fraction of the organic matter was consistently greater under pasture than in the equivalent soil layer under forest or arable land, which is in agreement with studies of Sparling (1992). In addition, a gradient of microbial colonization was found along the soil profile in the beech forest with maximal values in the litter layer, which is typical for fresh organic material at early decomposition stages (Dilly et al. 1997b). Generally, the microbial colonization corresponds to the current mineralization rates (Fig. 3.10).

Shifts in community structure of the microbial biomass between the soils are indicated by ratios between substrate-induced respiration (SIR) and fumigation-extraction (FE) in the A-horizon (Fig. 3.11). High values are typical of the earlier stages of decomposition and for agricultural in contrast to forest topsoils indicating a high metabolically responsive biomass in these soil biotopes. However, when land use was changed from monoculture to crop rotation, the portion of metabolically responsive microbial biomass was decreasing during the subsequent years and approached that in the adjacent soil under crop rotation.

The active portion of the microbiota and the contribution of fungi can be indicated by the ratio between bacterial colony-forming units (cfu) and direct (DAPI-stained) bacterial counts, and the ratio of fungal to bacterial cfu (Dilly et al. 1997b). The two ratios declined with decreasing soil depth and ongoing decomposition in the beech and alder forests (Fig. 3.12). The values decreased with soil depth by some orders of magnitude. The only exception was observed for the ratio of fungal-to-bacterial cfu in the nH horizon of the alder forest, most likely due to water saturation in this Histosol (see Chapter 2) and, thus, conditions favouring bacterial dominance.

The qCO_2 (metabolic quotient according to Anderson and Domsch 1990) values were high in the sandy soil under maize monoculture (Fig. 3.13). These high values

g SIR-C$_{mic}$ g^{-1} FE-C$_{mic}$

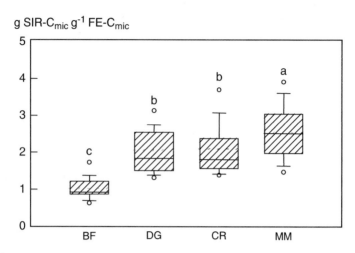

Fig. 3.11 Metabolically responsive biomass reflected in the ratio of SIR-C$_{mic}$ and FE-C$_{mic}$ in A horizons of different ecosystems from Jan 1992 to Oct 1993 ($n = 22$). *BF* beech forest (W1), *DG* dry grassland (A5), crop rotation field (A3), *MM* maize monoculture (A2). *Boxes* encompass 25% and 75% quartiles, the central and the *broken line* represent the median and the mean, and *bars* extend to the 95% confidence limits. *Open circles* indicate observations extending beyond the 95% confidence limit (after Dilly and Munch 1998)

indicate that current C mineralization of the soil microbiota liberated more C than expected on the base of the biomass content. The qCO$_2$ decreased during the decomposition of fresh organic matter indicating progressive efficiency. Thus, CO$_2$ loss from soil is high when the metabolically responsive biomass and the amount of readily biodegradable organic matter are high. Concurrently, the C availability decreases continuously in the course of decomposition as shown in Fig. 3.8 for beech and alder litter and oat straw when applying the 'C availability index – CAI' (Cheng et al. 1996). High ratios of basal-to-substrate-induced respiration indicated a high C availability, and this ratio was lowest in beech leaf litter.

Supplemental N favoured microbial growth in the arable soil under crop rotation more than in the beech forest soil and caused enzyme induction in the two soils in various ways (Fig. 3.14). The N-degrading urease activity was enhanced in the beech forest if available C such as glucose was present. In contrast, N addition was necessary to stimulate urease activity in the agricultural soil (Dilly and Nannipieri 2001).

Other enzymes were mainly affected in the beech forest. Nitrogen addition stimulated the glucosidase and phosphatase activities, and extra P addition repressed phosphatase activity and stimulated sulphatase activity.Thus, the soil microbial communities adjusted their ecophysiology to environmental conditions in the respective biotopes with particular reference to C supply. Such an adjustment seems to play a considerable role in N cycling, e.g. the in situ N$_2$O emission. The high C turnover may accelerate processes involved in N cycling such as N losses. High rates of nitrous oxides were determined on the crop rotation field and in the alder

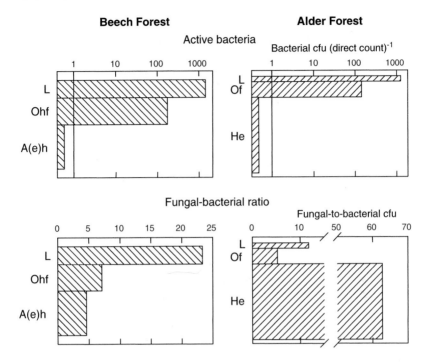

Fig. 3.12 Distribution of the microbial biomass along soil profiles of the beech (W2) and (dry) alder (W6) stands, defined as proportions of colony-forming units, direct counts and biomass estimated by fumigation-extraction (after Dilly et al. 1997b). Thickness of *bars* indicates the thickness of the respective horizons, being 10cm and 25cm in the beech forest and alder carr, respectively

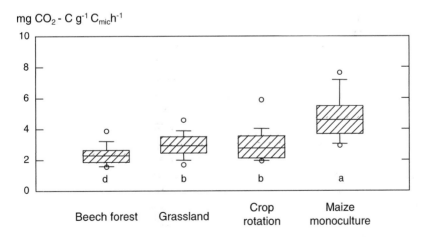

Fig. 3.13 qCO_2 values in the A horizons of ecosystems W1, A5, A3 and A2, measured from January 1992 to October 1993 ($n = 22$). *Letters* and *abbreviations* as in Fig. 3.11. *Boxes* encompass 25% and 75% quartiles, the *central line* represents the median, *bars* extend to the 95% confidence limits. *Open circles* indicate observations extending beyond the 95% confidence limits (after Dilly and Munch 1998)

Arable
soil
+C $\longrightarrow$ β-glucosidase ↓, sulfatase ↓
+C+N $\longrightarrow$ Urease ↑, sulfatase ↓
+C+N+P $\rightarrow$ Urease ↑, sulfatase ↓

Forest
soil
+C $\longrightarrow$ Urease ↑
+C+N $\longrightarrow$ β-glucosidase ↑, phosphatase ↑
+C+N+P $\rightarrow$ β-glucosidase ↑, sulfatase ↑

Fig. 3.14 Response of enzyme activities to substrate addition (glucose-C, glucose-C plus nitrate-N, glucose-C plus nitrate-N plus phosphate-P) in a field (A3) and a forest soil (W1; after Dilly and Nannipieri 2001). ↑ and ↓ indicate stimulation and repression, respectively

Table 3.4 In situ N_2O emissions (kg N ha^{-1}) from soils under tillage (A3, A2), grassland (A5, A8), and forest (W1, W6) use of the Bornhöved Lake District (after Mogge 1995; Dilly et al. 1997a)

Sampling period	Crop rotation	Maize monoculture	Dry grassland	Wet grassland	Beech forest	Alder forest
Jul/Dec 1992	0.39	0.21	n.d.	0.17	0.27	1.01
Jan/Jun 1993	10.04	3.18	2.29	n.d.	0.53	3.06
Jul/Dec 1993	2.80	2.12	3.19	n.d.	1.24	5.79

carr (Table 3.4). Lateral nitrate transfer, N input via N_2 fixation (Dittert 1992) and also the readily biodegradable litter may have favoured both in situ N_2O emissions and denitrification.

3.7 Discussion

Plants from agricultural sites such as *Secale cereale, Holcus lanatus, Alopecurus pratensis, Dactylis glomerata* and *Elymus repens* have a higher photosynthetic capacity than those found in the forest at the leaf level. They reached maximum photosynthetic capacities at values beyond 1 mmol photons m^{-2} s^{-1}. The highest photosynthetic capacity and, thus, the highest C sink efficiency was determined for the C_4 grass *Zea mays*. This holds for both leaf-area and soil-area based measurements since the maximum leaf area index was 4.5–4.8 for the trees (Herbst et al. 1999) and 7.2 for the Poaceae studied (Weisheit 1995).

In contrast, species from the forest floor such as *Galium aparine, Lamiastrum galeobdolon*, and *Corylus avellana, Alnus glutinosa* and *Fagus sylvatica* reached maximum photosynthetic capacities at lower photon flux densities ranging over 100–590 μmol m^{-2} s^{-1}. The higher quantum yields indicate higher C use efficiencies at low light levels.

Alder had the highest photosynthetic capacity and was more efficient under both light and shady conditions. The maximum net photosynthesis rate of *A. glutinosa* exceeded most values reported for deciduous tree species in temperate regions (Bazzaz and Carlson 1982; Heichel and Turner 1983; Koike 1987; Taylor and Davies 1988; Lyr 1992). Similar net photosynthetic rates were found for alder seedlings by Côté et al. (1988). Only *Populus tremuloides*, *P. fremontii* and *P. tristis* had higher assimilation rates, ranging over 20–30 μmol CO_2 m^{-2} s^{-1} (Roden and Pearcy 1993; Larcher 1994). The 1998 beech values exceeded those reported by Schulze (1970) and Schulte (1993) from the Solling Mts and this high capacity appears associated with high atmospheric nitrogen deposition rates (cf. Chapter 8).

High photosynthetic capacity (A_{max}) and low plasticity according to the light factor were determined for *Zea mays*, *Secale cereale* and *Alnus glutinosa* indicating both ruderal (Grime 1979) and pioneering strategies (Thomasius 1992). In contrast, *Fagus sylvatica*, *Corylus avellana* and *Lamiastrum galeobdolon* were able to acclimatize to low light conditions and low nutrient availability, which is typical of later successional stages (competitor or climax species). Thus, agricultural management via species selection induced higher C fixation potentials with high photosynthetic capacities. This means that human impact favours the dominance of early successional plants with correspondingly high C sink characteristics (cf. Chapters 4 and 8).

High photosynthetic capacities do not necessarily lead to high plant growth since C allocation and respiration patterns differ between the species. Plant C partitioning, soil C storage, and decomposition significantly varied between ecosystems and the degree of human impact in a landscape that hosted high biotic diversities. Ecophysiologically, alder that can be characterized as a pioneering tree (Eschenbach 1996b) which assimilates high amounts of CO_2 via high leaf area and higher photosynthetic fixation rates in comparison with beech. Similarly *Zea mays*, which is considered an effective agricultural crop, was not competitive because of its lower ground-area based annual GPP (see Chapter 4). In maize, however, the yield in generative organs is of major relevance. Beech and alder produced approximately 72 g and 33 g fruits m^{-2} year^{-1} (Lenfers 1994), whereas cereals and maize produced between 350 g and 900 g fruits m^{-2} year^{-1}.

Soil respiration resulting from both root and microbial activity exhibited a high proportion of the assimilated C transferred directly towards belowground, particularly in the alder forest at the eutric-dry site. The C amounts transferred above- and belowground varied site-specifically (Högberg et al. 2001; Janssens et al. 2001). The C partitioning towards belowground seemed related to the acquisition of limiting nutrients such as phosphorus. In contrast to alder, maize transferred low C amounts to soil. However, high mineralization of the scarce organic matter occurred at this site.

The decomposition proceeded rapidly close to the lake where pH value, water content and abundance of modulating fauna are high (Irmler 1995). The high root density (Dilly et al. 1999) and, thus, great amounts of root exudates may have additionally stimulated the microbial oxidation of the indigenous organic matter by 'priming effects' (Kuzyakov et al. 2000). The alder carr soil had frequently high water saturation and contained high densities of modulating macrofauna. Lateral input of nutrients and lake water buffering may have favoured the decomposition process. The comparison with the hillside part of the stand showed that a high water table did not necessarily restrict the decomposition rate and also O_2 supply may not be limited. In addition, low variation in microbial biomass content, low metabolic efficiency and high enzymatic potentials may have contributed to the rapid decomposition (Dilly and Munch 1996). The rates in the alder carr were similar to those of the agricultural site. In contrast, low decomposition rates were found in the beech forest due to lower carbon and nitrogen availability. Temperature as a general regulator (e.g. Kirschbaum 1995; Kutsch and Kappen 1997) seemed to play a minor role in controlling decomposition when the beech forest and the alder carr are compared.

The variations in basal respiration rates may be due to fluctuations in both microbial biomass and activity. In accordance, the biomass-specific respiration rates demonstrated the variable relationship between biomass and activity and lowest C use efficiency in fresh litter and in the field with maize monoculture. Both fresh litter and agricultural soil organic matter with simultaneous fertilization represent a labile C source. Concurrently, a metabolic-responsive microbial biomass favouring C depletion occurs in both fresh litter and agricultural soil (Mamilov and Dilly 2002). A depletion of soil organic C has to be expected in the long run particularly in soils with low C stocks such as the maize field and the respective management and fertilization system. Thus, plant selection modifies rhizo-deposition and partitioning, and also the structure of soil microbiota, and finally the C sequestration in soil.

3.8   Conclusions

The study area represents an array of ecosystems with differing C fixation strategies. *Zea mays* and *Alnus glutinosa*, representing agricultural crops and pioneering trees, had a high photosynthetic capacity. Plant C partitioning, soil C storage and decomposition varied substantially in ecosystems. In comparison with *Fagus sylvatica*, *A. glutinosa* assimilated high CO_2 amounts due to large leaf area (Eschenbach and Kappen 1996) and higher photosynthetic fixation rates; *Z. mays* was intermediate. The rapid decomposition of organic matter and leaf litter was typical of agricultural sites and the alder carr, particularly close to the lake, and provided an unstable C component. In the alder carr, high soil pH, favourable nutrient conditions and higher biotic activity certainly stimulated decomposition. In addition, anthropogenic lowering of the lake water level in the 1930s (see Chapter 2) and nutrient import from agricultural systems via the lake water may have promoted decomposition rate at the wet alder site. Rapidly mineralizing soil microorganisms, high allocation of readily available C, and the stimulation of exudates towards belowground obviously accelerated decomposition and enhanced the respiratory activity. Current decomposition rates varied between the biotopes in relation to the quality and ecophysiological stage of the soil microbial communities. High rates occurred particularly in fresh organic matter. The metabolically active biomass appears concurrently adjusted, being highly active in fresh litter and also in the maize field.

Agricultural practice modifies key processes of ecosystems via plant selection, soil management and fertilization and thus C assimilation (see Chapter 4). It may foster partitioning rates in favour of harvestable components. Soil management and fertilization stimulated metabolic-responsive biomass and microbial activities which regulated source-sink transitions in soil C pools. Future interdisciplinary research in other regions should provide information if ecophysiological adjustment will follow similar lines and exhibit patterns comparable to those found in the Bornhöved Lake District.

Chapter 4
Carbon and Energy Balances of Different Ecosystems and Ecosystem Complexes of the Bornhöved Lake District

Werner L. Kutsch, Georg Hörmann, and Ludger Kappen

4.1 Introduction

Energy and carbon cycles of ecosystems and landscapes are closely linked. The main source of energy in ecosystems is solar global radiation. Most of this energy is converted into heat and dissipated by long-wave radiation and convection or is used for the evapotranspiration of water. Only a small fraction of global radiation is used in the process of photosynthesis and changed into chemical energy stored in carbohydrates. Carbon assimilation initiates the biological energy cycle in which most of the carbohydrates produced by photosynthesis is lost by respiration in heterotrophic processes.

In agricultural landscapes parts of the biological energy budget (or carbon budget, respectively) are used for human necessities such as food or fibres. Agricultural landscapes, therefore, should primarily provide society with energy from farm products. However, in addition to local solar radiation today's agricultural management needs large amounts of energy, for instance for the production of fertilizers or pesticides and finally for field management and transport purposes. In addition, the population of rural areas consumes energy for households and mobility. Therefore the question is whether agricultural landscapes are net producers or net consumers of energy or, in terms of the carbon cycle, are sinks or sources for carbon dioxide and other greenhouse gases. This problem is a central issue for the sustainability of rural areas, and this Chapter explores the abiotic, biotic and anthropogenic carbon and energy budgets of single ecosystems, ecosystem complexes and farms of the Bornhöved Lake District. Finally, the energy balance of the whole region is deduced.

4.2 Carbon and Energy Fluxes and Balances at the Ecosystem Scale

4.2.1 Abiotic Ecosystem Energy Balances

Energy in ecosystems is transported by radiation, convection and conduction. The latter process is mainly relevant to soils. The main flux is transport of latent heat, which is

O. Fränzle et al. (eds.), *Ecosystem Organization of a Complex Landscape.*
Ecological Studies 202.
© Springer-Verlag Berlin Heidelberg 2008

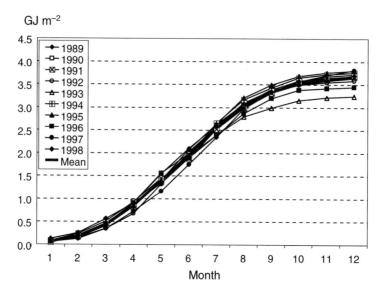

Fig. 4.1 Cumulative annual global radiation (GJ m^{-2} year^{-1}) of the study area (A3)

extracted from the environment during evaporation and released during condensation. Figure 4.1 shows the cumulative global radiation in the core area, which exhibits a low interannual variability in comparison with other climatic parameters.

The year 1993, with low values, is exceptional and can be explained by the considerably reduced irradiation during the period from July to September. Incoming short-wave solar radiation is split up into reflected short-wave radiation, long-wave radiation and the fluxes of latent heat, sensible heat, heat conduction in soil and metabolic heat, being mainly the energy utilized by the plant canopy for photosynthesis. The fluxes of soil heat and metabolic heat are small and therefore frequently neglected in the abiotic energy balance of ecosystems. Thus, the temperature of an ecosystem is taken as a function of latent versus sensible heat fluxes.

Complete abiotic energy balances over a period of seven years between 1990 and 1996 have been determined for the beech forest (W1) and the crop rotation field (A3). As shown in Fig. 4.2, the latent and sensible heat fluxes are approximately equal. In general, less radiation is reflected and a higher amount of energy is used for evapotranspiration in the beech forest.

However, there are several years when latent energy flux on the crop field is as high as in the forest. This situation is typical for years with dry summers and an early onset of the vegetation period. The inter-annual variations are exemplified by the energy balances for a moist (1994) and a dry (1995) year (Fig. 4.3). The forest has the capacity to absorb more energy than the agricultural plot, but the difference is less than expected if the albedo of grassland and forest is compared (see e.g. Monteith and Unsworth 1990).

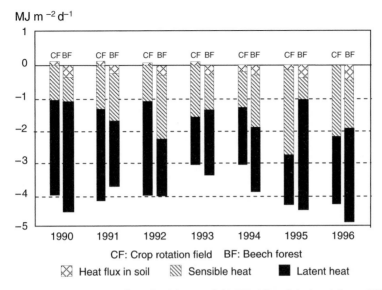

Fig. 4.2 Energy balances (MJ m^{-2} day^{-1}) of the crop field (CF, A3) and the beech forest (BF, W1)

4.2.2 Carbon Fixation, Primary Production and Biological Energy Consumption

Metabolic heat flux, although numerically only a marginal fraction of the ecosystem energy balance, is of fundamental importance. Reduction of CO_2 by photosynthesis is the major process which provides ecosystems with chemical energy essential for growth and maintenance of plants, animals and microbes.

Table 4.1 shows the calculated mean annual net income of three ecosystems of the core area (for details see Chapter 3). In the beech canopy (W1) the mean gross primary production is 1395 g C m^{-2} year^{-1}; in the alder carr (W6) it amounts to 2450 g C m^{-2} year^{-1}. This exceedingly high value can be interpreted as an edge effect because the alder carr forms only a small strip along the shore and is open to Lake Belau. The mean gross primary production of the maize stand (A2) is 1807 g C m^{-2} year^{-1}. Calculating 2874 kJ for 1 mol glucose, the beech canopy converts 1.59%, the maize field 2.06% and the alder forest 2.79% of the energy available from global radiation into chemical energy.

4.2.3 Soil Carbon Balances

As Dilly et al. (1996, 1997a) have demonstrated, the soil carbon relations of the Bornhöved ecosystems are mainly under the influence of three factors: (a) physical constraints such as soil temperature and soil moisture, (b) chemical constraints such

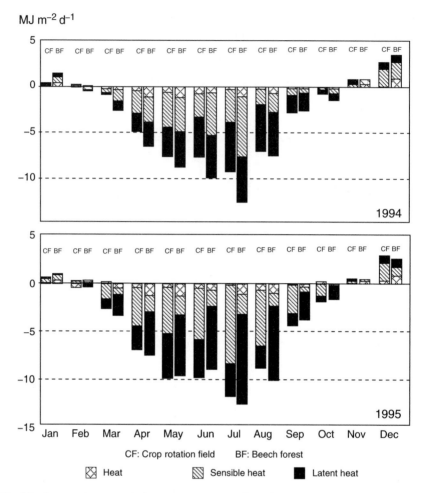

Fig. 4.3 Comparative energy balance estimates (MJ m⁻² day⁻¹) of the crop rotation field (CF) and the beech forest (BF) during the wet and dry years 1994 and 1995

Table 4.1 Mean annual photosynthetic gain of different ecosystems of the core area. Gross primary production (*GPP*) is the integral of the total leaf gas exchange processes including nocturnal respiration. *n.d.* No data

Ecosystem	Daytime GPP	Global radiation (%)	GPP	GR (%)
Beech forest	1395	1.59	1111	1.26
Wet alder forest	2450	2.79	2150	2.45
Crop field (maize monoculture)	1807	2.06	1492	1.91
Wet grassland	n.d.	n.d.	1230	1.40

as amount and quality of fresh organic matter and (c) the ecophysiology of plant roots and microbiota which adjust to the physico-chemical conditions in soil (see Chapter 3).

Figure 4.4 shows modelled soil carbon balances for four ecosystems of the core area. The carbon input is illustrated by two pathways: The 'litter pathway' (black

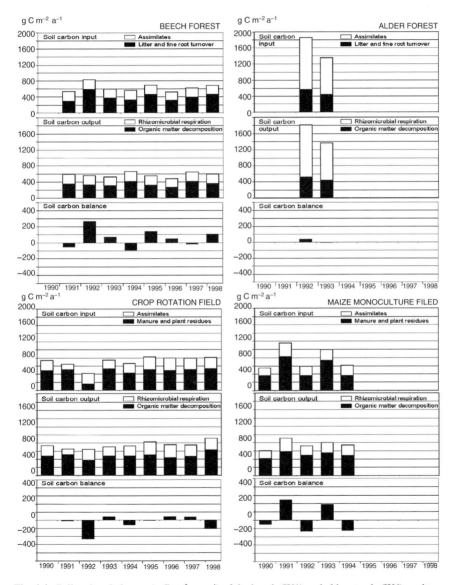

Fig. 4.4 Soil carbon balances (g C m^{-2} year^{-1}) of the beech (W1) and alder stands (W6), and two contiguous agrarian ecosystems under different land use (A2, A3). W6 has a peat soil, other soils are sandy

bars) comprises litter and fine root turnover which is mineralized by soil microbiota (heterotrophic soil respiration) whereas the other, the 'assimilation pathway' (white bars), shows the estimated amount of assimilates shifted belowground by the plants and used for the autotrophic part of soil respiration (see Table 3.3). The two pathways are also obvious in the soil carbon outputs. Black bars in Fig. 4.4 refer to heterotrophic bulk soil respiration, white bars to autotrophic soil respiration. In the sandy soils of the beech forest (W1) the average in- and outputs via the assimilate pathway were 233 g C m^{-2} year^{-1}, while the input via the litter pathway exhibited an average of 388 g C m^{-2} year^{-1} in eight years (1991–1998). High amounts of litter input (450–520 g C m^{-2} year^{-1}) were determined during years with high fruit production ('Mastjahre'). In other years the litter input range was 293–386 g C m^{-2} year^{-1}. Bulk soil respiration varied over 234–367 g C m^{-2} year^{-1} with a mean of 297 g C m^{-2} year^{-1}. On average 91 g C m^{-2} year^{-1} were stored in the beech forest soil during the reference period. This high value of soil carbon storage was the result of the extremely low pH value according to the negative correlation between litter accumulation and pH value found by Reiche et al. (2001).

In the peat soils of the alder carr (W6) carbon balances could be calculated only for 1992 and 1993 (Fig. 4.5, Table 3.3). The in- and output via the assimilate pathway (1113 g C m^{-2} year^{-1}) were more than four times higher than in the beech forest in the mean for the two years. The average input via the litter pathway comprised 551 g C m^{-2} year^{-1}. With a heterotrophic soil respiration of 487 g C m^{-2} year^{-1}, the alder carr soil sequestered an average of 64 g C m^{-2} year^{-1}. The total soil respiration of 1600 g C m^{-2} year^{-1} was extraordinarily high in this system. The microbiological properties of the soil and the litter in the alder carr (Dilly and Munch 1996; Dilly et al. 1999; Kutsch et al. 2001b) indicate that soil respiration was influenced by the

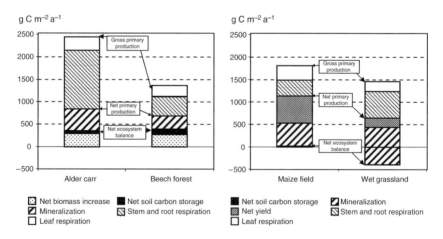

Fig. 4.5 Net ecosystem carbon balances (g C m^{-2} year^{-1}) of the alder (W6) and beech forests (W1) as compared with a maize field (A2) and wet grassland (A8)

large litter mass with a high nitrogen content which stimulated the activity of soil micro-organisms. This and the extraordinarily high rhizomicrobial activity were mainly responsible for the total soil respiration rates being much higher than in other forests.

On two crop fields (A2, A3) the litter pathway comprises fine-root turnover, crop residues, inter-cropping and organic manure. Therefore, the carbon balances of these systems highly depend on the farmers' management practices. During eight of nine years between 1990 and 1998 the crop rotation field (A3) was manured; only in 1992 was this treatment suspended. The mean input via this pathway was $454\,g$ C m^{-2} year^{-1}. The in- and output via the assimilate pathway were $256\,g$ C m^{-2} year^{-1}. Such input values are very similar to those determined for the beech forest. However, the chemical composition of the input via the litter pathway was very different from that of the beech litter. In addition, agricultural management practices such as harvesting, ploughing, fertilizing and pesticide application altered the ecophysiological properties of the soil microbial communities, resulting in higher mineralization rates in the crop rotation field compared with the beech forest (Kutsch et al. 1998; Dilly et al. 2000). The average carbon losses via the mineralization pathway were $486\,g$ C m^{-2} year^{-1} in the nine years under consideration. During this period the crop rotation field lost $31\,g$ C m^{-2} year^{-1} of soil organic carbon on the average.

The maize monoculture field (A2) got different treatments during the 1990–1994 period. During three years this field was only manured with cattle slurry so that the input via the litter pathway was $360\,g$ C m^{-2} year^{-1}. In 1991 and 1993 the field received unusually high carbon input by manure. In these years the inputs via the litter pathway added up to $830\,g$ C m^{-2} year^{-1} and $739\,g$ C m^{-2} year^{-1}, respectively. Bulk soil respiration varied between $412\,g$ C m^{-2} year^{-1} and $587\,g$ C m^{-2} year^{-1} with a mean of $506\,g$ C m^{-2} year^{-1}. It was increased in years of high inputs. In- and outputs via the assimilate pathway varied between $193\,g$ C m^{-2} year^{-1} and $325\,g$ C m^{-2} year^{-1} with a mean of $251\,g$ C m^{-2} year^{-1}. Consequently an average soil storage of $24\,g$ C m^{-2} year^{-1} resulted for the maize field within the reference period. However, this value does not reflect long-term management, because the high inputs by manure in 1991 and 1993 were exceptional. The mean annual C balance would have been negative according to a 30-year model simulation if the field had been manured with cattle slurry (Kutsch and Kappen 1997).

The carbon balance of the wet grassland (A8) soil was estimated from data of basal respiration rates measured in the laboratory (Dilly et al. 1997a). For details of the calculation of the carbon balances the reader is referred to Kutsch et al. (2001a). Peat mineralization caused high carbon losses from soil (see Chapter 3). Since the process depends mainly on the height of the water table (Kim and Verma 1992; Behrendt and Brüggemann 1993), it has to be taken into consideration that the water table of the wet grassland was lowered some 70 years ago (see Chapter 2). The chemical composition of the organic matter indicated that a high amount of peat had been mineralized since then (Wachendorf 1996). A mean annual peat loss of $400\,g$ C m^{-2} year^{-1} was calculated for the 1992–1994 period.

4.2.4   Net Ecosystem Carbon Balances

The two forest ecosystems (W1, W6) accumulated about 350 g C m^{-2} year^{-1} (Fig. 4.5). The major storage fraction was biomass increase. The alder carr with a gross primary production being twice as high as that of the beech forest had a lower net ecosystem carbon balance. The crop field with maize monoculture was a weak carbon sink during the reference period but only if the high manure inputs in 1991 and 1993 are included in the calculation. In addition, a net yield of 609 g C m^{-2} year^{-1} (gross yield – input by manure) was obtained for animal production or other human use. In the wet grassland (A8) only 205 g C m^{-2} year^{-1} were yielded, but due to peat mineralization the system was a source of CO_2.

4.2.5   Extended Energy Balances at the Patch Scale

Carbon or energy balances of agricultural ecosystems reveal two difficulties: the first is that high amounts of matter are transported across the ecosystem boundary as yield or manure. The fate of the material taken from the system as well as the origin of the material brought in should be included in the balance. The second difficulty is that the cultivation of the systems causes energy costs and the release of CO_2 (e.g. fuel for the tractor, energy for the production of mineral fertilizers or pesticides, etc.). Therefore, in the next step the carbon balances of the agricultural systems were converted into energy units and supplemented by the cultivation energy costs. Figure 4.6 shows this input for three agricultural systems in the core area, including all important factors except buildings.

Figure 4.6 shows that the temporal variation of energy input at the patch scale is high, even if similar crops are planted (e.g. corn in the years 1989 and 1993 on the crop rotation field A3). The annual input into this field varied between 1 MJ m^{-2} and 3 MJ m^{-2} mainly depending on different nitrogen inputs. The high input into the corn field (A2) was typical: a great part came from organic nitrogen which was applied in excess because the farmer had no other possibilities to dispose of the manure of his animals. The energy input into the grassland (A6) was only about one-third of that into the fields. This pasture was managed extensively; the only input was fertilizer.

The energy output and the output/input relation due to yield over a period of five years is illustrated for the crop rotation field (A3). In Fig. 4.7 the output varied between 4 MJ m^{-2} year^{-1} and 34 MJ m^{-2} year^{-1}.

The highest output was accomplished with maize as a C_4 plant in climatically favourable years. The normal output varied around 20 MJ m^{-2} year^{-1} if all parts of the plants were harvested. The low output of rye in the year 1992 was not typical for the region because of the extremely hot and dry summer. Cereals yielded energetic equivalents of 8 MJ m^{-2} year^{-1} up to 20 MJ m^{-2} year^{-1}. The relation between yield and total cultivation energy (or the output/input relation) varied between 4 and 12 in the five phases of crop rotation. The lower value is typical for high energy-consuming C_3 plants such as winter wheat, while the upper limit represents cereals with low input and low yield and C_4 plants such as corn with high yields.

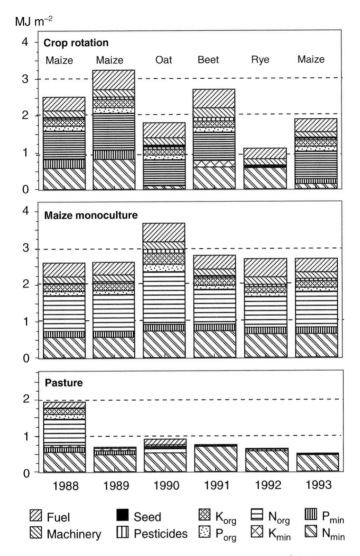

Fig. 4.6 Interannual variation of the anthropogenic energy input (MJ m^{-2}) in the agroecosystems A3, A2, A6

Figure 4.8 illustrates an integrated patch scale balance showing the extended balance of a crop field planted with maize in a climatically favourable year including anthropogenic energy inputs and processing of the biomass yield in the farm.

Gross primary production fixed 67 MJ m^{-2} year^{-1}, of which 32 MJ m^{-2} year^{-1} were harvested by the farmer. Assuming a demand for 10 kg dry matter per cow per day, it is possible to feed four cows with the yield of one hectare of the maize field. The biomass of the cattle was calculated to be 24 g C m^{-2} in carbon units and 1 MJ

Output [GJ ha⁻¹]

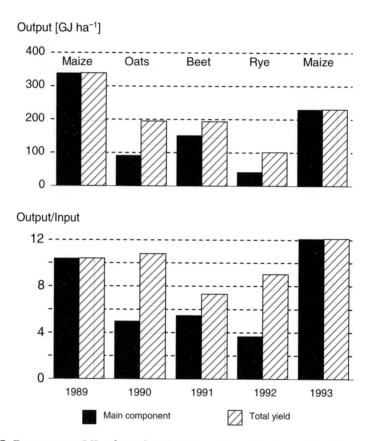

Fig. 4.7 Energy output (MJ m⁻² year⁻¹) and corresponding output/input ratios of crop rotation field A3 during the 1989–1993 period. The main component is cereals or beet, total yield comprises straw, too, but neglects the subterraneous organs

m⁻² year⁻¹ in energy equivalence, respectively (Kutsch et al. 2001b). Respiration of the cattle was estimated as 13 MJ m⁻² year⁻¹; about the same amount of energy was excreted by the animals and re-imported to the field as manure (Hayasaka et al. 1995; Jungbluth et al. 2001). Further, 5 MJ m⁻² year⁻¹ left the farmgate as products such as milk and meat. The total ecosystem respiration which also includes the respiration of the cattle and the anthropogenic energy inputs adds up to 63 MJ m⁻² year⁻¹ which is 4 MJ m⁻² year⁻¹ less than the GPP.

4.3   Farm Gate Balances

On the next level it was tested whether this balance reflected the usual agricultural practice in the Bornhöved Lake District. To this end a survey of farm gate balances was conducted at the study area level (see Fig. 1.2). A detailed questionnaire about

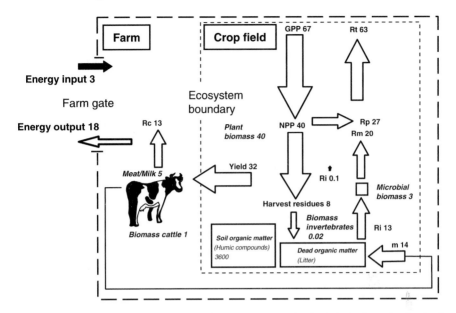

Fig. 4.8 Integrative energy balance (MJ m^{-2} year^{-1}) of maize field A2, including related animal husbandry and anthropogenic system inputs. *GPP* Gross primary production, *Rt* total ecosystem respiration (including the respiration of cattle which is not located at the site and the anthropogenic energy input), *Rp* plant respiration, *Rm* respiration of soil microbiota, *Ri* respiration of invertebrates, *Rc* respiration of cattle, *NPP* net primary production, *m* manure. For further explanations see text

fuel and fertilizer use, crop yield and the production of meat and milk was sent to all farmers. On this basis and including the results of interviews, eight farms were selected which can be considered representative of the different structural types of farmsteads of the study area (Fig. 4.9). Table 4.2 summarizes the acreages and different land use patterns of these farms.

The complete energy balance estimates of these farms, ranked in terms of growing output/input (O/I) ratios, shows that four farms attain only unsatisfactory O/I relations, although one of them has a slightly positive balance. By contrast, two farms specialized in crop production and pig fattening or pure high-yield crop production have high positive energy budgets which also correspond to distinctly positive O/I relations. It should be noted, however, that irrespective of the energetic situation described, not only the large crop II farm (extensively landed property) but every farm represents a microeconomically acceptable mode of agriculture under the present-day macroeconomic and agro-political boundary conditions (cf. Chapter 13).

On the basis of the whole data set four typical farms were selected for further analysis (Fig. 4.10a–d). 'Cropping' and 'Dairy' farming represented the best and worst O/I relations, whereas 'Cropping and Dairy' farming with intermediate values are indicative of the average farm type of the area. The high-yield crop farm (a) exports nearly 12 times the energy imported. Here the energy input is mainly due to fertilizer;

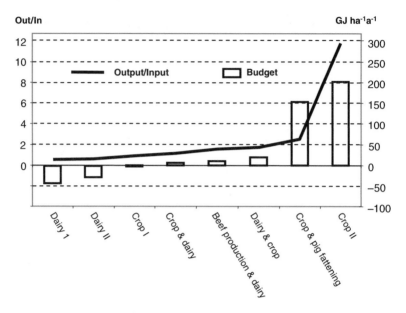

Fig. 4.9 Energy budget (GJ ha⁻¹ year⁻¹) and output/input ratios of eight representative farms of the Bornhöved Lake District. *Dairy I* Increased fodder supply (concentrated feed), *Dairy II* typical dairy farming, *Crop I* low yield farming, *Crop II* high-yield production

Table 4.2 Land use patterns of eight representative farmsteads of the study area during the 1987–1988 period

	Dairy I	Dairy II	Crop I	Crop and dairy	Beef production	Dairy and crop	Crop and pig fattening	Crop II
Arable land	86.6	49.3	62.9	21.0	45.0	11.0	23.0	520.0
Grassland	1.0	22.7	6.4	38.5	33.5	3.2	39.4	45.0
Wasteland and badlands	5.4	2.0	1.0	0.2	0.5		2.0	36.0
Forest		2.5						114.0
Episodic fallow						3.0		
Total acreage	93	76.5	70.3	59.7	82.0	14.2	64.4	729.0

other factors play a negligible role. The extremely inefficient dairy farm (d) with 60 ha grassland and fodder crop production exports only milk and meat. In addition to his own primary production the farmer has to import a considerable amount of energy as additional fodder. One of the mixed farms (b) has an O/I relation of 1, again with a high input rate due to fodder, fertilizer and fuel. The other (c) is selling milk and crops. Due to the high proportion of its own fodder production (grassland) the budget in farm b is balanced, whereas in farm c the preference of milk production necessitates an disproportionately high amount of fodder bought. Also the

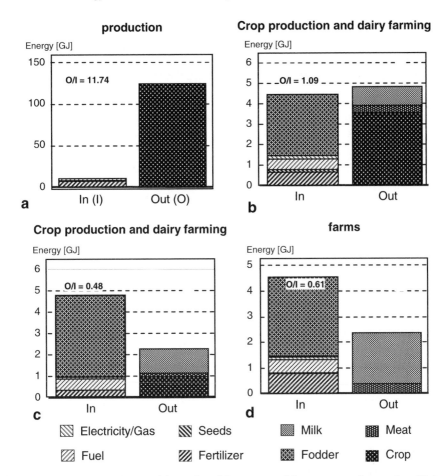

Fig. 4.10 Farm gate balances of four selected farm types of the core area of research. **a** High-yield cropping, **b** elevated cash crop production, **c** elevated proportion of animal husbandry, **d** inefficient dairy farming. Reference period is 1985–1988

dairy farm (d) has a high animal density and therefore a high energy demand for animal fodder. Although the owner's 40 ha of land provide an input of about 6–8 GJ year^{-1} the owner of this energy-consuming farm can gain by exporting only half of the energy bought on the market.

4.4 Energy Fluxes at the Landscape Scale

The actual land use of the Bornhöved Lake District, covering 89 km^2 with about 8700 inhabitants, displays the following pattern: 5% of the area are forests, 16% grasslands and 69% crop fields, the remaining 10% comprise lakes, fens, roads, buildings, etc. Since the energy balance of the latter landscape elements could not be assessed with a sufficient amount of accuracy, it is assumed to have remained constant in the framework of the scenario developed. This means a zero difference when comparing the energetic situation of the present-day land use pattern with that of the optimization scenario. In either case the (natural) advective component of the energy balance has been neglected.

Figure 4.11a shows that the mean annual net radiation of the above reference area amounts to 328.5 PJ year^{-1} of which nearly 60% are reflected or emitted as long-wave radiation. The metabolic heat flux, defined in terms of gross primary production (GPP) fixes about 2% of the global radiation, and only 1% is fixed by NPP (cf. Fig. 4.8). According to our calculations about 1000 TJ year^{-1} are supplied from the crop fields, with 450 TJ year^{-1} being exported as agricultural products; 350 TJ year^{-1} are used as fodder in the regional animal production, while the straw remaining on the fields contains about 200 TJ year^{-1}. Animal production in the region consumes about 750 TJ year^{-1} (350 TJ year^{-1} from fodder cropping, 200 TJ year^{-1} from the grasslands, 200 TJ year^{-1} from imports), and 50 TJ year^{-1} are exported in the form of animal products such as milk, meat and eggs. If the net energy production (NEP) of the forests in the area is either considered as a source of timber or as a CO_2 sink, it contributes to the exports by 70 TJ year^{-1}. Since anthropogenic imports comprise about 1400 TJ year^{-1}, an actual anthropogenic net energy import of 830 TJ year^{-1} would result for the region (Fig. 4.11a).

Figure 4.11b summarizes the result of an optimization scenario where anthropogenic energy input is reduced and energy efficiency is proportionally optimized. In addition, the land use pattern is changed. Based on 1998 prices, reduction of energy imports by about 30% appears possible due to more efficient energy use (e.g. better insulation of buildings, efficient heating, reduced use of cars, etc.). The technical reduction potential should be about 60% (Enquete-Kommission 1998). The use of renewable resources can further improve the energy balance. Thus, the scenario includes the use of wind power, solar energy (electricity), wood, straw, biogas and energy plants. The economic potential of wind power is about 100 TJ year^{-1}. The contribution of solar energy, however, is likely to remain low as long as the prices for solar generators remain on the present level. Even if enough area were available, the production of 25 TJ year^{-1} of electric energy would require investments of about 40–50 million euros. As the contribution of biogas is limited by the number of cattle the highest potential would come from the use of plant biomass for energy production. Consequently, the land use in this scenario has changed. Forests have increased slightly to 10% of the area, which involves an energy equivalent of 140 TJ year^{-1}. The proportions of grassland and 'other' elements remain at 16% and 10%, respectively. The area used for cropping purposes has decreased slightly and its use has changed. One quarter of the crop fields is used for fodder production, while the food and oil seed production is reduced to half of the cropping area. The remaining quarter is used for the production of C_4 energy plants. Together with straw used for energy production, 670 TJ year^{-1} can be exported for biomass burning, whilst the export based on food and oil seed decreases to 290 TJ year^{-1}. Also the export of animal products decreases because animal production is continued without any

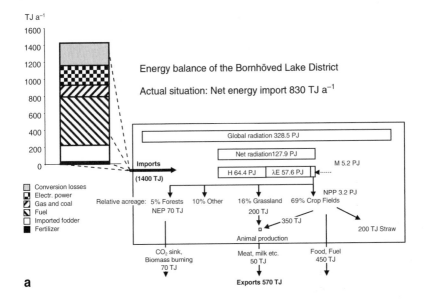

a

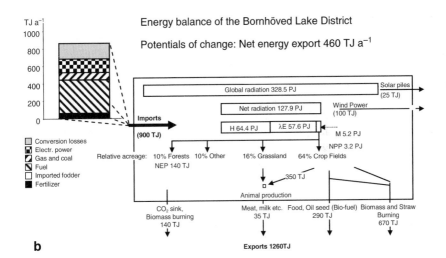

b

Legend: H: Sensible heat flux
λE: Latent heat flux
M: Metabolic heat flux
NPP: Net primary production
NEP: Net ecosystem production (in undisturbed ecosystems
= carbon sequestration)

Fig. 4.11 a Present-day energy balance of the study area. *H* Sensible heat flux, *λE* latent heat flux, *M* metabolic heat flux, *NPP* net primary production, *NEP* net ecosystem production (in undisturbed ecosystems = carbon sequestration). **b** Optimization scenario of the regional annual energy balance of the study area

imported fodder. In this scenario the energy input into the region is reduced from 1400 TJ year^{-1} to 900 TJ year^{-1}, whereas the exports are increased from 570 TJ year^{-1} to 1260 TJ year^{-1}. The scenario shows that it is possible to turn a rural landscape from a net energy sink with an actual O/I ratio of about 0.4 to a net energy source with an O/I ratio of 1.4. Theoretically, the O/I ratio could be raised above a value of 3.0, but this option would be liable to a set of ecological, economic and agricultural constraints.

4.5   Conclusions

According to the present inquiries into the carbon and energy cycles, forest ecosystems act as sinks, whereas agricultural systems are balanced or function as carbon sources, which is generally in accordance with the findings of other authors (cf. Goulden et al. 1996; Falge et al. 2002; Goodale et al. 2002; Smith 2004; Soussana et al. 2004). However, the results showed that the balances at the patch scale provided only limited information. Budgets at the farm level proved to be a better tool for controlling the sustainable development of rural areas. Compared with budgets at the patch scale and the regional scale farm-level budgets have the advantage that: (a) the data base is comparatively reliable because every farmer gave fairly good estimates for import and export of his products, (b) the results can easily be transferred into practice and (c) problematic parameters like the assessment of manure or the assignment of machinery to single fields and of the carbon in soils are avoided.

This kind of budget may be used as an indicative parameter for regional planning purposes, as will be discussed in Chapter 13. The results presented here offer possibilities to optimize the carbon and energy balances of farms and to create carbon sinks in rural areas (Vleeshouwers 2002). This implies:

- Investing in increased efficiency of energy use and the use of regenerative energy forms has the highest potential for CO_2 reduction.
- At the regional scale, forests are only able to compensate for a minor part of the CO_2 released. Afforestation leads to a temporary sequestration of carbon. If a forest were planted on a former crop field it would have a carbon sequestration potential of about 150 t C ha^{-1} for a period of about 100 years. Thereafter, the sink capacity of forest ecosystems is expected to decrease (Odum 1969; Nabuurs et al. 2003). It was set forth that even the most productive forest ecosystem of the region, the alder carr, had retained only a minor portion of the emitted CO_2.
- With respect to atmospheric conditions, it is necessary to develop management techniques that prevent losses of soil organic carbon (SOC) from agricultural soils. Intensive agricultural land use can lead to extensive losses of SOC in the long run (Kutsch and Kappen 1997), but crop rotation with intercropping and manure application would widely avoid carbon emissions to the atmosphere (Smith et al. 2001).

- Carbon release from peatlands can be diminished only if a high water table is restored. But such ecosystems have only a limited value for the farmer and may involve increased methane emissions (Meyer 2000).
- A strong source of CO_2 is imported fodder. Thus 'sustainable agriculture' can only be achieved by a reduction of fodder import. This, however, reduces the number of animals per unit area and the animal production.

Nevertheless, ecologically commendable strategies may conflict with economic goals. Producing crops for biomass burning may increase the income of farmers within a favourable framework of emission regulations, even if the implementation of emission trade mechanisms in agriculture is uncertain at present. However, the requirements of an energy optimization strategy may appear opposed to ecological requirements if one considers the straw on the fields an energy source for power plants but realizes that it gets lost as organic matter for the soil. Consequently, an administrative control system should balance both carbon ecology and carbon economy. Altogether, a change from net energy import to net energy export seems to be possible. A regional energy efficiency between 1 and 2 is indicated to make the region energetically sustainable.

Chapter 5
Water Relations at Different Scales

Georg Hörmann, Matthias Herbst, and Christiane Eschenbach

5.1 Introduction

The comprehensive data set of meteorological, hydrological, soil physical and plant physiological measurements permits a detailed inquiry into how ecosystems use water and how the water cycle is coupled with other cycles in agricultural and forest ecosystems. Since characteristic features of such cycling processes are distinctly scale-dependent and comprehensive water budgets are subject to detailed discussion in Chapter 9 of the present book, in the following the spatial scale is restricted to the range from leaves to stands of the ecosystems investigated in the core area of research (cf. Chapter 3). Thus, the first part presents water relations at the leaf scale, in the second the focus is on the ecophysiology of water exchange at the stand level, and finally there is a comparison of the empirical results of this enquiry with results of modelling approaches. The methodology underlying both measuring and modelling procedures is summarized in Section 1.3.

5.2 Water Relations at the Plant Leaf Scale

The water relations of plants depend on various interrelated physiological processes. Endogenous as well as various exogenous factors stimulate or inhibit the water turnover of plants. Among the interrelations between plant water and environment, the regulation of stomatal conductance is of crucial importance. It is influenced by microclimatic variables and in some instances directly by the leaf water potential (Jarvis 1976; Raschke 1979; Nonami et al. 1990). Increased stomatal resistance limits transpiration water loss, but is always accompanied by decreased CO_2 assimilation (see Chapter 3). To test the physiological adjustment of a species at leaf level, conductance, transpiration rate and water potential were investigated in situ on maize and rye fields, grasslands, hedgerows, beech and alder stands and reed swamps.

O. Fränzle et al. (eds.), *Ecosystem Organization of a Complex Landscape.*
Ecological Studies 202.
© Springer-Verlag Berlin Heidelberg 2008

5.2.1 Diurnal Course of Leaf Water Relations

From the data sets collected in different species and at different sites, diurnal courses of leaf water relations of *Alnus glutinosa* were chosen as examples. Conductance, transpiration rates and water potentials of black alder leaves were investigated at two neighbouring sites with different water regimes: alder trees in the occasionally water-logged alder carr (W6; Fig. 1.4) and alder shrubs in a nearby, much drier hedgerow between A2 and A5 (Eschenbach and Kappen 1999).

During the 1992 growing season the soil water regimes at the two sites differed in a typical and expected manner. The stand at the lake shore was influenced by the water level of the lake with the water table being 20–50 cm belowground in summer, which involves soil water tensions around −30 hPa. Near the hedgerow soil water tensions were strongly influenced by precipitation and decreased during normal summer conditions several times to values below −1000 hPa at soil depths of 30 cm and 60 cm. At 160 cm depth however the variation was small and soil water tension did not fall below −60 hPa (Bornhöft 1993).

Differences between 'predawn' leaf water potentials of trees at different sites were small and sometimes the values were even less negative in the supposedly 'dry' hedgerow than in the alder forest. No site showed any seasonal trend for the 'predawn' leaf water potential.

On all days in 1992, except 21 October, the leaf water potentials of the wet alder carr and the 'dry' hedgerow followed the identical diurnal trend with higher values in the early morning and late evening and lowest values at noon (Fig. 5.1; Eschenbach and Kappen 1999). On 21 October the weather was overcast, wet and cool and the water potential stayed at about −0.3 MPa throughout the day.

Water potentials of the leaves at different positions in the hedgerow (N-side, S-side, middle) were always similar but in the canopy of the alder carr the leaves at the top of the trees had markedly lower values in July and August. The diurnal minimum of the leaf water potential of the dry hedgerow alders declined from −1.3 MPa in May to −2.2 MPa in September whereas in the alder carr the diurnal minimum, after an initial decline from May to July, remained constant at about −1.7 MPa during August and September. Exposed leaves from the tree top reached even minima of −2.3 MPa in July and minimal water potentials of the leaves from the inner parts of the crown ranged between −1.1 MPa and −1.5 MPa throughout the season.

Exemplary diurnal courses of leaf conductance, transpiration and leaf water potential of black alder are shown in Fig. 5.2 (Eschenbach and Kappen 1999). 18 August was a typical sunny summer day, the irradiance increased to 1.2 mmol photons $m^{-2} s^{-1}$ and fluctuated due to clouds. Air temperatures were 12 °C at sunrise and reached a maximum of 20 °C whilst leaf temperatures differed from these by less than 1 K. Leaf–air vapour pressure differences showed almost bell-shaped curves with low values in the morning and in the evening (about 2 mbar bar^{-1}) and a maximum of 12 mbar bar^{-1} at noon. The black alder trees in the carr and the hedgerow had nearly identical climatic conditions during the day. However, the

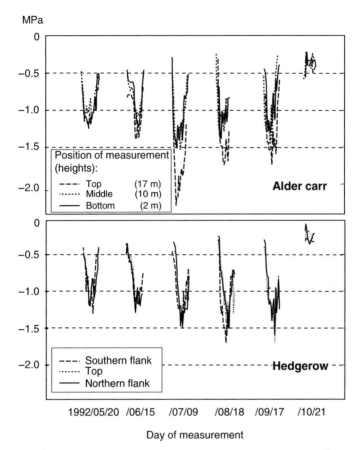

Fig. 5.1 Diurnal course of leaf water potentials (MPa) of *Alnus glutinosa* during the 1992 growing season. Top: leaves at three different levels in the canopy of the alder carr (W6), bottom: alder leaves at three different locations in a hedgerow (near A 5)

peripheral sun-lit leaves had a much higher PPFD and there were few differences in temperature and saturation deficit (ΔW).

The diurnal courses and absolute values of leaf conductance and leaf water potential were strikingly similar at both sites of the alder tree. Leaf conductance was highest in the morning (peripheral leaves about 360 mmol m^{-2} s^{-1}), when irradiance was high and ΔW was low, and decreased steadily during the day. Conductance and transpiration rates of the leaves of the inner crown were lower due to reduced irradiance (Fig. 5.3). The diurnal course of leaf water potentials reflected the actual transpiration rates but with an inverse pattern. Typically, water potentials became gradually more negative during the morning, decreasing from −0.3 MPa at sunrise to −1.7 MPa at noon, before recovering again in the afternoon.

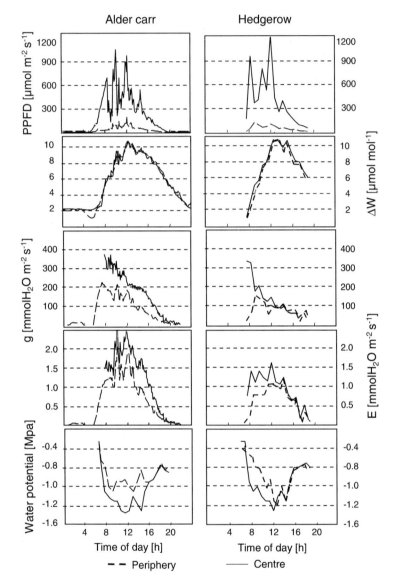

Fig. 5.2 Exemplary diurnal course (August 18, 1992) of water relations of *Alnus glutinosa* in the alder carr (left) and the hedgerow (right): leaf conductance (g, mmol m⁻² s⁻¹), transpiration (E, mmol m⁻² s⁻¹) and leaf water potential (MPa) measured on leaves from the sun-lit periphery (solid line) and from the inner part of the crown (dashed line). The microclimatic parameters are irradiance (PPFD, µmol m⁻² s⁻¹) and leaf-air vapour pressure difference (ΔW, mbar bar⁻¹)

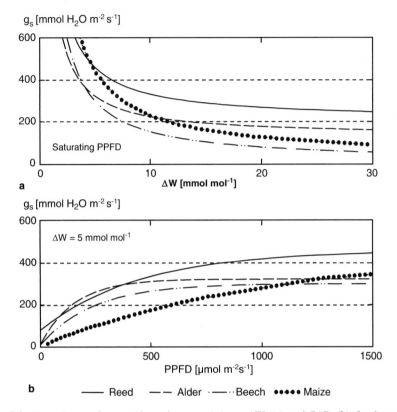

Fig. 5.3 Dependence of stomatal conductance (g_s) on ΔW (**a**) and PAR (**b**) for leaves of *Phragmites australis, Alnus glutinosa, Fagus sylvatica* and *Zea mays* at a ΔW of 5 mmol mol^{-1}. All measurements were taken in sun leaves during the peak of the growing season in the absence of water stress. The measured data were fitted to an exponential function (ΔW) and a saturation type curve (irradiance)

5.2.2 Dependence of Leaf Conductance and Transpiration on Irradiance and Saturation Deficit

The relationship between leaf conductance and ΔW was hyperbolic at saturating irradiance (Fig. 5.3a) in alder trees and in species of other ecosystems studied, i.e. lake ecotones and sites W6, W1, A2. Stomata were opened maximally at low evaporative demands (low ΔW) but leaf conductance decreased rapidly with increasing ΔW. Concerning the levels of maximum conductance at different ΔW values and the capabilities of regulation, indicated by the steepness of the decrease, the investigated plant species differed in a typical way. All the measurements were taken in sun leaves during the peak of the growing season, but in the absence of water stress. Reed showed the smallest capability for regulating water turnover at leaf level which is due to the litoral habitat of this species. Also the differences between the two tree species were related to water availability at their habitats: *Alnus glutinosa* is less capable of limitation than *Fagus sylvatica*. *Zea mays* as a water-saving C_4 species exhibited a strict stomatal response to increasing ΔW. At high irradiance however, and without soil water shortage, leaf conductances were slightly higher than those observed in beech.

The light dependence of stomatal aperture of the investigated species followed a typical saturation curve as PPFD increased (Fig. 5.3b). The initial slopes of the curves, however, differed among the species. A steep increase in the sun leaves of *F. sylvatica* and *A. glutinosa* indicated a sensitive reaction to ambient irradiance. The separated curve fitting of the data sets from both the beech and the alder leaves revealed that the response of conductance to PPFD was slightly less pronounced in beech than in alder. In both trees, the stomata of leaves in the inner shaded parts of the canopy reacted more strongly to irradiance than those of peripheral leaves. Stomata of *Phragmites australis* and *Z. mays* showed a smoother reaction to increasing light.

5.2.3 Interrelations between Leaf Water Potential, Leaf Conductance and Transpiration

In *A. glutinosa* the relationship between leaf water potential and leaf conductance was linear over all measured water potentials between nearly zero to −1.75 MPa (Eschenbach and Kappen 1999). There was an even stronger linear relationship between leaf water potential and transpiration. Neither in black alder nor in beech was any transpiration-limiting threshold of leaf water potential detected (Fig. 5.4), as described by Hacke and Sauter (1995) for beech trees. Under comparable site

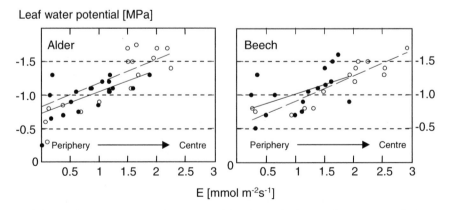

Fig. 5.4 Relationship between leaf water potential and transpiration rate for mature, sun-lit leaves of *Alnus glutinosa* and *Fagus sylvatica*.

conditions, the dependence of the leaf water potential on transpiration water loss was linear, but the transpiration rates of alder leaves were higher than those of beech leaves (Eschenbach et al. 1996).

5.2.4   Acclimation of Leaves to Environmental Conditions

Leaves can change their response to environmental factors like PPFD, T, ΔW due to seasonal and spatial acclimation processes as demonstrated for the Bornhöved ecosystems by Herbst and Kappen 1993. Besides phenologically based variations, a reversible mid-term acclimation process of stomatal response to air humidity following the weather patterns has been detected (Vanselow 1997; Kutsch et al. 2001b). Because these acclimation responses are species-specific, a realistic parameterization of the leaf and canopy resistances to transpiratory water loss covering all investigated plant stands, all seasons and different years is complicated. The measured leaf scale data had to be sorted and grouped with respect to species, crown position season, year, PPFD and ΔW, before appropriate types of mathematical functions describing the influence of the environmental factors on stomatal conductance were chosen and curve-fitting procedures, based on least-square analysis, could be applied to the sub-sets of data. The resulting estimates of canopy conductance for four plant stands during four subsequent years are summarized in Fig. 5.5.

5.2.5   Measurement and Parameterization of Interception and Soil Evaporation

Patch-scale water relations and water turnover are determined by plant physiological properties as well as by vegetation structure. Most crucial for an estimation of the magnitude of interception evaporation is the interception storage capacity of the vegetation (S). Following Gash and Morton (1978), S can be determined by plotting net precipitation versus gross precipitation for individual rainfall events (Fig. 5.6). As a consequence, S becomes the negative intercept of a regression line through the data points on the y-axis. The slope of that line equals $1-p_t$, with p_t being the proportion of rain diverted to the shoot surfaces. Considering the rainfall data collected above and below a fully developed maize canopy (LAI = 2.5) between July and October 1995, S was determined as 0.95 mm and p_t as 24.2% (Fig. 5.6). Assuming a linear relationship between leaf area index (LAI) and the size of S and p_t, the observed values can be normalized to S = 0.38 mm and p_t = 9.7% LAI (Herbst and Vanselow 1997). A similar analysis of data collected in the beech forest yielded S values of 1.28 mm in summer and 0.84 mm in winter (Herbst and Thamm 1994). For an alternative approach treating S as variable and influenced by wind speed, the reader is referred to Hörmann et al. (1996).

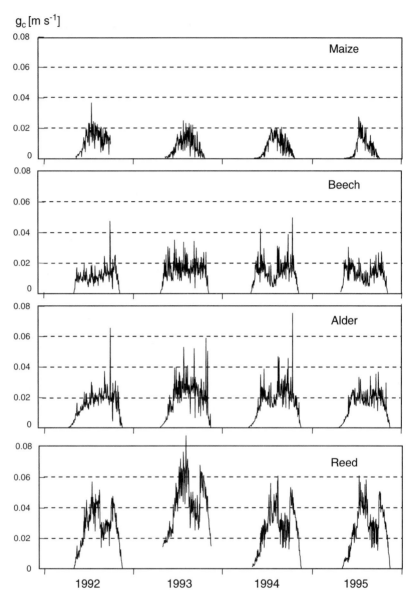

Fig. 5.5 Modelled noon values of canopy conductance (g_c) for four years in stands of maize (A2), beech (W1), alder (W6), and reed (east of W6)

The determination of the role of soil surface conductance (g_s^s) in water vapour transfer from soil to atmosphere is based on the simplifying assumption of a dry soil layer topping a saturated zone (Fuchs and Tanner 1967; Shuttleworth and Wallace 1985). A decrease of g_s^s corresponds to a theoretical increment of the dry

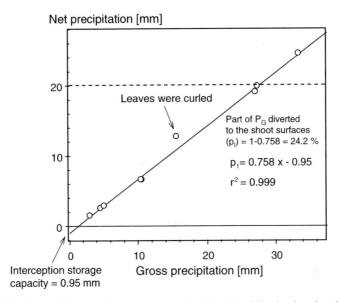

Fig. 5.6 Calculation of interception storage capacity (S) of the fully developed maize canopy (A2) from measurements of gross and net precipitation (see text for further explanations). Redrawn from Herbst and Vanselow (1997)

layer. Based on mean values of g_s^s for the daylight period in the maize field (A2), a relation was obvious between g_s^s and the time which had passed since the end of the last rainfall event. Individual rainfall events were separated from each other if interrupted by at least four rainless hours. In Fig. 5.7 the daylight means of g_s^s are plotted versus the time period after the last rainfall for different seasons. For a wet soil surface ($t = 0$), g_s^s values varied between $3\,cm\,s^{-1}$ and $4\,cm\,s^{-1}$ during most of the year, except in winter before ploughing, when surface water could accumulate on the closed soil surface; then g_s^s became $7\,cm\,s^{-1}$. The curvature of the regression lines varied significantly between the four seasons. It corresponds to the desiccation rate of the uppermost soil layer and may reflect the mean vapour pressure difference (ΔW) of the time periods for which it was fitted (Fig. 5.7, inset): With increasing ΔW there was a linear decrease of the coefficient representing the curvature of the g_s^s versus time relationship, corresponding to a more rapid drying of the soil surface.

5.3 Water Turnover at the Stand Level

The above findings on leaf conductance, soil surface conductance and rainfall interception capacities were used to parameterize a two-layer evaporation model of the Shuttleworth–Wallace type which is based on the Penman–Monteith equation (Shuttleworth and Wallace 1985; Herbst and Kappen 1999). By means of standard

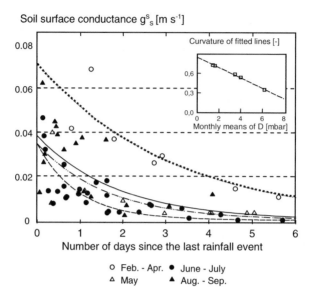

Fig. 5.7 Dependence of soil surface conductance (g_s^s) on the time since the last rainfall event in the seasonal course. *Inset* Dependence of the coefficient representing the curvature of the regression lines on monthly means of saturation deficit of the air (D). Low values of the coefficient correspond to a quick decrease of soil surface conductance after the end of a rainfall event

weather data, transpiration, soil evaporation and interception evaporation of the maize field, the beech forest, the alder carr, the reed swamp and the open water surface of Lake Belau were estimated from this model for two meteorologically different years (Fig. 5.8).

Transpiration of maize plants never exceeded 4 mm day^{-1} and was controlled by both energy supply and stomatal behaviour, corresponding to decoupling coefficients (W) between 0.2 and 0.6. Over the whole year, soil evaporation was the largest evapotranspiration component in the maize field, because transpiration lasted only during five months per year and was sometimes suppressed by soil drought.

Table 5.1 compares these values as annual totals with gross precipitation and net radiation input over a four-year period. Total evaporation amounted to 57% of the gross precipitation (PG), providing for a high amount of percolation into deeper soil layers (Herbst 1997). In the 105-year-old beech forest (W1: tree height 29 m, maximum LAI 4.5) total annual transpiration (Tr) varied between 326 mm and 421 mm (mean 389 mm or 50% of PG) and annual evapotranspiration (ET) between 567 mm and 665 mm (mean 617 mm or 79% of PG). In the 60-year-old alder stand (W6: tree height 18 m, maximum LAI 4.8) the respective values were 375 mm and 658 mm (mean 538 mm or 69% of PG) for Tr and 612 mm and 884 mm (mean 768 mm or 99% of PG) for ET. The daily maximum of transpiration was around 5 mm in the beech stand but up to 8 mm or exceptionally even 10 mm in the alder carr (Fig. 5.9), and W was between 0.2 and 0.3 in both stands (Herbst et al. 1999). In years with

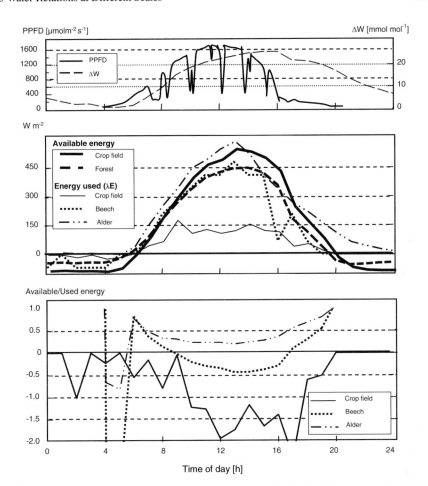

Fig. 5.8 *Upper panel* Microclimatic conditions (PPFD, μmol m^{-2} s^{-1}; ΔW, mmol mol^{-1}). *Middle panel* Available energy (–X, W m^{-2}) of crop rotation field (A3) and beech forest (W1) on 20 May 1993 and energy used (ΔE in W m^{-2}) for transpiration by the different ecosytem types crop rotation field (A3), beech forest (W1) on 20 May 1993 and alder carr (W6) on 27 June 1992. *Lower panel* Ratio of advective energy to energy used for evapotranspiration (–X/ΔE)

high radiation input, ET in the alder stand (along a lake shore with unlimited water availability) exceeded both PG and net radiation. The higher interannual weather-dependent variation of transpiration in alder stands corresponds to a lower capacity of stomatal regulation as compared with beech stands (Herbst et al. 1999).

For comparison purposes some figures from other beech stands are given. An evaluation of pertinent literature by Brechtel and Lehnhardt (1982) provides estimates of the annual evapotranspiration from 209 mm to 497 mm according to site charac-teristics, climate, structure and age of stands. A 90-year-old beech in the Schönbuch (Baden-Württemberg) had an average summer evapotranspiration rate of 466 mm

Table 5.1 Annual totals (mm year^{-1}) of transpiration (Tr), interception evaporation (EI), soil evaporation (ES), evaporation from the water surface (E) and total evapotranspiration (ET) in four plant stands and Lake Belau as compared with annual gross precipitation (PG)

		1992	1993	1994	1995
PG		703	827	878	698
RN		738	638	788	797
Maize	T	189	139	160	115
	EI	43	84	42	37
	ES	174	254	256	278
	ET	405	477	459	430
Beech	T	421	326	389	419
	EI	116	141	129	126
	ES	103	100	79	120
	ET	640	567	597	665
Alder	T	658	375	525	596
	EI	111	133	124	115
	ES	115	104	82	135
	ET	884	612	731	846
Reed	T	901	453	701	827
	EI				
	E				
	ET	1324	824	1024	1254
Lake Belau	E	676	524	662	731

during the 1979–1982 period with a transpiration of 263 mm. During the winter months (November to April) evapotranspiration amounted to 71 mm only, while the transpiration rate was negligible (Fleck 1986). Transpiration of the reed canopy (adjacent to W6) was remarkably high if compared with the other ecosystems and sometimes exceeded 10 mm day^{-1}. The rates corresponded to local advection or downward fluxes of sensible heat in the reed belt. Table 5.1 shows that the annual total evapotranspiration (ET) from the reed belt generally exceeded the annual evaporation of the open lake surface (E). Depending on meteorological conditions, ET ranged between 824 mm year^{-1} and 1324 mm year^{-1} and the ratio of ET vs E ranged between 1.5 and 2.0. In three of four years of investigation, ET also exceeded annual gross precipitation, which underlines the general significance of reed transpiration for the water balance of lakes and wetlands (Herbst and Kappen 1999).

5.3.1   Evaporation and Advective Energy Supply

During the period from mid-June to mid-September, modelled and measured daily sums of ET exceeded the daily sums of net radiation sometimes in the beech canopy and in the alder stand. For instance, this phenomenon can be demonstrated for a clear day at the end of June (Fig. 5.9). In the middle panel of the figure, net radiation is compared with the amounts of energy supplied for evapotranspiration in the different vegetation types or ecosystems alder carr W6, beech forest W1 and the crop rotation

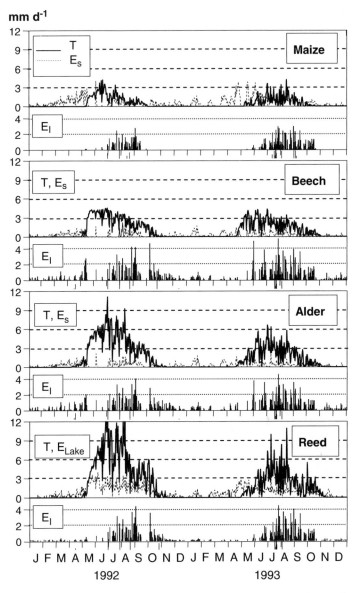

Fig. 5.9 Modelled daily sums of transpiration (T), interception evaporation (EI), and soil evaporation (E_s) for maize (A2), beech (W1), alder (W6) and reed in 1992 and 1993

field A3. Due to ecosystem properties, the available energy over the forests was higher than that over the crop rotation field on this clear day. Whilst under these conditions only a fraction of the available energy was used for transpiration by the crop plants, most of the available energy was kept within the forest systems. Especially in the alder stands, the energy used for water turnover was temporarily even higher than the energy available by net radiation. Regarding the proportion between advective energy and total energy used for evapotranspiration ($-X/\Delta E$; lower part of Fig. 5.9) the following sequence was found: alder forest > beech forest > crop field. Hence it follows that the water consumption of small forest areas cannot be sufficiently explained by the locally available energy alone. Depending on the physiology of the trees, the properties of the site, the degree of wetness of the canopy and the atmospheric vapour pressure deficit, advective energy in form of sensible heat and wind contributes to a large extent to forest evapotranspiration.

5.3.2   Comparison of Water Vapour Flux Measurements with Different Modelling Approaches

The investigations described above are time-consuming and require expensive equipment. Such preconditions are rarely met in practice where water budgets are needed at a daily, seasonal or annual time scale, e.g. in order to evaluate water and related nutrient fluxes in climate scenarios (Herbst and Hörmann 1998). Usually, only climatic data from standard weather stations and some soil physical data are available. Water budgets are calculated with more or less complicated computer models, the choice depending on the experience of the modeller and often on limits imposed by available staff and funding. To bridge this gap between science and practice, the above results were compared with the output of two models representing completely different levels of complexity. The first is the SOIL/COUP model (Jansson and Karlberg 2001), a state-of-the-art Richards-type soil water model based on the Penman/ Monteith evaporation approach. The second model is called 'SIMPEL' and is a collection of several spreadsheets which can be combined like a toolbox. SIMPEL stands here for the low end of hydrologic computing (Hörmann 1997).

The data base for this comparison is deliberately as diverse as possible to cover the whole range from science to daily practice. The parameterization of the above Shuttleworth/Wallace model is based on plant physiological measurements and hourly micrometeorological data to calculate the fluxes of heat and water based on the Sverdrup/Bowen ratio method. In contrast the parameterization of the other two models is based on daily mean values.

The SOIL model uses an energy balance approach which is similar to the Shuttleworth/Wallace (SW) procedure. Interception is calculated on the basis of LAI and Penman/Monteith ET values; storage capacity is a function of LAI during summer and is set to zero in winter. SIMPEL uses a bucket model with variable storage capacity which is a function of LAI during the growing season and set to a constant value (0.83 mm) in winter.

Despite the completely different computation methods, the interception sums turned out to be nearly identical for the maize field (A2) and the beech forest (W1). The annual difference between the SIMPEL bucket method and the SW method is only 2 mm for maize (Fig. 5.10) with an annual mean value of 51 mm (1992–1995) and 6 mm for the beech forest with an annual sum of 128 mm.

mm month^{-1}

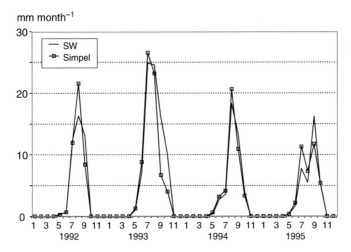

Fig. 5.10 Modelled monthly interception of a maize stand (A2) during the 1992–1995 period. *SW* Shuttleworth/Wallace model, *SIMPEL* Simple bucket model

mm month^{-1}

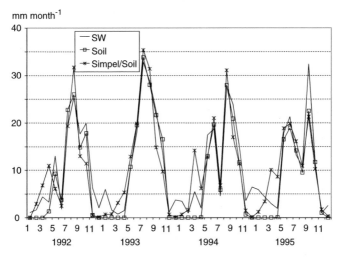

Fig. 5.11 Modelled monthly interception of a beech stand (W1) during the 1992–1995 period. *SW* Shuttleworth/Wallace model, *SOIL* COUP model/SOIL, *SIMPEL/SOIL* SIMPEL model with evaporation from COUP model

Figure 5.11 shows a closer look at the calculated values for the beech forest. The predicted values by the three models were close together in the summer season, but deviated from each other in winter and spring when the SOIL model set LAI to zero because of the absence of foliage.

To describe the actual evapotranspiration (ETa) four different approaches were used, covering the range from simple to complex: for the agricultural plot the Haude evaporation, for the beech forest the PM-based SOIL model and SIMPEL with potential evapotranspiration (ETp) copied from SOIL.

5.3.2.1 Maize Field

Although the models applied yielded consistent values for interception, the evapotranspiration estimate by the Haude approach was distinctly less satisfactory. In the maize field the mean ETa is 443 mm year^{-1} according to the SW model and 305 mm year^{-1} according to the Haude method. The potential ETp for Haude is only 11 mm year^{-1} higher than the ETa calculated by the SW model. A further analysis of the monthly values (Fig. 5.10) shows that the summer evaporation is underestimated and the growth of the plants starts earlier than predicted by the Haude method.

5.3.2.2 Beech Forest

Evapotranspiration of the beech forest was calculated by means of three models. The corresponding curves in Fig. 5.12 show the results of different strategies of modelling: the 'COUP-model/SOIL' line as calculated with the default values recommended for beech forest by Jansson and Karlberg (2001) and the 'SIMPEL'

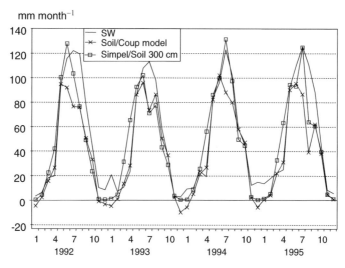

Fig. 5.12 Modelled monthly sums of the actual evapotranspiration (ETa) of the beech stands (W1) during the 1992–1995 period. *SW* Shuttleworth/Wallace model, *SOIL* COUP model/SOIL, *SIMPEL/SOIL 300 cm* SIMPEL model with evaporation from COUP model and a rooting depth of 300 cm

curve show the result of the model run with deep roots (3 m) and a high stress tolerance of the tree, i.e. the tree transpires fully down to 7% available soil water with a wilting point at 5% soil water content. The evaporation calculated with the SW method is as high as the potential evaporation calculated with conventional methods (Penman/Monteith approach of the COUP model). The high values obtained can only be explained by a high advection of sensible heat.

The simulation runs in Fig. 5.12 demonstrate that two factors are important: the reaction of plants to soil water stress and the available water content. It ensues from this figure that the high evaporation rates measured by the Sverdrup/Bowen ratio method are only possible if the trees have full access to soil water of a 3 m soil column and if the plants transpire at full rate until nearly all of the soil water is depleted.

5.4   Conclusions

The integration of results from plant physiological and meteorological measurements revealed a rather high evaporation for the beech stand (W1) compared with results from similar forests. Thus, even in very dry years expected stress symptoms did not appear. The only plausible explanation is that the trees drew water from much deeper soil layers than expected.

The consequences for modelling are significant: most models have used in practice a standard default configuration. The results of these model runs revealed evaporation rates far beyond the measured ones. Only two factors remain as an explanation: a much higher soil water storage combined with a high stress tolerance, i.e. the tree roots are deeper than usually expected and they transpire at full rate even at low soil water contents. The consequence for calculating the local water budgets is that meteorological and plant physiological measurements have to be combined for a valid parameterization of the models. The default values for rooting depths should be regarded with care.

Chapter 6
Site-Related Biocoenotic Dynamics

Ulrich Irmler, Oliver Dilly, Joachim Schrautzer, and Klaus Dierssen

6.1 Introduction

The dynamics of biota are largely dependent on the available habitats in a landscape while the heterogeneous distribution of resources is an essential reason for the constitution of ecological niches and for the structure and composition of communities. Vertical structures are marked within forest ecosystems with several strata, i.e. floor ground, herb, bush, and tree layers, which are reflected in the stratification of the animal community (Overgaard Nielsen 1987). Dead logs in the treetops offer other living conditions than dead wood lying on the ground. Animals that use different heterogeneously distributed resources must exhibit a high amount of spatial flexibility (Weidemann 1986).

The spatial distribution of habitats and microhabitats is the result of local relief and related soil texture, water and nutrient cycles, and the influence of light. A heterogeneous environment has different effects depending on the size of the organisms. Thus, it is better suited to small species than to large ones since it fosters the coexistence of small species more than that of larger ones (Levin 1974). Giller (1996) focused on fractal geometry for elucidating the differences in the niche structure of soil animals. Besides the environmental factors organisms themselves produce a plenitude of further structures.

A major part of the biocoenotic investigations, therefore, focused on the influence of the irregularly distributed habitats and on the dynamics of organisms (cf. Chapter 7). The dynamics of biota follow mostly seasonal dynamics, e.g. the leaf development of plants and the generation cycles of species, or long-term non-seasonal fluctuations. Animals in particular show seasonally controlled movements. Many animal species living in the vegetation during summer retreat into the soil in autumn or winter. Others migrate from agrarian ecosystems to forests for hibernating purposes. Seasonal movements are particularly wide-spread in a heterogeneous landscape. Long-term fluctuations may depend on successional changes of resources, i.e. the degradation of woody debris or leaf litter, or the succession of the vegetation after changing management practices.

6.2   Vertical Distribution Patterns of Fauna
and Microbiota in Beech and Alder Stands

6.2.1   Vertical Stratification in the Soil Layer

Within the soil profile the microbial biomass decreases and the physiological strategies change towards less metabolically active and glucose-responding microbes. This holds for the L, O and Ah horizons in the beech forest (W1, W2) as biomass determinations on the basis of a complementary methodology show (cf. Section 1.3). Thus, biomass values estimated by substrate-induced respiration and fumigation–extraction methods could be compared (Vance et al. 1987). The substrate-induced respiration seems to be related to the activated fraction of the microbial community while fumigation–extraction yields total estimates of the microorganisms susceptible to fumigants (Dilly and Munch 1998).

The density of microorganisms generally decreases in the forest soils with increasing depth (Fig. 6.1). This is evident by relating microbial data to soil dry mass, but also by referring to potentially available organic C. Only if the bulk density of a fresh litter horizon of the beech forest is low does the biomass of microorganisms appear to be higher in the deeper horizon. Carbon and nitrogen mineralization rates exhibit generally similar trends, but relating the transformation rates to the size of the microbial biomass provides additional information on the microbiota. The biomass-specific activity rate is high in the presence of less decomposed organic matter. The vertical pattern in microbial biomass content and biomass-specific activity rates correspond to shifts in enzyme activities (Schulten 1993; Dilly 1994) and to the character of microbial communities (Bach 1996). When applying cultivation techniques, r-strategists frequently dominate in the litter horizon, whereas populations with low diversities are found in the acid mineral horizon.

Also the soil fauna generally decreases with increasing depth (Peterson and Luxton 1982). It is obvious from comparative studies on mull and mould that the vertical gradient differs according to the humus type (Schaefer and Schauermann 1990). According to hand sampling (macro-fauna), heat extraction (mesofauna), and wet funnel technique (Enchytraeidae, Nematoda), soil biomass is higher in mull than in moder. Biomass is determined directly as fresh mass or indirectly by especially determined coefficients for different soil faunal groups (Irmler 1995). In relation to soil carbon mass the major proportion of the soil faunal biomass occurs in the litter layer of the beech and alder stands. The alder carr (W6) generally contains a greater part of the biomass in the mineral horizon when compared with the litter (cf. Section 2.2.3)

6.2.2   Vertical Stratification of Animals in the Vegetation Cover

The vertical stratification of animals was investigated in the beech forest (W1, W2) and the alder carr (W6) with regard to Empidoidea (Diptera: Empididae, Hybotidae, Dolichopodidae) and beetles (Coleoptera), using intercept traps exposed from 26 April

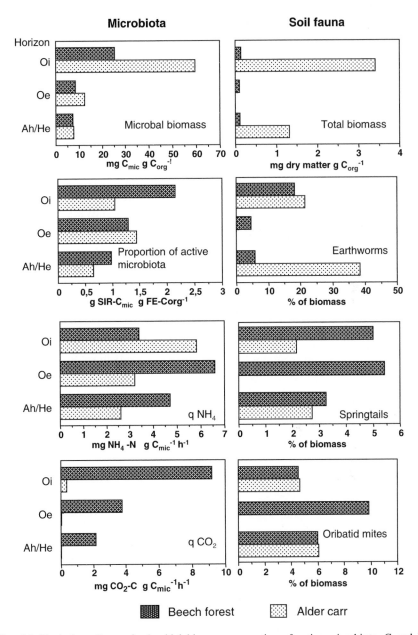

Fig. 6.1 Vertical gradients of microbial biomass, proportion of active microbiota, C and N efficiency, and biomass of the total soil fauna or different soil faunal groups (mg g^{-1} C$_{org}$) in the Arenic Umbrisol of the beech forest (W1) and the Histosol of the drained alder carr (W6). SIR and FE indicate complementory determination procedures: *SIR* substrate-induced respiration, *FE* fumigation–extraction

to 14 October 1992 at 1.5 m, 9 m, 18 m, 27 m, and at 1.5 m, 5 m, 17 m height (Raabe et al. 1996; Irmler 1998a).

The species richness and flight density of empidoid flies and beetles decreased with increasing altitude (Fig. 6.2). The dipteran families developing mostly in the soil showed recognizable differences in vertical zonation. Altogether, Empididae, Hybotidae and Dolichopodidae were found with 35%, 28% and 14% of the species in all layers. The preference of the near-ground layer was especially pronounced with the Dolichopodidae. More than 50% of the species occurred only in this stratum, while it was preferred by only 15% of the Empididae.

Flight activity was still more restricted to the ground layer than species richness. In both alder and beech forest 76% and 61% of all individuals occurred in the bottom layer. This means that between 7 and 19 individuals m^{-2} day^{-1} and only 2 individuals m^{-2} day^{-1} flew in the bottom stratum and the canopy, respectively. Beetles were also most frequently caught in the ground layer of the forests (1–2 beetles m^{-2} day^{-1}) during the vegetation period.

The characteristic vertical distribution of Empididae (Fig. 6.3) is linked to the displacement between different biotopes by many species during the mating phase (Delettre et al. 1992) and may be described as a diffusion process with permanent replacement of the individuals (Johnson 1957). For their swarming behaviour during mating the species require additional habitats that are protected against wind and form prominent structures. In contrast, the hygrophilous Dolichopodidae that have no swarming behaviour display a predatory running activity in the vegetation layer on the forest ground, where they hunt pollen-consuming insects (Chvàla 1983).

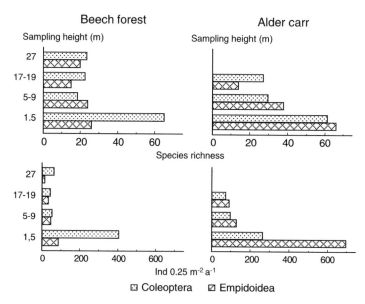

Fig. 6.2 Species richness and abundance of three Empidoidea (Diptera) families and beetles (Coleoptera) in the flight intercept traps of the beech forest (W1) and the alder carr (W6)

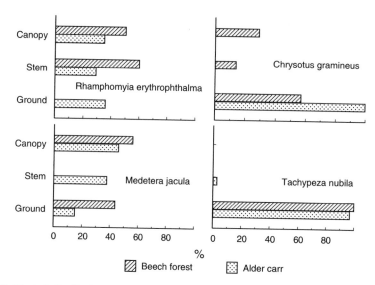

Fig. 6.3 Vertical distribution (% of individuals) of some Empidoidea species in the beech forest (W1) and the alder carr (W6)

Therefore, most Dolichopodidae are found within 0–10 cm above ground (Pollet and Grootaert 1991). The low density of flying Diptera in the canopy is caused by the influence of the wind in the trunk region (Overgaard Nielsen 1987). Seasonal changes of the use of strata are also due to the influence of wind, because the major proportion of the insects shifted from the wind-protected canopy region in summer to the bottom region with continuing leaf fall in autumn; during spring the opposite trend was recognized (Taylor 1958; Lewis 1966). The different use of strata might also depend on the nutritional conditions of the vegetation stands. In the alder carr a rich ground fauna of Diptera is available as a prey, in particular diverse midge families (Irmler 1995) which are absent in the beech forest.

At least five types of stratification can be distinguished (Irmler 1998a). About 33% of the beetle species in the intercept traps belong to the ground fauna. This was essentially due to the high species richness of Staphylinidae. Some 23% or 24% of the species were saproxylic or habitants of herbs, trees, or flowers, respectively. Most species of the ground fauna occurred in the bottom layer, but even at 27 m altitude in the beech forest. On average an increase of species living on dead wood was noticed with increasing altitude, i.e. 26% at the bottom and 30% in the upper-most layer. The beetles living on herbs and trees were equally distributed in the vertical. A detailed analysis of the trophic guilds of saproxylic species showed that the carnivore guild comprised the most active animals among the saproxylic species with an abundance of 16 individuals m^{-2} to 100 individuals m^{-2} (Table 6.1).

In the intercept traps mainly species were found that live in small, irregularly distributed habitats, e.g. fungi, dead wood, nests, carrion, and faeces, while the species of the litter layer were largely absent. This refers not only to flightless species of

Table 6.1 Vertical distribution of ecological beetle guilds in terms of catches in intercept traps (%) in the W1, W4, and W6 stands

Forest layer	Saproxylic				Ground	Herb/ tree	Nest	Fungi	Carrion, faeces
	Xylophag.	Xylosaprophag.	Zoophag.	Sum					
Beech									
1.5 m	1.5	0.6	14.4	17.2	10.3	41.0	1.8	10.9	2.4
9 m	2.5	0.4	5.9	8.9	6.8	49.8	3.4	21.9	0.4
18 m	4.4	0.4	16.3	24.2	3.5	34.4	2.6	13.7	0.4
27 m	5.4		19.1	25.1	2.4	36.4	1.5	10.1	
Alder									
1.5 m	0.9	1.4	5.6	7.9	27.4	45.3	3.2	3.1	5.2
5 m	2.0	0.2	2.0	4.5	14.0	74.6		1.0	1.7
17 m	1.6	1.0	7.9	11.2	16.8	28.6	22.7	9.5	0.7

Carabidae or Staphylinidae, e.g. *Othius myrmecophilus*, but also to flying species, e.g. *Othius punctulatus* which is frequently found in the litter of the beech wood (Irmler 1995). Also a number of Staphylinidae living on dead wood was not registered in the intercept traps, though they contributed to the most abundant saproxyclic species on the ground (Irmler et al. 1997). Other species which were exceedingly frequent in dead wood and litter occurred in the intercept traps only with low abundances.

Comparing the species composition of dead wood on the ground with that in the intercept traps leads to the conclusion that several saproxylic species inhabit mainly or exclusively the canopy region (Irmler et al. 1996, 1997). In the dead wood of the ground layer 44 saproxylic beetle species were determined. Only 20 species were identical with the 71 saproxylic species of the intercept traps. Even some of the most abundant and frequent species in the intercept traps were not found in the dead wood on the ground. *Rhinosimus planirostris* that occurred in the intercept traps in high abundance was only registered in young sections of dead wood on the ground. Probably most saproxylic species exclusively found in intercept traps live in young dead wood of the trunk and canopy layers. The high flight activity of these species can be attributed to the search for breeding habitats.

6.3   Spatial Patterns of Microbiota, Vegetation and Fauna in Beech Forests

6.3.1   Microbiota and Decomposition

Available substrates are of great importance for microbial growth and the accumulation of soil microbial biomass in ecosystems. In addition to the litter high amounts of energy are supplied via exudates of the roots to the heterotrophic microbiota (Grayston et al. 1996). Other microorganisms are also concentrated in the space around the roots exhibiting great differences in density in comparison with the bulk soil (cf. Sections 2.2.3 and 2.3).

For the Bornhöved beech forest (W1) Stork and Dilly (1998, 1999) found patterns of microbial biomass and microbial respiration rates in different soil compartments. Either related to soil mass or organic C content, the highest variability of microbial parameters occurred at the 10 m scale (Fig. 6.4). Thus, the distance to the tree trunks and the variations between trees may have controlled the pattern of microbial communities and activities. Griffiths et al. (1996) found that the presence of host trees is important for ectomycorrhizal mat patterns. They showed that mats of different morphology did not physically overlap and mat distribution was influenced by: (a) the proximity of one mat to another, (b) the distance from the mat to the closest living tree, (c) the density of living trees in the stand, and (d) the successional stage of the stand. The importance of the forest structure for the spatial distribution of microbiota may result from the pattern of litterfall and root growth and, thus, from considerable variations of exudates. Acid runoff from tree trunks is also reported to affect the soil organisms, particularly in heavily polluted forests (Baumgarten and Kinzel 1990).

The activity of the soil fauna was investigated by means of the bait-lamina test (Irmler 1998b). The method can be used as a screening test and is supposed to reflect the feeding activity of the mesofauna (Törne 1990). Two 10×10 m quadrats and a smaller subplot (1×1 m) were examined in the beech forest (W1) with 100 sampling points each. Each point contained five bait-lamina strips with 16 pores each filled with standard agar-cellulose mixture.

The results of the statistical analysis showed a higher variance in the upper layer (pores 1–8) than in the lower layer (pores 9–16). The variance was higher on the 10×10 m scale than on the 1×1 m scale of the subplot, while the mean feeding activity was not different between the two 10×10 m quadrats. The pH value and thickness of litter had no significant effect on the feeding activity. The distance from adjacent trees was also tested as a factor for the pattern, but the slight increase of feeding activity with decreasing distance from the tree was not significant. Only the feeding activities of the upper and lower layers were significantly correlated (Irmler 1998b).

6.3.2 Vegetation

The floristic diversity of the beech forest (W1, W2) was investigated at different spatial scales. Three factors were considered to be responsible for the spatial heterogeneity: random, endogenic and exogenic parameters (Eber 1981, 1989; Dierschke 1989). Eber (1981) stressed the vegetative and generative potential for dispersal of the species as endogenic parameters. The spatial distribution of the species is scale-dependent and is also influenced by exogenic factors such as soil reaction, trophic status, light availability, and forest management.

The characterization of the herb layer was based upon a 230-grid map, each grid measuring 10×10 m² (Fig. 1.5). Seven grids were randomly selected and

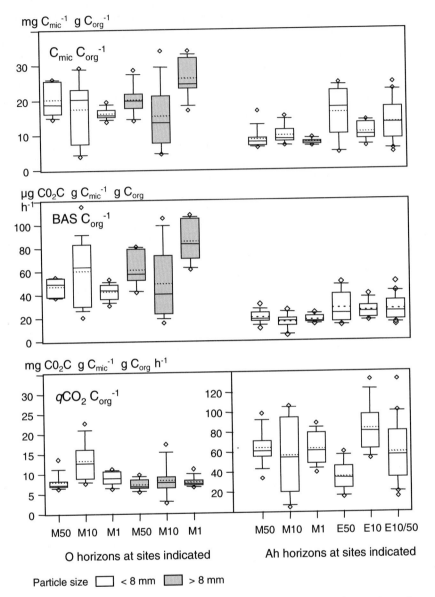

Fig. 6.4 C_{mic} content, microbial respiration rate and qCO2 per unit organic substance in two layers of the beech forest W2. M50, M10, M1 and E50, E10: both mixed and individual samples, respectively, from 50-m, 10-m, and 1-m grids. *E10/50* Combined individual samples, *boxes* indicate 25% and 75% quartile, *solid lines* and *dotted lines* are mean and median values, respectively, *columns* reach the 90% confidence intervals, *open rhomboids* indicate the tenth and nintieth percentiles

subdivided into 100 plots of 1 m². Furthermore, all 10×10 m² grids were grouped into areas of 250 m². The vegetation was classified using the Braun–Blanquet (1964) method.

Three sociological groups were differentiated (cf. Chapter 2): group 1 (*Asperulo–Fagetum typicum*) located in the centre of the area, group 2 (*Asperulo–Fagetum*, subassociation of *Circaea lutetiana*) near the forest track, and group 3 (*Asperulo–Fagetum*, subassociation of *Deschampsia flexuosa*) near the fringe to the intensively cultivated field A1.

The species composition and abiotic parameters of the subassociations were related to contents of N_{tot}, K, and Mg (Fig. 6.5). Group 1 was significantly different (Man–Whitney U-test with $P < 0.05$) from the two other groups that were more similar to each other, reflecting a vegetation on soils with high K, Mg, and N content. Both soil reaction and potassium content decreased from north (forest-path) to south (field) and from south to north, respectively. Light availability seems also to exert a considerable influence on the floristic distribution as is indicated by *Milium effusum*, which is significantly more abundant on clearings.

Species richness was significantly higher at the 10×10 m² and the 50×50 m² scales (Mann–Whitney U-test; $P < 0.01$) than at the 1 m² scale (Dierssen 1990).

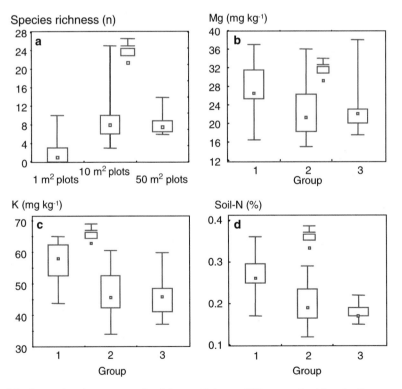

Fig. 6.5 Comparison between species richness of plots at different scales (**a**) and soil parameters (**b–d**) of subassociations of the beech stand (W2)

6.3.3 Fauna

Ground beetles (Carabidae) and the robin (*Erithacus rubecula*) were investigated to analyse, in a deliberately exemplary way, the spatial distribution of the fauna in the beech forest. Carabids were registered using 98 pitfall traps in the southern part of the beech forest (W2; Bortmann 1996). Three different grids were used: a 25-m grid in the major part, a 50-m grid in the western part and a 10-m grid along a transect from the field to the centre of the forest (Fig. 6.6). In addition, two adjacent habitat margins were investigated to analyse potential edge effects, namely, the southern maize field and the eastern forest area with a high proportion of spruce, oak, and hazel (cf. Fig. 1.4). A total of 44 ground beetle species was registered, with 35 species occurring in the beech forest only. The grid points of the area adjacent to the maize field, the eastern hillside forest, and the spruce forest formed separate clusters resulting from the average-linkage-cluster-analysis of the dominance identity (Renkonen 1938) at a level below 60% similarity. Within the beech wood three clusters were separated: (a) The central area subdivided into four further clusters on a level of 70% similarity, (b) a windfall area of four grids, and (c) several grids at the edge of the forest, particularly on the eastern hillside with its mixed forest cover.

The separation of the six clusters can be attributed to different distribution patterns of the carabid species. The characteristic species of cluster 3 at the edge of the forest area was *Carabus hortensis* that is a typical species of the forests on sandy soils (Irmler 1999). The windfall area (W1.1; cluster 2) was characterized by the euryoecious species *Pterostichus niger*, whereas the stenoecious species, i.e. *Abax parallelepipedus* and *Pterostichus oblongopunctatus*, dominated in the interior forest (cluster 1).

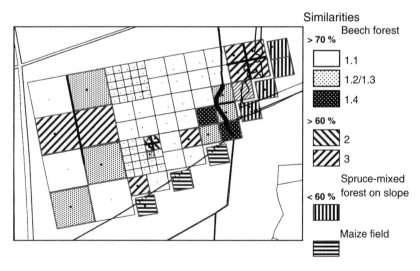

Fig. 6.6 Distribution of ground beetle assemblages in terms of different similarity levels of the dominance identity at the beech stand (W2) and its vicinity

Table 6.2 Relationship between abundance of ground beetles and various habitat parameters in the beech forest (W2) and the adjacent ecosystems (A1, W4), expressed using the Kendall τb-correlation coefficient: * $P<0.05$, ** $P<0.01$ (Bortmann 1996)

Species	Distance from:			
	Forest edge	Maize field	Spruce forest	Herb layer
Carabus coriaceus	−0.21**		−0.16*	
Carabus hortensis	−0.20**		−0.19**	
Carabus nemoralis		0.15*		
Harpalus latus		−0.16*		0.20*
Leistus rufomarginatus	0.17*		0.23**	
Nebria brevicollis			−0.25**	−0.20*
Notiophilus biguttatus			−0.16*	−0.30**
Pterostichus niger	−0.16*	−0.24**	−0.23**	0.20*
Pterostichus oblongopunctatus	0.31**	0.20**		
Synuchus vivalis	−0.26**	0.20*	−0.32**	−0.29**

The further subdivision of cluster 1 could be explained by the edge effect to adjacent ecosystems; thus, for instance, cluster 1.4 is predominated by *Nebria brevicollis*, being a typical species of the contiguous, intensively cultivated grassland.

A close relationship was found between the distribution of ground beetle species and the distance from the forest edge (cf. Table 6.2). In particular, the typical forest-dwelling species *Pterostichus oblongopunctatus* was positively correlated with the distance from the forest fringes. Further relationships were determined between the development of the herb layer and some ground beetles. No relationship was found with regard to soil parameters, e.g. soil texture, pH, both thickness and density of the litter, and structural parameters like distribution of dead wood and tree stumps. In fact species richness varied quite irregularly between nine and 18 species.

The potential effects of heterogeneity on the territorial quality of the robin sites were determined within the 50-m grid in the beech forest. Both mean species richness of birds and annual quantity of fully fledged robin squabs were estimated in every quadrat (Fig. 6.7). The highest annual reproductive rates were determined adjacent to forest tracks or smaller streets. The lifetime-reproduction differed significantly in the quadrats. Males showed a high site fidelity. During six years males with territory in specific quadrats always attained high reproductive rates, whereas males in other quadrats had low reproductive success. The higher lifetime-reproduction of high-grade territories was linked to both high mating and annual reproduction rates.

The body mass of the nestlings showed the same relationship, while the species richness of breeding birds had just the opposite trend (Fig. 6.8). The reproductive success of the robin was highest in quadrats with low structural diversity and species richness of breeding birds. It may be the result of interspecific food competition with other bird species, e.g. tits, that used predominantly the same food source, i.e. caterpillars, for rearing. Caterpillars were available only during a short time and thus constituted the main regulator for the development of nestlings. In the beech forest (W1, W2), a high species diversity of birds is mainly found along the fringes, i.e. adjacent to hedgerows or other forests which also have a high structural diversity.

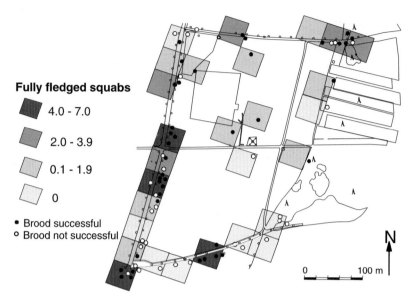

Fig. 6.7 Reproductive success of robins in different quadrats of the beech forest (W1,W2; mean fledged squabs year^{-1}; 50 m^2 quadrat) in the years 1992–1996

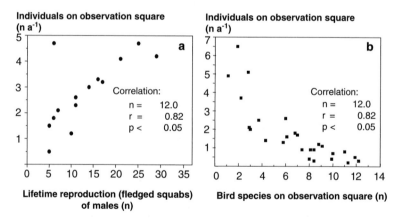

Fig. 6.8 Lifetime reproduction of individual males of the robin in the beech forest (W1, W2) in relation to the fledged squabs year^{-1} quadrat^{-1} (**a**) and correlation between production of fledged squabs and species richness in the quadrats (**b**)

However, the breeding success of the robin is also strongly influenced by predation on eggs and nestlings. The nest failures resulting from predation were highest in quadrats with a high species richness of birds, which can be explained by the particular attraction of these areas for predators.

6.4   Seasonal Dynamics

6.4.1   Seasonal Changes in the Animal Community

The seasonal change of the faunal biomass is influenced by the vegetation strata or soil layers (Figs. 6.9, 6.10). The animals of the vegetation layer attained maximum numbers during the vegetation period. In the beech forest (W1, W2) the maximum biomass of the herbivorous animals of the ground vegetation occurred in September, though higher phytomass was available in April/May rather than in autumn. In contrast, the maximum faunal biomass of both canopy and marginal area of the alder carr (W6) was reached in June. The early biomass maximum of the leaf-feeding insects on alder was due to the 'early season feeders' that profited from the better food quality of the young leaves (Ambsdorf 1996).

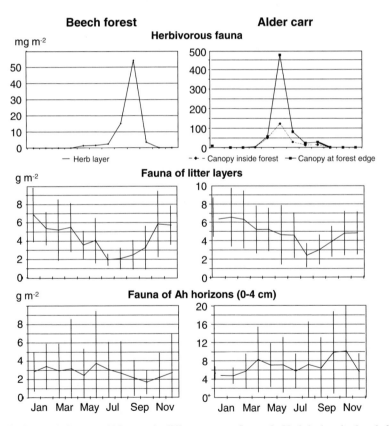

Fig. 6.9 Seasonal changes of biomass in different strata of a sandy Umbrisol under beech forest (W1) and the Histosol under alder carr (W6). Herbivore fauna in the beech forest as in 1988, in the alder carr as in 1990; values of the litter and the soil layers are averages of the 1988–1995 period

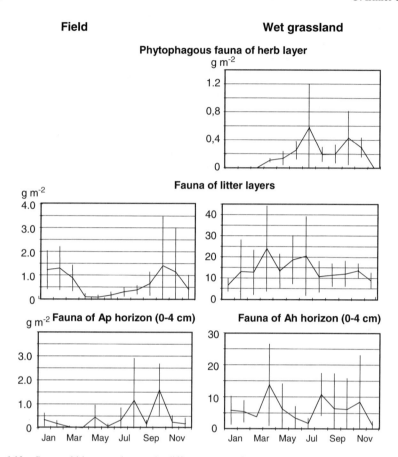

Fig. 6.10 Seasonal biomass changes in different strata of a sandy soil of field (A3) and a peaty soil under wet grassland (A9). Averages relate to the 1988–1991 period; no data were available on the phytophagous fauna in the field

The maximum abundance of the litter fauna was found in winter, with exception of the wet grassland (A8) where the maximum biomass occurred during the vegetation period. In the forests the yearly minimum biomass in the litter layer was always found in July or August, and the minimum biomass in the field was in April. The cultivation of maize (A2) between 1988 and 1990 and subsequent ploughing in March or April was most likely responsible for these findings. The seasonal changes of the biomass in the litter layer of the two forests may depend on leaf fall and soil moisture that was highest in autumn.

No seasonal change in the soil faunal biomass was found in the deeper soil layer (Ah, nH, Ap). The deviations of the monthly averages were essentially higher than the seasonal fluctuations, indicating a more or less constant biomass during the whole year.

Land use has a distinct effect on the seasonal pattern of the fauna by selection of phenological types, as can be demonstrated by the Sciaroidea midges (Heller 1996). In both woods and agrarian systems univoltine and bivoltine or polyvoltine (i.e. one, two, or more generations/year) species predominated (Table 6.3). The different hatching periods of the cranefly *Ericonopa trivialis* (Limoniidae) depended on the grassland use (Rief 1996). The minimum larval period of the species was 123 days on the fallow and 84 days on both the meadow and the intensely used pasture with higher irradiation on the soil. In the average of four years (1988–1991), the juvenile stages of spiders were most abundant in May and June, whereas the adults attained their maximum in winter (Fig. 6.11). Only on the field the lowest density of juvenile spiders was found in May/June, probably due to ploughing at that time. On the field juvenile spiders first appeared in July with increasing density, indicating an immigration from the surrounding habitats. The 3/2 relation of juvenile/adult spiders also indicated that the indigenous reproduction was at least reduced on the field.

Table 6.3 Species with known number of generations in four ecosystems of the core area of research

	Univoltine	Bivoltine	Polyvoltine
Beech forest	5	5	1
Alder forest	6	6	1
Field	2	5	4
Wet grassland	1	3	–

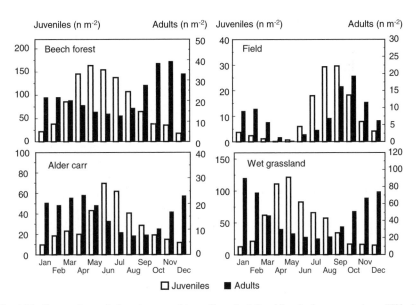

Fig. 6.11 Seasonal population changes of juvenile and adult spiders in four ecosystems (W1, A3, W6, A9) of the core area

6.4.2   Seasonal Change of Habitats

Many animal species require coherent habitats. Species changing between strata form an ecological guild, passing the larval period in the soil and requiring pollen as food for egg development in the adult stage, which is found in blossom-rich forest margins. Noticeable habitats for mating, e.g. dancing places, are also necessary. Several animal groups of the tree layer execute seasonal peregrinations between separated canopy habitats; thus for instance, the Psyllidae (*Psylla alni*) migrate between margin and canopy of the alder stand (W6) during early spring (Ambsdorf 1996). The species occurred in May mainly in the margin of the alder carr at the lake and subsequently migrated into the canopy between June and August.

The pollen beetle *Meligethes aeneus* hibernated mainly in the beech forest, while the larvae developed in the field or grassland systems (cf. Fig. 1.5) on rape or other cabbage species (Fig. 6.12). Between August and February the pollen beetle was found in the litter of the beech wood, hatching in the winter quarters in April and subsequently spreading over the adjacent agrarian systems. After the development of larvae in early summer they spread immediately, which can be deduced from the maximum flight activity in the early half of July, followed by choosing their winter quarters in August.

Unsuitable living conditions in an ecosystem induce a seasonally alternating occurrence of different generations in contiguous ecosystems of a landscape. This phenomenon was noticed for the bibionid midge *Dilophus febrilis* (Heller et al. 1991). The species had two generations per year with the larval stages of the first

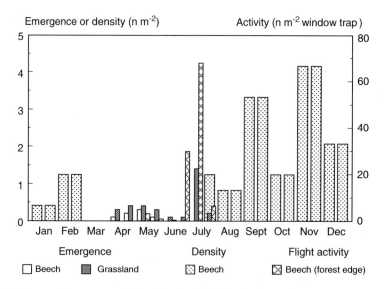

Fig. 6.12 Seasonal appearance of the pollen beetle *Meligethes aeneus* in emergence traps and in the litter layer of various ecosystems (average of the 1989–1990 period)

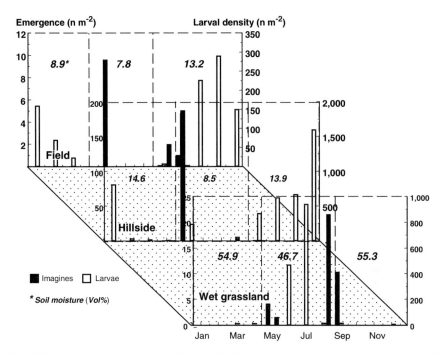

Fig. 6.13 Mean abundance of adult *Dilophus febrilis* in emergence traps and larvae in the soil of the three ecosystems of the agrarian catena (A3, A6, A9) during the 1988–1990 period

generation in winter from September to March and of the second generation in June and July (Fig. 6.13). The adults of the early generation hatched in May, the adults of the second generation in August/September. The different living conditions of the two generations were largely due to soil moisture conditions. The winter generation did not develop in the wet grassland as a consequence of a long inundation period, whereas the summer generation did not develop on the summer-dry areas of the hillside grassland (A5, A6) or the field, indicating that larvae develop in a range of 13–50% soil moisture. Thus the species requires for its development a landscape mosaic with contiguous winter-moist and summer-moist areas.

6.4.3 Seasonal Changes of Food Resources

Like most passerid birds the robin (*Erithacus rubecula*) has a predatory lifestyle during the breeding period. Most invertebrates then occur only for a short time within the action range of the bird, which compels switching between different prey. The seasonal use of food resources by the adult robin was determined by faeces samples (cf. Davies 1977; Bryant 1978). The food of nestlings was recorded using photographs of the adult birds carrying food to the nest (Lille 1996, Grajetzky 1992). The adult birds hunted almost exclusively in the litter, preferring small running invertebrates in the range of 2–8 mm body length, which could be rapidly discovered and swallowed without further treatment. Food requirements ranged between 100 kJ day^{-1} and 120 kJ day^{-1}, corresponding to a daily consumption of 700–1000 animals.

Occupation of territories started in April. In that early period the rove beetle *Lesteva longoelytrata* dominated in the faeces sampled (Fig. 6.14). In the middle of April *Lesteva* declined and the green weevil *Phyllobius argentatus* increased in the diet. During two weeks the green weevils were substituted by small Hymenoptera.

For the female robins the first hatching period of *Phyllobius argentatus* in April was of decisive importance. In this period they invested energy in egg production and their intensive use of weevils obviously depended on the high energy requirement of the females (Bezzel and Prinzinger 1982).

The high energy demand of the nestlings induced the feeding adults to switch to larger and energy-rich animals, e.g. caterpillar and craneflies, originating from both bushes and stinging nettle stocks of the beech wood margin. As soon as the transport for the nestlings food stressed the time and energy budgets of the adults this prey was more profitable than the small animals on the ground which they collected for their own diet.

During the first rearing period from 25 May until 10 June, the food selection for the nestlings was as follows: (a) high dominance of caterpillars during the early 10 days of the rearing period subsequently decreased to about 15% at the end of the rearing period, (b) an increase in cantharid beetles and later tipulid midges in the food, (c) increasing diversity of the food. This seasonal pattern was observed in every year investigated ($P < 0.01$; χ^2 test). The decrease of caterpillars as a food resource in June was a result of the synchronous development of many species from the larval stage to the pupal stage in soil.

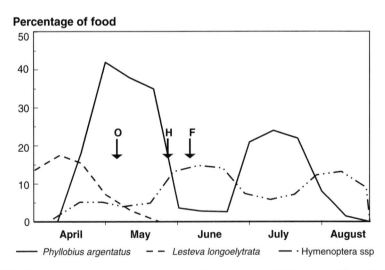

Fig. 6.14 Food resources of the robin during the breeding period 1993 in the different habitats within the beech stands (W1, W2) based on the dominance of prey remnants in the 216 faeces samples. *O* Start of oviposition, *H* hatching of nestlings, *F* flying out of young birds

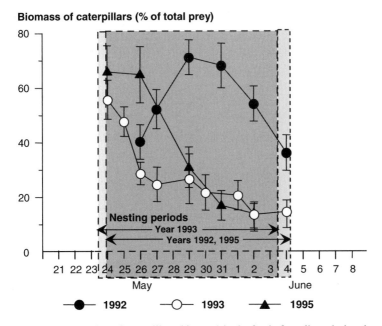

Fig. 6.15 Average proportion of caterpillars (biomass) in the food of nestlings during the rearing period of three early breeding periods of the years 1992, 1993, and 1995 in the different habitats of the beech stands (W1, W2)

The portion of the three prey-groups in the food and the period of the caterpillar decline varied in different years. In 1992 the caterpillars contributed most and for a long time to the diet, in 1993 and 1995 the caterpillar portions in the food had already decreased by the end of May (Fig. 6.15), resulting in high quantities of cantharid beetles and tipulid midges as a compensatory food for the first brood nestlings which received 62.3 mg day^{-1} during the early 3 days. For at least 4-day-old squabs the prey mass increased varying between 115.4 mg day^{-1} and 123.7 mg day^{-1}. Sclerotization of invertebrates did not correspond to both the age of nestlings and the development of the digestive tract of the robin as it did in the stonechat (*Saxicola torquata*). The prefered prey was smooth-membranous, which could be digested at every stage by the nestlings.

The nestlings of the first rearing period consumed between 740 g and 960 g food mass (FM). The squabs of the following period got between 531 g and 609 g FM, although they were more frequently fed and received twice as many animal prey-items. The correlation between the number of animals in the food and the total food mass was negative (Fig. 6.16). Although the frequency of feeding increased with decreasing biomass of the prey, squabs obtained less biomass. This feeding behaviour compensated for the small biomass offer and indicated the sub-optimal food conditions in the later rearing periods.

Biomass per feeding (mg)

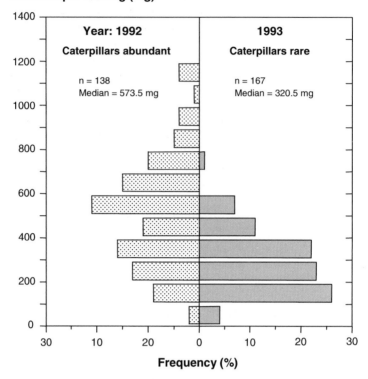

Fig. 6.16 Influence of caterpillar availability on food biomass in the beech forest (W1, W2) and adjacent ecosystems, in particular A1. Indicated is the prey biomass per feeding of the nestlings in the contrasting years 1992 (many caterpillars available) and 1993 (caterpillars rare)

6.5 Long-Term Dynamics

6.5.1 Effects of Climatic Change

Climatic changes affect the fauna directly and indirectly through the food web, e.g. by modified predator/prey relationships or competition. Pertinent data sets from the years 1988–1995 obtained by different methods are included in the present analysis. The macrofauna of the soil (W1) was recorded from four replicate $0.1\,m^2$ samples of both the litter and the upper soil layer down to 4 cm depth by hand sorting and heat extraction. The meso-fauna was obtained by heat extraction only. During the years 1988–1990 and in 1992 the samples were taken monthly, during the years 1991 and 1993–1995 samples were taken every second month. Spiders and ground beetles were additionally recorded using pitfall traps which were replaced every second week. Exponential curves were generated for time series; and relationships between curves were identified by cross-correlation at the significance level $P < 0.05$.

During the study period, low fluctuations were observed for the spider *Coelotes terrestris*, the rove beetle *Othius punctulatus*, and the ground beetle *Abax parallelepipedus* (Fig. 6.17). *C. terrestris* passes through a two-year generation cycle with maximum activity in August and September (Weidemann 1986). The abundance fluctuates between 18–24 individuals trap^{-1} month^{-1} on average. *O. punctulatus* passes through a one-year or two-year generational cycle with oogenesis in the winter months and oviposition in spring (Kasule 1970). The population density fluctuated between 2–4 indiviuals m^{-2}. Both the minimum and maximum abundance in 1989 and 1993, respectively, may be due to the precipitation pattern. The activity density of *A. parallelepipedus*, passing through a generation cycle of a few years, did not fluctuate during the first years. Starting in 1993, activity density of the species preferring dry living conditions decreased continuously, however, perhaps a consequence of the humid climatic conditions of that year.

Remarkable fluctuations were found for microarthropods. The springtail *Isotoma tigrina* attained maximum densities in May and June with particularly high and low abundances in the dry year 1990 and the humid year 1993, respectively (Fig. 6.17). Most Collembola and oribatid mites, e.g. *Lepidocyrtus lanuginosus*, preferring humid conditions, reacted with decreasing or increasing density in the dry years 1989, 1990, or the humid year 1993, respectively. From 21 collembolan species investigated only five species displayed no significant relationships between precipitation and abundance. These five species were colonizers of the vegetation

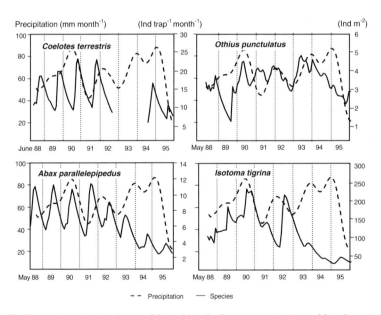

Fig. 6.17 Fluctuations in abundance of the spider *Coelotes terrestris* (Araneida), the rove beetle *Othius punctulatus* (Staphylinidae), the springtail *Isotoma tigrina* (Collembola), and the ground beetle *Abax parallelepipedus* (Carabidae) in the beech forest (W1) in comparison with the monthly rainfall rates of the 1989–1995 period

layer or the litter surface, e.g. *Orchesella flavescens, Dicyrtoma saundersi* or *Sminthurus fuscus*. For the other species significant time lags of 0–15 months were determined. Euedaphic species, e.g. *Isotomiella minor* and *Tullbergia sylvatica*, and large species, e.g. *Tomocerus flavescens,* have lags of 13–15 months. Colonists of old litter layers, e.g. *Friesia mirabilis* and H*ypogastrura denticulata*, show lags of 2-10 months and colonists of young litter layers, e. g. *Lepidocyrtus lanuginosus* and *Entomobrya muscorum*, have lags of 0–2 months.

6.5.2   Influence of Temperature on the Generation Cycle

The development of herbivorous arthropods was distinctly related to the temperature conditions of the different years (Hanssen and Irmler 1995). The grasshopper *Chorthippus albomarginatus* has an univoltine generation cycle with four larval stages. Oviposition is in autumn and eggs pass through a prediapause before dormancy in the winter. Egg development continues in the spring of the following year and larvae hatch in early summer.

During 1990 and 1992 the first larval stage was observed in the twenty-first and twenty-second weeks and in 1991 between the twenty-third and twenty-fifth weeks of the year (Fig. 6.18). In 1992 the larval development finished after approximately seven weeks, while 1990 and 1991 the development took ten weeks. The integral temperature required for the development was estimated taking as a basis a starting point of 12 °C, which was derived from field data. At this temperature no development was detected in the field, though van Wingerden et al. (1991) found a starting point of development at 10.8 °C. Using this estimation the species needed an integral temperature of 330 °C per generation.

The leaf hopper *Arthaldeus pascuellus* is a bivoltine species. The development of larvae requires five larval stages. *A. pascuellus* hatches in May and finishes the first generation in June; during summer the second generation develops. Eggs pass an eudiapause stage in winter.

Comparing the years 1990–1993, all larval stages were already found in the nineteenth week of 1990, whereas only the first larval stage occurred at that date in 1991 and 1992. The high temperatures in summer 1992 contributed to a fast development with the result that for the second generation no difference was recognized compared with 1990. Based on a starting point of 0 °C (Müller 1980; Dobler 1985) *A. pascuellus* required an integral temperature of 1450 °C for development per generation.

6.5.3   Succession from Grassland to Alder Carrs

The secondary succession of the fauna or the vegetation of abandoned wet grassland (A8, A9) was investigated for periods of seven and 12 years (1988–1999), respectively, and compared with the climax stage of the alder carr (W6). Vegetation dynamics of

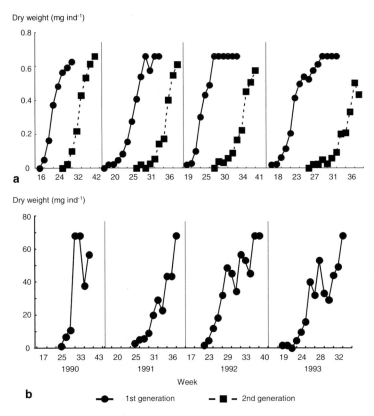

Fig. 6.18 Growth of the leaf hopper *Arthaldeus pascuellus* (**a**) and the grasshopper *Chorthippus albomarginatus* (**b**) in the abandoned dry grassland on the slope (A6) during the 1990–1993 period. *Figures* on the *x*-coordinate indicate current weeks of the year

the grassland were most intensively studied after abandonment on permanent plots recording species composition at 14-day intervals. The faunal succession of ground beetles and spiders was analysed by collecting specimens in three replicate pitfall traps between 1989 and 1995 or 1996. According to the changes during that period, species were classified into the following succession guilds: (a) disappearing, (b) abundance decreasing, (c) abundance constant, (d) abundance increasing, (e) abundance with maximum in a few years, (f) sporadic. Permanent species were determined as indicators of the intensity of change by counting the species occurrence in successive years.

The land use of the pasture changed between extensive cattle grazing, abandonment and mowing, resulting in short-term changes of plant and animal species composition (Fig. 6.19). The endurance of plant species shows no succession changes, but species richness slightly decreased and tall-growing grasses dominated probably as a result of the reduced grazing intensity.

In contrast to pastures, the species composition of the fallow changed distinctly. Plant species richness increased in succession stage I until 1990, remained constant

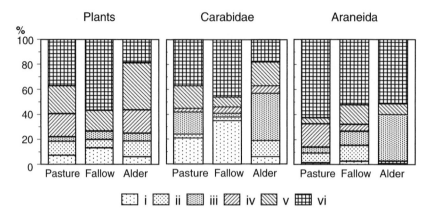

Fig. 6.19 Composition of successional guilds in the wet grassland pastured (A8), and in fallow land or the alder carr, respectively (species: *i* disappearing, *ii* decreasing, *iii* constant, *iv* increasing, *v* with maximum, *vi* remaining)

in the following stage II until 1993 and decreased in the last stage (Jensen and Schrautzer 1999). In stage I the vegetation was dominated by light-demanding species of the Calthion, e.g. *Caltha palustris* and *Lychnis flos-cuculi*. In stage II tall-growing species, e.g. *Alopecurus pratensis* and *Carex acutiformis*, invaded and became dominant while light-demanding species disappeared. In the last stage species richness decreased from 20 species to 12 species. Canopy hight increased distinctly mainly due to the dominance of *Carex acutiformis*. Further herbaceous species included in particular nitrophilous species, e.g. *Urtica dioica* and *Galeopsis tetrahit*.

The alder carr can be defined as the final stage of the weakly drained grassland in the secondary progressive succession. The species richness is on a constant level and species composition remains more or less the same, in particular with respect to ground beetles and spiders.

6.5.4 Changes during Decomposition Processes

The decomposition of the leaf litter takes only a few years (Irmler 1995; cf. Section 3.5), whereas wood decomposes over several decades (Harmon et al. 1986). Changes in the community during the different decomposition processes are related to the successive alteration of the substrate (Takeda 1995; Irmler 2000), the change of competition or food conditions (Anderson 1975; Dilly and Irmler 1998), and eventually the seasons and the different climate of the years (Irmler 1996).

The changes in the community during litter decomposition were investigated from January 1992 until October 1994 using mesh bags with 5 mm mesh size each filled with 10 g dry weight autumnal leaf litter which was harvested every month (alder carr) or every second month (mixed forest, beech forest). The fauna was determined by heat extraction of four replicate mesh bags/date (Irmler 1996; Dilly and Irmler 1998; Irmler 2000; cf. Section 1.3.3).

Collembolan and oribatid mite assemblages changed differently during the breakdown process (Irmler 1996). At the beginning, the oribatid assemblage was similar in the three stands. It successively changed during three years and was significantly different in the forests at the end of the experiment. In contrast, Collembola displayed a more seasonal change in species composition. The two animal groups included species of the early, medium and late stages (Fig. 6.20). An ordination of the species by means of a cluster-analysis (merging rule: unweighted pair-group-average, distance measure: 1-Pearson r) resulted in a cluster for an early and a late assemblage (Table 6.4). Both the early and the late assemblages appeared

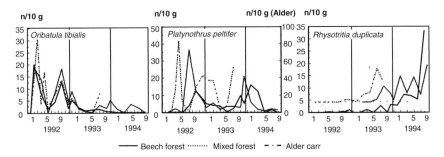

Fig. 6.20 Changes in the normed abundance of collembolan or oribatid species during breakdown litter in three forests of the core area (W1, W5, W6)

Table 6.4 Mean occurrence of the collembolan (*Coll.*) and oribatid mite (*Orib.*) assemblages, minimum and maximum occurrence of species and species richness of the two successive assemblages during the breakdown process of leaf litter in the four forest types (W1, W5, dry and wet facies of W6) of the core area

	Beech		Mixed		Dry alder		Wet alder	
	Coll.	Orib.	Coll.	Orib.	Coll.	Orib.	Coll.	Orib.
First assemblage								
Mean occurrence (month)	10±2	10±4	9±3	12±3	6±2	7±2	9±2	8±2
Mean occurrence (monthly min.–max.)	7–13	7–15	5–15	7–18	3–9	4–9	4–9	6–12
Species richness	6	5	17	12	13	10	18	9
Diversity (Hs ln)	0.9	0.8	1.5	2.3	1.6	1.4	1.7	2.0
Second assemblage								
Mean occurrence (month)	21±6	21±5	22±2	22±3	12±3	13±3	12±4	14±2
Mean occurrence (monthly min.–max.)	12–32	12–30	17–24	16–29	7–18	9–20	6–12	9–16
Species richness	21	19	10	28	18	22	8	18
Diversity (Hs ln)	2.0	2.5	1.2	2.6	1.2	2.5	1.4	1.7

after one year and two years in the beech (W1, W2) and the mixed forest (W5), respectively. In the two alder facies the two communities used to occur after 0.5 years and 1.0 year. On average the oribatid mites were found slightly later than the Collembola.

Dominant collembolan species in the early assemblage were *Lepidocyrtus lanuginosus* and *Isotoma notabilis*. *Isotomiella minor, Onychiurus armatus*, and *Folsomia quadrioculata* contributed to the dominant late collembolan colonizers. In all forests the oribatid mites *Oribatula tibialis* and *Adoristes ovatus* were early colonizers, while *Nothrus silvestris* and *Hypochthonius rufulus* were late colonizers. Many species occurred later in the beech forest than in the mixed or alder forests (Irmler 2000).

According to Anderson (1975), seasons scarcely influence the succession of the litter colonization by animals. Changes in the springtail community were related to the N-dynamics during litter breakdown (Tanaka 1995). In the forests late colonizers were more frequent among oribatid mites than among springtails as was also found by Beck et al. (1988) and Anderson (1975). Early colonizers were similarly represented in both animal groups. The colonizing process is also determined by the available food resources (Hasegawa and Takeda 1996). In particular the changing fungal colonization results in a change of food requirements for soil animals (Anderson 1975). In the alder carrs, primarily cellulose and lignin are transformed by bacteria and fungi within a densely connected food web at the beginning of decomposition (connectivity 0.22). At later decomposition stages (connectivity 0.19) fat and starch decomposition became more important (Wachendorf et al. 1997b; Dilly and Irmler 1998).

During the decomposition of dead wood three stages can be distinguished (Simandl 1993): an early stage with fresh withering wood, a median one with hard dead wood, and a late stage with soft dead wood. As a result of the long decomposition process, the different participation of microorganisms, the environmental variability and the different tree species bring about a very heterogeneous decomposition pattern, which is reflected in a high arthropod diversity (Schimitschek 1952; Kaila et al. 1994; Siitonen 1994).

In the Bornhöved stands a more differentiated age classification appears indicated with regard to the analysis of beetles and the midge families Mycetophilidae and Sciaridae (Irmler et al. 1996; 1997): (a) the bark is fixed firmly to the solid wood, (b) the bark is fixed weakly to the less solid wood, (c) the form of the soft wood is still recognizable, (d) dead wood has become amorphous. The animals were registered using 50 emergence traps with a total volume of $1.7\,\mathrm{m}^3$ of wood.

A total of 228 beetle, 38 sciarid, and 55 mycetophilid species were found. Species richness of beetles, sciarid and mycetophilid midges increased with the age of the dead wood from 19 species/sample, 7 species/sample and 2 species/sample in age class 1 to 33 species/sample, 12 species/sample and 8 species/sample in age class 4, respectively. The composition of the species community changed during wood decomposition. Species of the early stages were the beetles *Anomognathus cuspidatus* and *Cerylon histeroides* and the mycetophilid midge *Apolephthisa subincana*. Typical species of old stumps and logs were the beetles *Corticaria*

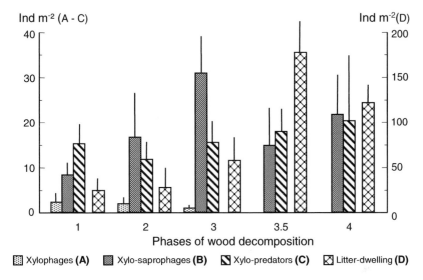

Fig. 6.21 Abundance of beetle guilds in different age classes on and in dead wood of W1, W4, W6 stands

elongata and the sciarid *Lengersdorfia detriticola*. During the late stages species also occurred which normally colonize the litter layer, e.g. the beetle *Tachyporus hypnorum* and the midges *Ctenosciara hyalipennis* and *Phronia basalis*.

Both mycetophilid and sciarid species were exclusively found in dead wood, in the litter layer and in both substrates with 46%, 32%, and 22%, and with 30%, 45%, and 25%, respectively. About 39% of the beetle species were saproxylic. Xylophagous beetle species were relatively frequent at the beginning of decomposition, e. g. Cerambycidae and Scolytidae (Fig. 6.21). During intermediate decomposition stages mainly saproxylic species occurred which were replaced in the later stages by ground-living species. Predatory species contributed to the faunal density during all decomposition stages, with high abundance figures.

6.6 Conclusions

Biotic dynamics is largely influenced by spatial heterogeneity. Stratification in the soil profile depends on changing environmental parameters within the transect from the lake to the hill top, while the dynamics in the different horizons is mainly affected by the land use pattern. Within the vegetation layers interactions of animals are well developed. The strata are differently used by the same species in correspondence with the distance from the lake, which may be due to the variable pattern of food resources on the ground. Thus, many animal species of the soil layer, e.g. dipteran species, use the whole system from the soil layer to the canopy, while another part of the fauna, e.g. wood-living beetles, only use separate layers of the tree strata.

Ecosystem size is an important parameter influencing biotic dynamics. As all ecosystems of the study area are comparatively small, marginal effects on the structure of the biota are quite remarkable. Furthermore, the beech forest is additionally intersected by several forest tracks which also affect the biotic dynamics, e.g. the vegetation cover, the distribution of invertebrates and birds and thereby the dynamics of food resources. As a result of the small-sized ecosystems several species can easily subsist on different ecosystems in correspondence with the seasosonal or interannual fluctuations of environmental conditions. Thus, many species of the agrarian systems use the forest to hibernate, populations migrate according to moisture conditions between dry and wet sites. The changing land use results in succession processes and alters the composition of the community.

Biotic dynamics is, therefore, difficult to analyse on the landscape level. They cannot be explained by the normal seasonal changes or the overall climatic differences between successive years alone. The biotic interactions within the complex landscape, in particular the environmental changes from the lake to the hill top and from intensively tilled agrarian systems to little-managed forests, substantially influence ecosystem dynamics.

Chapter 7
Biocoenotic Interactions between Different Ecotopes

Ulrich Irmler, Franz Hölker, Hans-Werner Pfeiffer, Walter Nellen, and Hauke Reuter

7.1 Introduction

In comparison with a relatively uniform landscape organisms in a complex one have to cope with a distinctly higher number of problems. On the one hand, the irregular habitat pattern may be inducive to higher biodiversity, on the other isolation of habitats may result in a decrease in species numbers, if extinction is not compensated by immigration. In fractal landscape models, the number of species coexisting in microsites increased up to the fractal dimension of 2.75, but decreased with higher dimensions (Palmer 1992). Studies on ground beetles in the Netherlands showed a significantly higher proportion of species with high migration potentials in small habitats than in large ones (Vries et al. 1996). In stressed habitats an opposite effect might develop. Thus, Fahrig and Jonson (1998) found an increasing species richness in alfalfa fields with increasing isolation.

In diverse landscapes the quality of habitat fringes is, therefore, important for the interactions of habitats through species and the species richness of the habitats. Species interacting between different habitats can also influence ecosystem functions, e.g. seed dispersal or fructification of plants. For the pollinator–plant interaction Steffan-Dewenter and Tscharnke (1999) found that the proportion of large bees and bumblebees was higher in small habitats than in large habitats, which was attributed to the better flying ability of the larger species.

Borders between ecosystems (ecotones) may either serve as guidelines for migrating animals or as barriers that inhibit the dispersion of species. Corridors between ecosystems are generally regarded as structural prerequisites to optimize the migration into isolated habitats. However, corridors with suboptimal living conditions are only used by a few species (Collinge 2000) and only over short distances (Haas 1995). Furthermore, the interactions between habitats depend on their sink or source qualities (Cornelsen et al. 1993; Irmler et al. 2000).

In the present inquiries different borders between habitats were investigated: the transitional zone between terrestrial ecosystems and Lake Belau, the ecotones between the beech forest and the arable land, between the beech forests and the mixed forest on the slope, and between the littoral and pelagic of Lake Belau (cf. 1.5). Research focused on the quantification of movements across the habitat boundaries

O. Fränzle et al. (eds.), *Ecosystem Organization of a Complex Landscape.*
Ecological Studies 202.
© Springer-Verlag Berlin Heidelberg 2008

for different groups of animals and the function of boundaries for the orientation of animals. The interaction processes between habitats and at habitat margins were additionally simulated by models to analyse ecotone effects on the animal populations. Furthermore, model simulations should help to predict the influence of a future landscape development on biodiversity and ecosystem functions.

7.2 Interactions between Different Ecosystems

7.2.1 Interactions between Terrestrial Ecosystems

7.2.1.1 Ground Beetles

The permeability of habitat borders was investigated with different groups of animals representing different life forms: ground beetles, flies and small mammals. A variable number of ground beetle species along the gradients between the central beech forest (W1, W2) and both the arable land (A1, A2) and the mixed forest on the slope (W5; Fig. 1.4) shows the high influence of the arable land on the species composition of the beech forest (Table 7.1). The species richness of ground beetles varied between 9 and 10 species trap^{-1} in the central beech forest, increased to 14–15 species trap^{-1} at the border of the field and amounted to 12 species trap^{-1} on arable land. Between the beech forest (W2) and the mixed forest (W5) species numbers also increased to 12–14 species trap^{-1}, but were slightly lower than between beech forest and arable land (A1). In particular, species turnover was different between the two gradients. In the central part of the beech forest (W2) species turnover between traps was 0.14 to 0.06 on the average. In the 70-m marginal belt adjacent to the field species turnover was significantly higher, with 0.42 to 0.14 (U-test: $P < 0.01$). In the transition zone between the beech forest and the mixed forest on the slope no significant increase in species turnover was found. Considering species numbers and turnover rates within the beech forest, a marginal belt between 70 m and 100 m can be defined that is influenced by the arable land, while no pronounced border can be defined between beech and mixed forest (Fig. 7.1).

Table 7.1 Species richness at borders between different systems. The direction and number of species moving between habitat 1 and habitat 2 are given in the columns labelled > and <

Habitat 1–Habitat 2	Total Species richness	species richness of ground beetles (%)			
		Habitat 1	>	<	Habitat 2
Beech (W 2)–field (A 1)	45	48	42	84	86
Beech (W 1)–alder forest (W 6)	32	53	52	42	79
Arable land(A1)–grassland (A 6)	51	96	30	90	36

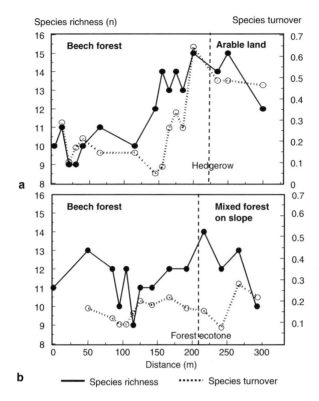

Fig. 7.1 Species richness (*n*, annual number of species per trap) and species turnover: [T= (number of species lost + number of species added)/(species in trap$_x$ + species in trap$_{x+1}$)] for ground beetles along two gradients in the beech forest: **a** from central beech forest (W1) to arable land (A1), **b** from central beech forest to mixed forest on slope (W5)

Transfer rates of species were determined along the borders between the beech forest and the adjacent arable land, between beech forest and alder carr, and between arable land and grassland (Table 7.1). Species number was highest in the arable land with 36 species. Of the 45 species found in the arable land and the beech forest 34% occurred in both systems. Also, between the other adjacent habitats about 30% of the species were common to both habitats. Low transfer rates were determined from beech forest to arable land, from alder carr to beech forest and from arable land to grassland.

The ground beetle *Abax parallelepipedus* is almost exclusively restricted to woody habitats (Irmler 2000). In the beech and alder stands and on arable land an average of 9.4 individuals trap^{-1} were caught in 1988 (Table 7.2). Some 82% of the individuals originated from the beech forest (W1, W2) and its transition to the mixed forest (W5) on the slope. Only 8.5% of the individuals were involved in exchange processes between other habitats. Interactions at the border to the alder carr were very small. According to the ANOVA test only the abundance of ground

Table 7.2 Abundance of three ground beetle species in different habitats and at their borders in 1988. The direction and number of species moving between habitat 1 and habitat 2 are given in the columns labelled > and <. Standard deviation are given in brackets, superscript letters denote significant differences

Habitat 1–Habitat 2	Habitat 1	>	<	Habitat 2
Abax parallelepipedus				
Beech (W1)–slope forest (W5)	58.0 (11.1)[a]	37.0 (5.6)[a]	21.0 (12.1)[a]	1.3 (0.6)[b]
Slope forest (W5)–alder forest (W6)	1.3 (0.6)[b]	2.3 (2.5)[b]	4.0 (4.0)[b]	1.7 (2.1)[b]
Beech (W2)–arable land (A1)	58.0 (11.1)[a]	0.3 (0.6)[b]	3.0 (1.0)[b]	0.7 (1.2)[b]
Arable land (A2)–hedgerow (A3.1)	0.7 (1.2)[b]	1.0 (0.0)[b]	0.5 (0.7)[b]	0.7 (1.2)[b]
Carabus hortensis				
Beech–slope forest	306.7 (11.9)[b]	69.7 (22.4)[c]	59.7 (15.0)[c]	396.3 (37.1)[a]
Slope forest–alder forest	396.3 (37.1)[a]	29.0 (12.5)[d]	33.3 (14.6)[d]	15.7 (12.0)[d]
Beech–arable land	306.7 (11.9)[b]	62.7 (40.1)[c]	67.3 (21.8)[c]	21.7 (2.9)[d]
Arable land–hedgerow	21.7 (2.9)[d]	55.3 (18.2)[c]	49.7 (6.7)[c]	50.3 (8.0)[c]
Pterostichus niger				
Beech–slope forest	9.0 (6.0)[b]	14.3 (4.7)[b]	6.7 (4.9)[b]	59.0 (27.9)[a]
Slope forest–alder forest	59.0 (27.9)[a]	3.3 (1.5)[b]	5.3 (1.2)[b]	6.7 (3.2)[b]
Beech–arable land	9.0 (6.0)[b]	9.3 (4.2)	16.7 (4.7)[b]	1.3 (1.5)[b]
Arable land–hedgerow	1.3 (1.5)[b]	4.0 (1.7)[b]	3.7 (2.5)[b]	1.3 (0.6)[b]

beetles in the beech forest and the transition to the mixed forest were significantly higher than in the other habitats and border zones. *Abax parallelipedus* used only the beech forest and its border zones to adjacent forests. Interactions between the other habitats were extremely low. The species was absent on arable land, except field A1. A second ground beetle species, *Carabus hortensis*, occurred predominantly in forests, but was also frequently found in other habitats (Irmler 2000). Within the core area 46% of the individuals were caught in the beech forest and its border zones, 2% of the individuals reached the field adjacent to the beech forest and only 0.4% of the individuals reached field A3, about 250 m from the beech forest. This was nearly as much as was found for *A. parallelepipedus* in the field adjacent to the beech forest. ANOVA with post hoc test yielded the following four groups, for which abundances differ significantly: the group from mixed forest on the slope with the highest abundance, the group from beech forest, the group from the forest margins, and the group from habitats with the group from more remote habitats. Some 24% of the individuals were involved in interactions between habitats outside the beech and mixed forests. *C. hortensis* can be regarded as a species which strongly participates in the exchange between habitats. The third ground beetle species, *Pterostichus niger*, inhabits various terrestrial habitats of Schleswig–Holstein (Irmler 2000). Within the core area the highest abundance was found in the mixed forest on the slope. All other habitats could not be differentiated according to abundance figures (ANOVA with post hoc test). 30% of the individuals were involved in interactions between habitats outside the mixed forest.

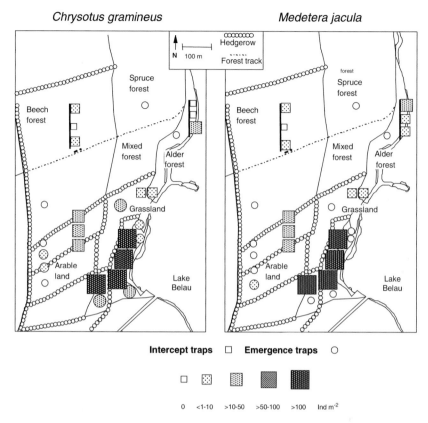

Fig. 7.2 Distribution of two dolichopodid flies in flight intercept traps and emergence traps in the core area

7.2.1.2 Flies

In addition to the ground beetles empidid, hybotid and dolichopodid flies were investigated, representing animals which use the air for dispersion. In total 31 species of Empididae, 39 species of Hybotidae, and 36 species of Dolichopodidae developed in the different habitats of the core area, according to emergence trap catches. However, these were only 52%, 34%, and 53%, respectively, of the species caught in flight intercept traps. For nearly half of the species no evidence for the development in the core area was found. A comparison of three zones differing in distance from the beech forest showed that the beech forest margin adjacent to the arable land (A1) exerted the highest attraction to the flies. All investigated 13 dipteran species occurred in a higher abundance at the beech forest margin than in the centre of the adjacent arable land. Only four species were found more frequently in the hedgerow opposite the beech forest than in the centre of the arable land (Table 7.3).

Table 7.3 Abundance of empidid fly species in flight intercept traps at the beech forest margin (FB; W2.2), in the central field (CF; A2), and at the adjacent hedgerow (H; A3.1) between two fields. *n.s.* Not significant, *** $P<0.001$, ** $P<0.01$, * $P<0.05$, according to U-test

Species	Individuals m^{-2} year $^{-1}$			Significance	
	FB	CF	H	FB > H	H > CF
Rhamphomyia crassirostris	360	1	93	***	***
R. longipes	290	8	18	***	*
Empis chiroptera	202	4	15	***	n.s.
E. nuntia	174		5	***	*
E. aestiva	124	14	23	***	n.s.
Hilara maura	94		11	***	*
Rhamphomyia sulcata	39	1	3	***	n.s.
H. interstincta	34	1	5	***	n.s.
R. plumipes	35		2	***	n.s.
H. intermedia	28	1	3	***	n.s.
R. tarsata	26		1	**	n.s.
R. sulcatella	15			***	
E. tesselata	13		1	*	n.s.
Total abundance	1434	28	180		
Species number	13	7	12		

The most frequent species was *Rhamphomyia crassirostris*, which occurred with 0.3 individuals m^{-2} in the grassland. Therefore, nearly 18 000 individuals emerged in the grassland covering 6 ha. Along the 400-m margin of the forest the catches in a 1 m^2 flight intercept trap yielded more than 450 individuals. Although the trap itself might have attracted the flies and extrapolations would therefore appear biased, it is evident that the beech forest margin appears particularly attractive to the flies. The reason is the swarming behaviour of the species which prefers conspicuous landmarks sheltered from strong wind (Tréhen 1977; Delettre et al. 1992). In addition, also the southerly exposure of the wood margin seems to play a role, in particular in comparison with the northern flank of the hedgrow.

Even among the dolichopodid flies which have no swarming behaviour, several species had a high dispersal potential. *Chrysotus gramineus* and *Medetera jacula* developed in the arable field and in the grassland (Fig. 7.2). *M. jacula* emerged with 14 individuals m^{-2} year^{-1} in the field, *C. gramineus* with 1 individual m^{-2} year^{-1} in the field and with 5–8 individuals m^{-2} year^{-1} in the grassland. From these habitats they flew into the beech forest and the alder carr, where they were found even at heights of 19 m and 27 m (cf Chapter 2). In particular *M. jacula*, developing only on arable land, was more frequent in the canopy region of the alder carr than *C. gramineus*. *M. jacula* seems to disperse mainly along the margin between the forest and the open terrestrial habitats or between the alder carr and the lake, respectively.

Table 7.3 specifies the dispersal behaviour of 11 species. Six of them reached adjacent habitats, out of which four originated from the grassland, one species from the arable land, and one from woody habitats. The four species which developed in the grassland dispersed along the ecotone between the lake and the alder carr. Even *Hybos culiciformis* [which develops in the beech forest (W2) and the adjacent hedgerow] used to disperse in this ecotone. The species dispersed directly through the forest and the high canopy layer, but also along the beech forest margin and the hedgerow. Thus, flies appear much involved in biocoenotic interactions within the research area.

7.2.1.3 Mammals

Small mammals are the third group which contributes essentially to habitat interactions. In the beech forest the Yellow Necked Mouse (*Apodemus flavicollis*) is the most abundant small mammal species, with more than 50% of the total abundance of small mammals. In summer the largest part of the population used to feed on the adjacent arable land. For 46% of the individuals (11 individuals from 24 investigated mice) oat accounted for more than 50% of the intestinal food remnants. In July oat was the preferential diet, amounting to more than 40% of all remnants, on average, although the oat field was 300 m distant from the beech forest (Table 7.4). After harvest oat remnants (mainly grains) decreased in the diet and beech remnants, in particular beech nuts, were the main food resource. The corn field directly adjacent to the beech forest was of low importance, because corn was not ripe in July. Also in late summer corn was not used, since the food supply seemed to have been sufficient in the beech forest during this season.

7.2.2 Interactions between Terrestrial and Aquatic Ecosystems

With a few lepidopteran species short migrations between terrestrial and aquatic stands of reed could be observed. The inhabitation of the noctuid moth *Archanara* sp. mining in reed blades is related to the diameter of reed blades and the growth of the animals. Only reed stands with more than 30 blades m^{-2} and a basal diameter

Table 7.4 Diet composition (vol% of the total intestinal content) of the Yellow Necked Mouse (*Apodemus flavicollis*) in the beech forest (W1, W2) during the July to September period. Standard error is given in brackets

Month	July	August	September
Number of investigated individuals	24	20	3
Beech remants	1.7 (8.1)	54.8 (43.7)	83.7 (14.3)
Remaining plant remnants	33.5 (39.1)	26.3 (39.6)	11.7 (16.9)
Oat remnants	43.3 (41.0)	11.1 (29.8)	
Undefined remnants	13.0 (26.1)	3.2 (11.5)	
Arthropods	4.4 (9.0)	4.6 (6.1)	4.7 (7.2)
Corn remnants	3.2 (15.1)		
Fungi	0.9 (3.3)		

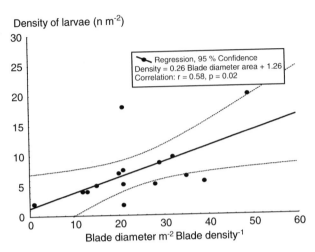

Fig. 7.3 Density of *Archanara* larvae in relation to blade diameter (blade diameter m^{-2} vs blade density of the reed stands adjacent to W6)

of at least 3 mm were inhabited. Reed stands with an average basal blade diameter of more than 6 mm and a density above 90 blades m^{-2} were preferred (Grabo 1991). Thus, a positive correlation between the blade diameter area (blade diameter m^{-2} blade density) and the density of *Archanara* larvae could be defined (Fig. 7.3).

The larvae were conspicuously abundant in semi-aquatic stands. Young larvae hibernate in the blades of terrestrial reed stands and migrate to young sprouts in spring. They migrate on the ground, because leaves can only be used for migration if the blades are fully developed. Since the larvae cannot cross inundated ground, they need a non-inundated spot or a thick litter layer during this season. Using marked blades, it was estimated that one individual larva needed three to six blades for pupation. Blade changing is therefore necessary for the larvae. A supply of four to five blades was determined in cultivation experiments in the laboratory.

During the course of the year the growing larvae migrated from central stands to aquatic stands, because aquatic stands had a higher average blade diameter. This is obvious for the number of blades *Archanara* larvae feed on (Fig. 7.4). Adult moths emerged at the beginning of August predominantly from the lake-side reed stands, but oviposition was performed in central reed stands with wet terrestrial conditions.

7.2.3 Interactions within Aquatic Ecosystems

Interactions within the limnetic, littoral and profundal zone of Lake Belau were investigated for adult fish (>2 years old) and 0-group fish (<1 year old).

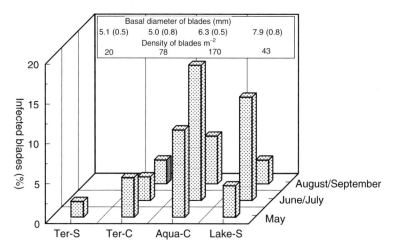

Fig. 7.4 Spatial and seasonal distribution of reed blades infected by *Archanara* sp. moths within a transect in front of the wet grassland (A9) of the core area from the terrestrial side (*Ter-S*), central stand with wet terrestrial conditions (*Ter-C*), central stand with aquatic conditions (*Aqua-C*) to a riparian stand (*Lake-S*)

7.2.3.1 Adult Fish

Fish behaviour and density were determined by a sonar technique between March and October during the 1989–1991 period (Pfeiffer 2000). Species composition and structure of fish populations were analysed by beach seine and electric fishing.

From April to October the fish occurred mainly in the pelagic zone of the main basin during the daytime and also at night when they avoided the littoral (Fig. 7.5a). At dusk and dawn fish migrated into the littoral feeding there, both at the bottom and near the water surface. This diurnal migration behaviour was observed directly and by horizontal echo sounding. In the open water fish stayed mainly at a depth of 2 m to 12 m but occurred also down to 21 m, i.e. below the thermo- and oxycline, respectively. In October/November the fish disappeared gradually from the pelagic water column and started to form dense winter aggregations on the deep bottom (Fig. 7.5b). During the warm season large parts of the fish population were thus feeding in the littoral while defecation and excretion took place in the limnetic zone. So nutrients were transferred from the littoral into the limnetic zone.

This nutrient transfer by fish was investigated in the two dominant species: roach (*Rutilus rutilus*) and bream (*Abramis brama*). These two accounted for 60% and 13.5%, respectively, of the total fish abundance and for 39% and 36%, respectively, of the total fish biomass. The latter was estimated at 204 kg ha^{-1} (fresh weight) and the former at 2360 fishes ha^{-1} (Pfeiffer 2000). The intestinal content of 336 breams and 229 roaches caught near the shore and in the deeper part of the lake were analysed (Table 7.5). Gastropods, chironomid larvae, other insects, and Turbellaria

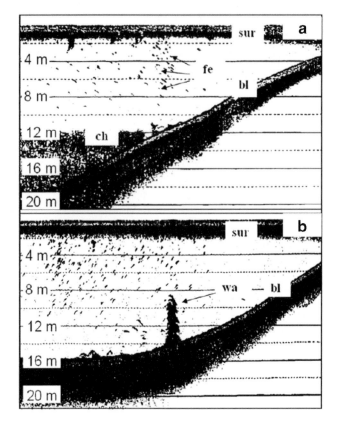

Fig. 7.5 Sonic depth diagrams illustrating the distribution of fish on two different days: **a** 26 September 1990, **b** 28 October 1991; both diagrams produced at 1100 hours. *sur* Surface, *bl* bottom line, *fe* fish echoes, *ch* Chaoboridae, *wa* winter aggregation of fish

Table 7.5 Seasonal variation of food composition (% of food mass) and food consumption of bream and roach in Lake Belau (kg fresh weight ha^{-1}). Periods: *I* January–March, *II* April–June, *III* July–September, *IV* October – December

Periods	Bream				Roach			
	I	II	III	IV	I	II	III	IV
Number of intestines	54	56	131	95	55	50	97	27
Gastropoda	4.4	1.4	1.0	–	5.1	–	32.1	5.1
Insecta	–	–	–	–	–	15.5	31.4	10.7
Chironomidae	42.1	62.9	64.2	22.8	2.1	27.8	16.8	1.8
Chaoboridae	36.8	-	25.9	66.1	–	–	–	–
Turbellaria	15.2	34.7	0.2	5.2	–	–	–	–
Copepoda	0.1	0.9	2.3	3.0	0.3	36.4	0.1	3.2
Cladocera	0.1	0.1	4.2	2.8	1.4	19.9	4.4	61.5
Total consumption	77.6	251.1	456.3	161.0	64.8	167.3	315.3	128.2

were consumed in the littoral zone, whereas in the open water the diet comprised chaoborid larvae, copepods, and cladoceras. Bream fed mainly on chironomid and chaoborid larvae, while roach had a wider food spectrum, additionally characterized by zooplankton and gastropods. During summer, zooplankton accounted for 45% in abundance, but only for 2% in biomass of the food of bream. Roach mainly fed on snails in spring and more heavily on zooplankton in summer when zooplankton accounted for more than 60% in abundance and 36% in biomass in the stomachs of this species. For both fish populations it was estimated that throughout the year 70% of the food biomass originated from the littoral (July–September 78.6–79.7%) and 30% from the open water. Thus, the fish populations accounted for a considerable nutrient transfer from the littoral into the pelagic zone, where the primary production increased in the limnetic zone.

It was investigated how important this process actually was in contributing to primary production in the lake and, furthermore, how much the fish could influence the stocks of their food organisms in the littoral zone. To this end, the food consumption of the adult fish population was calculated by means of a bioenergetic model, based on growth and related physiological data and also allowing the determination of nutrient remineralization rates. The total annual food consumption of bream and roach amounted to 946 kg ha^{-1} and 676 kg ha^{-1} (fresh weight), respectively (Table 7.6). Q/B (consumption/biomass) ratios of 13 for bream and 8 for roach were attained. These relatively high values resulted mainly from high losses of energy and mass during the cold season. Bream lost 73% fat, 26% energy, 21% carbon, and 8% nitrogen of its last summer maximum. For roach these losses were even higher, with 87% fat, 43% energy, 41% carbon, 27% nitrogen, and 10% phosphorus.

On an annual basis nutrient remineralization by bream and roach was of minor importance (Table 7.6); during summer, however, when phosphorus became the minimum factor, it became relevant. Thus, in August 1991 a rate of 28% of the total phosphorus remineralization was caused by fish activity in the sub-littoral and profundal zones. In October 1991 it amounted to 21%, and in June/July to nearly

Table 7.6 Biomass, food consumption and total nutrient remineralization of bream and roach in Lake Belau

Parameter	Units	Bream	Roach	Total
Biomass	kg ha^{-1}	74.1	79.4	153.5
Consumption	kg ha^{-1} year^{-1}	946.1	675.6	1621.7
Consumption/biomass ratio (Q/B)		12.8	8.5	10.6
Q pelagic species	kg ha^{-1} year^{-1}	295.0	192.0	1135.0
Q littoral species	kg ha^{-1} year^{-1}	651.0	483.0	487.0
Nutrient remineralisation				
Phosphorus	kg year^{-1}	159.0	118.0	276.0
Nitrogen	kg year^{-1}	1490.0	771.0	2260.0
Littoral phosphorus	kg year^{-1}	111.0	82.0	193.0
Littoral nitrogen	kg year^{-1}	1043.0	539.0	1581.0

6% and 10%. Thus, the role of the fish for phosphorus release was of some importance and was linked with the high phosphorus demand for primary production at a time when the phosphorus content of the epilimnion was not detectable ($<5\,\mu g\ l^{-1}$ phosphate-phosphorus). In August, the daily phosphorus demand for phytoplankton production was $7.8\,g\ m^{-2}$ whereas the daily total remineralization rate of fish, phyto-, and zooplankton together amounted only to $4.7\,g\ m^{-2}$. Additional but minor phosphorus-releasing processes may be due to other members of the pelagic community such as Rotifera, Protozoa, and bacteria.

Nearly no phosphorus loss or gain was found in August 1991 when the primary production was limited by phosphorus remineralization of the pelagic community (Landmesser 1993). The amount of phosphorus released by the two dominant fish populations corresponded to 17% of the total primary production or an increase in algal productivity by 21%. This was made possible mainly by a nutrient transfer from the littoral zone into the pelagic water column where most of the planktonic primary production takes place. In the past such rates of fish phosphorus remineralization during phosphorus limitation in summer were likely to have been underestimated for the following reasons: (a) phosphorus remineralization by phyto- and zooplankton was overestimated due to decreasing phosphorus contents in algal cells and zooplankton excretion in summer, (b) sedimentation of algae and lysis and total remineralization had been calculated only for the epilimnion, whereas sedimentation and nutrient transport into the hypolimnion had been neglected, (c) phosphorus remineralization of algal cells was partially taken into account twice, namely in the form of autolysis and of zooplankton grazing and excretion. Furthermore it must be taken into consideration that fish also transported nutrients from the pelagic water column into the littoral zone in spring and early summer, as most species deposited their eggs in shallow waters. This nutrient transport was assessed at $27\,kg\ ha^{-1}$ (fresh weight), $4.4\,kg\ ha^{-1}$ carbon, $1.1\,kg$ ha^{-1} nitrogen, and $0.07\,kg\ ha^{-1}$ phosphorus (cf. Chapter 9).

The importance of fish predation on the population of food organisms was estimated by comparing fish consumption and production of prey biomass (Table 7.7). Fish consumption came up with 90% and 200% of the mean biomass of the taxa Ceratopogonidae, Mollusca, Copepoda, and Cladocera. These values are equivalent to 13% and 33% of the annual production of these groups.

The fish preyed especially hard on Trichoptera and Chironomidae. The consumption exceeded 8–10 times the mean biomass of this prey, which was more than its calculated annual production. From May to August the fish reduced the biomass of Chironomidae and Trichoptera drastically, while the effects of fish predation on Mollusca, Cladocera, and Copepoda were less severe. Thus, the breakdown of the crustacean peak in June–July could not be caused by adult fish. They may only reduce the mean size of specimens in the respective populations of their food organisms.

On an annual basis nutrient remineralization by fish was not extremely important. Only in summer were phosphorus and nitrogen remineralized to a relatively high degree in the pelagic zone by the dominant fish species of the lake. Analysis findings are reported, for instance, by Vanni and Findlay (1990), Carpenter et al.

Table 7.7 Fish consumption in relation to biomass and production of prey taxa in Lake Belau

Feeding taxa	Biomass	Production	Emergence	Bream and roach consumption		
					% of prey	
	kg ha^{-1}			kg ha^{-1}	Biomass	Production
Cladocera	74.29	621.85		150.78	202.94	24.25
Copepoda	78.20	654.49		83.21	106.41	12.71
Chironomidae	71.30	534.72	25.73	623.59	874.65	116.62
Trichoptera	9.48	28.44	4.34	93.42	985.50	328.50
Ceratopogonidae	0.95	3.34	?	0.87	91.66	26.19
Mollusca	246.51	728.43		242.91	98.54	33.35

(1992), He et al. (1993), Kraft (1993). Contrasting data are presented by Paulson (1977), Nakashima and Leggett (1980, Brabrand et al. (1990), Kraft (1992), and Schindler et al. (1993). But Kraft (1992) described a possible methodological overestimation of phosphorus remineralization by zooplankton in the study of Nakashima and Leggett (1980); and the same may apply to the study of Brabrand et al. (1990). Thus, these fish populations could in fact have played a more important role in nutrient remineralization than stated by those authors.

The transfer of nutrients by fish from the littoral into the open water increases the primary production in the lake at a time when phosphorus is a limiting factor. This confirms the thesis of Shapiro and Carlson (1982) which says that primary production is increased above the level supported by allochthonous nutrient influx when benthivore fish make daily feeding migrations from the offshore to the inshore areas. Also Schindler et al. (1993) and He et al. (1993) provided evidence that the structure and feeding modes of the fish population are of considerable importance for nutrient remineralization in a lake. Nearly no transfer effect in the above sense is to be expected, for instance, when pelagic and planktivorous fish are the dominant groups, as they keep off the near-shore areas and rather stay in the epilimnion where nutrients are then recycled only (Shapiro and Carlson 1982; Brabrand et al. 1990).

Similar effects of fish predation on littoral invertebrate populations, especially Chironomidae, were also described by Hayne and Ball (1956), Kajak (1968, 1988), Morgan et al. (1980). Chironomids and other benthic organisms are always preferred as food by bream and roach whereas zooplankton is only a substitute (Bauch 1966; Mann 1973). Altogether the above results illustrate a typical feature of a mesotrophic to eutrophic lowland lake (Landmesser 1993; Barkmann 1998) where the forgoing habits of first-year (or O-group) fish are of particular importance for the development of zooplankton communities (cf. also Chapter 10).

7.2.3.2 0-Group Fish

0-group fish preyed mainly on zooplankton. Whether they were able to control the zooplankton populations could not be answered definitively because a really satisfactory technique to quantify the abundance of <1-year-old fish in lakes is still lacking. Usually catches are extremely variable regardless of the kind of gear used (Bertram 2002) and therefore both 0-group fish consumption and zooplankton production, can be assessed only roughly.

Birth and mortality rates of pelagic *Daphnia* populations were determined and compared with the estimated consumption of 0-group fish (Fig. 7.6). Simplifying, a constant abundance of 1.5 fish m^{-3} was assumed which corresponds to the highest concentration found so far. The data show population declines of *Daphnia* in June 1991 and May 1992, respectively. These happened before the juvenile fish were able to forage on *Daphnia*. In July the daily predation on this zooplankton amounted to a maximum of three specimens, whic was close to the estimated mortality rate of the *Daphnia* population. Thus, the 0-group fish did not cause an early collapse but rather prolonged the summer depression of this prey population.

With calanoid copepods inhabiting the pelagial the situation is similar, i.e. the decline of their population started in the second half of May. It was linked to a clear water phase, but was not due to fish larvae predation. The latter did not exceed 0.3 mg m^{-3} day^{-1} (dry weight) at that time of the year. In June and July, however, 27 mg m^{-3} dry weight of copepods or 15% of their standing crop were devoured by fry every day.

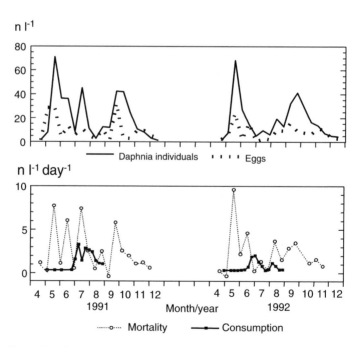

Fig. 7.6 Comparison between the estimated mortality rate and the estimated daily consumption of *Daphnia* sp. by 0-group fish (Bertram 2002)

The critical density of 0-group fish being able to control population dynamic processes of zooplankton in the pelagial is discussed controversially. Mehner et al. (1995, 1997) mentioned that in the Bautzen reservoir a biomass of 50–60 kg fresh weight of 0-group percoids per hectare are needed to reduce the *Daphnia* population substantially. For other lakes threshold values of 20–50 kg fresh weight of 0-group fish were estimated (Mills and Forney 1981; Dunkan 1997). In Lake Belau the highest biomass of 0-group fish found was 3 kg ha^{-1} (fresh weight).

In the littoral zone both 0-group fish and zooplankton had an extremely patchy distribution which made it difficult to obtain reliable data on their quantity. In 1992 a reduction in the abundance of *Bosmina* was observed at the end of May, at a time when 0-group cyprinids were not yet able to forage on these crustaceans. Two weeks later, in early June, fry densities of >100 specimens m^{-2} were found which probably caused a predation pressure of 40% of the *Bosmina* biomass. It is likely that fish larvae were responsible for the reduction in the concentration of this zooplankton from about 11 mg m^{-3} (dry weight) down to only 50 mg m^{-3} (dry weight) until to the middle of June.

The biomass of *Daphnia* remained permanently below 240 mg m^{-3} (dry weight) in the littoral zone during spring and summer while fish larvae consumed approximately 80 mg m^{-3} (dry weight) by the end of June. This means that more than 30% of the standing crop of *Daphnia* was grazed up by 0-group fish daily. Such an amount would have exceeded the regeneration rate of the population. Therefore, it is assumed that the *Daphnia* found in the stomachs of the fry had been consumed in the offshore area.

7.3   Modelling Species Interactions between Habitats

In order to extend the empirical investigations to other ecological situations several individual-based models were developed (Breckling et al. 1997; Hölker 2000; Reuter 2001). These models serve to support a better understanding of the empirical results and to extrapolate the results in time and space to other environmental situations and species with comparable properties and permit to integrate spatial heterogeneity and processes on different organisational levels (Reuter and Breckling 1999). Furthermore, the models may help to identify relevant details of behavioural dynamics (Schmitz 2000). Two models dealing with interactions within terrestrial and aquatic ecosystems are presented.

7.3.1   Modelling Interactions between Terrestrial Habitats

The terrestrial model 'Faust' (from German: *Faunenaustauschmodell*) was developed to analyse the dispersal of ground-dwelling arthropods in correspondence with the landscape structure and the biological properties of the species investigated (Reuter

2001). For the core area the model was used to investigate the conditions of population survival, the dispersal to suboptimal habitats, and the factors responsible for distribution.

The 'Faust' model describes the dispersal of wingless ground beetles on the individual level (Fig. 7.7). Each model individual has the same set of variables describing its properties, e.g. position in space, age, and reproduction state, and a set of rules (activity repertoire), which determines how the variables change in correspondence with the local environment and the inner state of the individual. The activity repertoire is limited to the behavioural processes which are essential for dispersal. The motility characteristics are described by speed, distribution of step length, turning angles and their distribution, seasonal and daily activity time, temperature dependence, behaviour at habitat boundaries, and perception of the environment. As dispersal processes at the landscape scale usually exceed the lifetime of an individual carabid beetle, habitat dependent mortality and reproduction processes are included in the model. Stochastic processes in the implementation of the behavioural repertoire and the local environmental situation lead to different motility patterns and fertility of model individuals.

The model environment is implemented in a two-dimensional grid of different habitat types such as forest, hedgerows, pasture, and arable land. As the habitat-specific parameterization of the carabid behaviour is supplied by independent parameter sets, the number of habitat types may be extended easily. Further spatial information connected to the grid map concerns, for instance, the spatial resistance and the translucency of the vegetation as an important orientation factor for carabid beetles (Neumann 1971). Spatial attributes may be changed dynamically

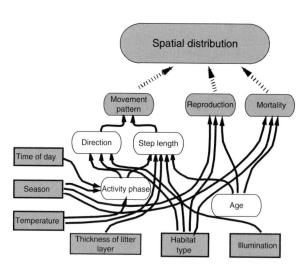

Fig. 7.7 Biological and environmental parameters included in the Faust model. The long-term spatial distribution of carabid beetles results from environmental characteristics and biological properties such as motility characteristics and reproduction rates

during a simulation, e.g. factors affected by seasonal changes in the vegetation cover. Simulation runs were performed with a high temporal resolution (15 min per time step) to represent behaviour at habitat borders correctly, which is important to understand the dispersal process. A detailed model description is given by Reuter (2001).

Model specifications were developed for the carabid beetle *Abax parallelepipedus*, which is a dominant species in the forests of the core area (cf. Section 7.2.1). In addition to the data resulting from the present investigations, the model parameterization is based on data published by other authors on beetle abundance in different habitats (Burel and Baudry 1994; Petit and Burel 1997), their physiology and reproduction motes (Symondson 1994; Chaabane et al 1997), and their motility patterns and dispersal power (Loreau and Nolf 1993; Charrier et. al. 1997; Butterweck 1998).

A. parallelepipedus is a middle-sized ground beetle (2 cm), often recorded as dominant in temperate European forests, where it is recorded at 1000–4000 individuals ha^{-1}. In northern Germany the species is strongly restricted to woody habitats or hedgerows. The seasonal activity phase in northern Germany extends from April to October with a maximum before mid-August. Both reproduction rate (10–18 eggs female^{-1}) and walking distances (1.8 m on average in forests) are low (Loreau and Nolf 1993; Charrier et al. 1997; Butterweck 1998). The model distribution of step length is based on a hyperbolic function, which is typical for many carabid species (Baars 1979; Riecken and Ries 1992). In suboptimal habitats the implemented walking pattern changes into a directed walk with larger step length and lower turning angles (Fig. 7.8).

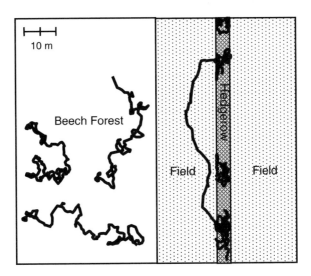

Fig. 7.8 Modelled random-walk characteristics of ground beetles in the beech forest (*left*) and the field–hedgerow complex (*right*)

The simulated movement was evaluated on a small spatial scale and for single individuals: (a) step length and distribution of turning angles for different habitat types follow data from empirical investigations, (b) the fractal dimension has typical values between 1.21 and 1.62 depending on habitat-dependent movement characteristics (Wiens et al. 1995; Butterweck 1998). The fractal dimension describes properties across scales and thus gives important information on long-term dispersal abilities.

An application of the model compared empirical data from pitfall traps with results from model traps at identical places and exactly the same operation schedule. The results from this analogous data analysing process will answer questions concerning empirically determined results on abundance and population density and provide for a feedback on the quality of the simulations.

In order to establish a basic distribution pattern for the core area, simulations started with an initial random beetle distribution in the beech forest. A stable abundance configuration was reached after simulating 20 years and provided the initial situation for all other simulations for the core area, because it strongly corresponds to the empirical data from the beech and slope forest and the hedgerows. The results of the trap simulations confirm the quality of the underlying motility assumptions. The overall distribution of trap catches is very similar between the empirical data and the simulation. Mean yearly catches were 23.4 individuals trap^{-1} in the real pitfall traps and 24.1 individuals trap^{-1} in the simulated traps. Variance and standard error, however, were lower in the simulated pitfall traps, with 92.8 and 9.6; in the real pitfall traps they were 184.9 and 13.6. This indicates that the variability of the distribution in the model is lower and that additional factors determine the local occurrence.

7.3.2 Modelling Interactions between Aquatic Habitats

The aquatic model was used to investigate size-dependent scaling effects in the simulation of diurnal habitat shifts and their role for food consumption of roach (*Rutilus rutilus*) in Lake Belau (Hölker 2000; Hölker and Breckling 2005). The individual-based modelling strategy allows to study habitat shifts with relevance to prey populations and enhances the spatio-temporal resolution of the empirical results on roach in the lake. Parameterization allows to define the swimming pattern of the fish in correspondence with the habitat and the light conditions.

Two swimming modes can be distinguished according to Wieser (1991): (a) a low-cost activity corresponding to a linear swimming mode, and (b) a high-cost swimming pattern which consists of irregular movements, e.g. starting, stopping, turning, or accelerating. This is comparable with the routine swimming pattern in the water column, e.g. during food search. The energy costs for both swimming modes were determined by respiration measurements in the laboratory (Hölker 2000).

Pfeiffer (2000) and the local fisherman observed that the roach population is active at day with a preference for the pelagic water column, and at dusk and dawn with a preference for the littoral. Roach are not active during the night and favour the pelagic water of the lake during this time; at twilight they prefer the littoral zone. These preferences were adopted to the model. The internal physiological status is controlled by Boolean variables. The status 'not hungry', for example, is triggered by fish getting the maximum daily ration. These variables affect the corresponding habitat selection and swimming mode (Table 7.8).

Roach are facultative shoalers and move frequently in groups of 3–50 individuals (Haberlehner 1988). According to observations of the local fisherman and our own inquiries, roach form groups of individuals of similar length. For roach fry Manteifel et al. (1978) observed that the fish swim and feed in shoals, when light intensity exceeds 0.1 lux. At lower light intensity the shoals disperse. The diurnal behaviour was also adopted to the model where the direction of an individual was calculated in relation to all other visible specimens. Each individual of the shoal adjusts its swimming direction to the average of all visible specimens. Swimming distances are calculated according to body length, resulting in different swimming speeds for fish of different size. Thus, only specimens of similar length join a shoal for a longer time.

Figure 7.9 shows that the preference of the littoral area at twilight is more pronounced for the simulated adult roach of age class VI than for age class I. However, adults were rarely found in the shallow southern bay during twilight, which may be explained by the long distance and the short period of twilight. During the longer daylight period this effect was not observed. As a result of the correlated random walk and the shoaling behaviour, adult roach were frequently located over the whole lake, although they preferred the pelagic area. Due to their higher swimming speed mainly larger fish swam into the southern bay, which had a longer time-cost in returning to the preferred pelagic area. The roach behaviour was identical for young and old fish. However, abundance in the shallow southern bay was lower for the simulated young roach than for the old roach during twilight, which is also explained by the long distances between the pelagic area and the large littoral area of the southern bay.

Table 7.8 Design of the habitat selection mode of roach in Lake Belau (modified after Hölker and Breckling 2001)

Physiological status	Light conditions	Habitat preference	Swimming mode
Hungry	Night	Pelagial	Low-cost swimming pattern
	Day	Pelagial	High-cost swimming pattern
	Twilight	Littoral	High-cost swimming pattern
Not hungry	Night	Pelagial	Low-cost swimming pattern
	Day	Pelagial	Low-cost swimming pattern
	Twilight	Pelagial	Low-cost swimming pattern
Spawning	Night	Pelagial	Low-cost swimming pattern
	Day	Littoral	Spawning behaviour
	Twilight	Littoral	Spawning behaviour
Winter	Night	Pelagial	Low-cost swimming pattern
	Day	Pelagial	Low-cost swimming pattern
	Twilight	Pelagial	Low-cost swimming pattern

Twilight   **a**   **b**   Daylight

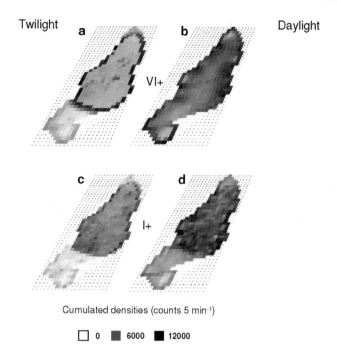

Cumulated densities (counts 5 min⁻¹)

☐ 0   ▨ 6000   ■ 12000

Fig. 7.9 Cumulated densities of 100 roach specimens during a simulated standard year (computed for 5-min intervals): **a** VI roach at twilight, **b** VI roach at daylight, **c** I roach at twilight, **d** I roach at daylight (modified from Hölker and Breckling 2005)

Length, weight (November 1991, 1992), and food composition (autumn 1991) of 0-group roach in Lake Belau were used to calibrate and to connect the spatial behaviour and the energetic needs of roach in the presented model with the growth and food composition of roach in Lake Belau (Hölker and Breckling 2001). The diet of the simulated 0-group roach consists of about 60–90% zooplankton and only 10–40% insect larvae in the first year.

Most of the ontogenetic changes in the diet composition of fishes are probably caused by morphological changes, in particular the increase in mouth size and the improvement in locomotory ability (Wootton 1990). An emergent attribute of the modelled foraging strategy is the decrease of zooplankton in the food of the first years, which is a consequence of ontogenetic changes in the scaling of spatial behaviour (Hölker and Breckling 2001). Though the feeding strategy did not change during growth, large roach exhibit a higher propensity to search for food in the littoral zone at dusk or dawn. Consequently, simulated adult roach ingested a higher proportion of littoral food during twilight than during daylight. However, the predation pressure is higher on the comparably small littoral area than on the larger pelagic zone, which is a response of roach to the higher abundance of food in the littoral zone (Hölker and Breckling 2005). In accordance, a significant top-down effect was calculated for roach and bream on the insect larvae in Lake Belau and a high nutrient transfer from littoral to limnetic zone (see Section 7.2.3).

7.4 Conclusions

The small-scale ecosystems of the study area lead to strong interactions. In particular the fauna of the beech forest is influenced by the adjacent agrarian systems, while typical species of woods rarely invade the agrarian systems. Ecosystem interfaces are well defined between forest and agrarian systems and between the lake and terrestrial systems. Only a few eurytopic species or species with a high motility interchange between forests and agrarian systems, e.g. mammals, use the agrarian systems as food resource. The forest borderline, furthermore, operates as an optical guideline for the dispersion of many animal species. Dipteran species with swarming mating behaviour concentrate there or disperse along this guideline. Interactions between terrestrial and aquatic systems are limited to the small littoral zone. In contrast, interactions within the aquatic system are well developed. It can be deduced from the motility patterns of fish that interactions between the two zones of the lake are basically comparable with the interactions of forest and agroecosystems in the light of the activity patterns of mammals. The littoral zone is the main food resource of fish in dependence on season and development status.

The modelling results illustrate that spatially explicit individual-based models may be very useful in representing and integrating complex empirical investigations. In order to analyse dispersal and interaction processes between different habitats the models emphasize that motility patterns have to be combined with a set of further factors. Thus, in the fish model bioenergetic processes have to be included in order to reproduce realistic interactions between the different habitats of a lake, whereas the beetle model must integrate life history factors and habitat-dependent mortality to simulate long-term colonization processes. Based on such complements, the models provide a commendable instrument for inter- and extrapolation purposes and a wide set of possibilities for adaptation to other environmental situations.

Chapter 8
Element Fluxes in Atmosphere, Vegetation and Soil

Otto Fränzle and Claus-Georg Schimming

8.1 Introduction

The juxtaposition of element sources and sinks in a landscape is clearly the major determinant of overall element fluxes. Connections between landscape components can be complex and counter-intuitive; for example, groundwater flow can bypass riparian zones, as the following Chapters will show. Therefore, quantitative landscape-scale evaluations of element fluxes are difficult. They require models capable of depicting internal and input/output element dynamics in diverse ecosystems, as well as landscape-scale hydrologic processes (Chapter 10). These requirements are a challenge to modellers who have to balance the need for detail in their models with the need to be comprehensive; this involves linking simulation models of flux rates with geographic information systems to carry out landscape-scale assessments. As Chapters 9–11 may show, this combination of various approaches to quantitatively define element fluxes through interrelated terrestrial and aquatic ecosystems of the study area on a hierarchy of temporal and spatial scales provides reciprocally new and deeper insights into structure and function of ecosystem complexes.

8.2 Atmospheric Deposition and Leaching Processes of the Vegetation Cover

8.2.1 Medium-Scale Deposition Patterns of the Study Area

The small-scale deposition networks of the Bornhöved research scheme with their high temporal and spatial resolutions constitute reference systems for the state-run medium-scale Schleswig–Holstein network. Consequently the comparative evaluation of deposition values at two complementary spatial scales permits one to reliably differentiate between local and regional influences on the chemical composition of the solutes collected in geostatistically validated arrays of bulk and wet-only

O. Fränzle et al. (eds.), *Ecosystem Organization of a Complex Landscape.*
Ecological Studies 202.
© Springer-Verlag Berlin Heidelberg 2008

samplers, or to distinguish local terrestrial sources of solute components from farther-distant marine ones, respectively (Jensen-Huß 1990, 1992).

The comprehensive data sets of these complementary networks show in comparison with the other countries in Germany that Schleswig–Holstein is characterized by distinctly higher deposition rates of the air-borne marine components sodium, magnesium and chloride, while both the nitrate and sulfate rates are average. Owing to intensive emissions from agriculture ammonium deposition is considerably increased, i.e. exceeding the federal average. Potassium and calcium deposition prove largely determined by local sources, so generalized regional distribution patterns cannot be identified.

In the above-canopy precipitation marine solute components attain substantial percentages; chloride, for instance, normally amounts to 40% of the total anions, sodium represents about 35%, ammonium another 30% of the total cations. Regionalized estimates of the air-borne marine elements and correlation analyses show that sodium and chloride are almost exclusively of North Sea origin, while the marine proportion of magnesium is about 90%, that of sulfate varies between 10% and 30%; potassium and calcium normally have marine components of $<20\%$ (Jensen-Huß 1990).

In greater detail the analysis of monthly concentration values in wet and dry atmospheric deposition discloses a great deal of broad adjustment to the annual course of precipitation in the sense of an inverse relationship, i.e. in general months with little rainfall (cf. Section 2.2.2) are characterized by increased solute concentrations since for low-volume rain events of short duration, wet deposition due to intensified wash-out processes represents a more intense element flux to vegetation and soil than does dry deposition during the intervening dry periods. The correlation does not hold for elements of marine origin, however; their maximum concentrations occur in winter. With regard to increased ammonium and proportionately elevated sulfate concentrations, which are linked by gas-phase reactions, the seasonal pattern of agricultural activities with highly variable emissions from point and non-point sources is important (Fränzle 1993). Wet deposition accounts for more than 85% of the inputs of sodium, copper, magnesium, ammonium, chloride, sulfate and nitrate into the ecosystems of the study area, while about 50% of the potassium and calcium inputs are due to dry deposition. As a consequence of these interrelationships and corresponding to rainfall intensities, the deposition rates of all of the foregoing components, with the exception of the marine ones (and also the heavy metals cadmium and lead), are maximum in summer (Lalubie 1991; Lenfers 1994; Branding 1996).

8.2.2   Small-Scale Atmospheric Deposition, Canopy Throughfall and Litterfall of Beech Stands

A core programme of deposition and interception measurements provided weekly data of wet deposition by means of wet-only samplers and bulk deposition in permanently open samplers at the agricultural sites (A3.1), while throughfall and stemflow measurements were made under different forest canopies (W1). They were supplemented by analyses of relevant atmospheric trace gases and leaf extraction experiments to assess leaching from beech trees. The results served as inputs to a canopy interaction model and a refined resistance model of pollutant dry deposition. In addition, the importance of element transport by litterfall was assessed in the different forest ecosystem types of the study area in selected years representing normal and exceptional conditions of growth and productivity.

8.2.2.1 Ozone Concentrations

The formation of tropospheric ozone is on the one hand linked to sufficient concentrations of the precursor substances NO_x, CO, CN_x and volatile organic compounds, on the other specific meteorological conditions are required. For the study area these comprise high-pressure situations with easterly to southeasterly winds, high insolation and atmospheric stability. Under these circumstances O_3 concentrations above the A3.3 field as a reference unit frequently exceed both the short-term and long-term critical levels, as Fig. 8.1 shows. When O_3 occurs in combination with other pollutants like SO_2 or NO_2 adverse effects are not simply the sum of the individual component effects, but the relative concentrations of the components or the specific sensitivity of the plants to one of them may determine the response. Thus, a mixture of gaseous pollutants may display antagonistic, additive or synergistic effects (cf. Guderian 1985).

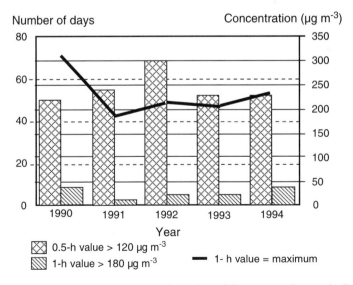

Fig. 8.1 Frequencies of critical levels and 1-h maxima of O_3 concentrations at the Bornhöved field site A3 (after Orth 1996)

8.2.2.2 Bulk Deposition of Cations and Anions

Deposition rates were assessed by means of bulk samplers distributed in the eco-systems of the core area (W1, W5, W6, A3, A6, A8) in compliance with geostatistical requirements (Fränzle 1993). Bulk samplers however, can only provide estimates of the wet deposition and (an unknown part of) the dry deposition of gases and par-ticles. Thus, apart from free acids, bulk deposition measurements yield only mini-mum values of the total deposition and, in combination with wet-only samplers, of the dry deposition, which assumes a particular importance in the framework of cali-brating elemental flux models. Therefore, complementary estimates of total deposi-tion were made on the basis of the Ulrich (1983) and van der Maas et al. (1991) approaches in combination with resistance models for dry deposition estimates (Wesely 1989; Erisman 1992).

Estimates of the mean annual deposition rates of the above chemical species during the 1989–1997 period, based on the combined assessment strategies men-tioned, are summarized in Fig. 8.2.

In unpolluted air a major source of H-ions is carbonic acid in precipitation; another important source is the CO_2 production by soil organisms and plant roots, and H-ions are also produced when nitrogen and sulfur compounds are mineralized. Under the influence of agriculture, industry and traffic the background H-ion concentrations and the resultant wet deposition rates are distinctly elevated in the study area. Nevertheless, both in bulk and throughfall deposition there is a clear trend to lesser rates during the 1989–1997 period as a consequence of various measures reducing directly or indi-rectly H-ion emissions or release.

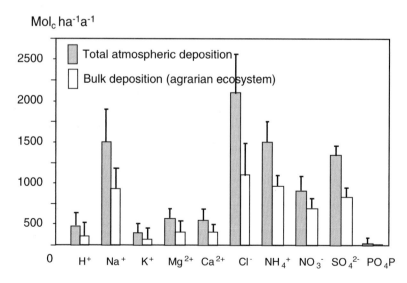

Fig. 8.2 Average annual deposition rates in the beech stand and the agrarian ecosystems of the core area (W1, W5, W6, A3, A6, A8) during the 1988–1998 period (after Wellbrock 2002, modified)

Neither sodium nor chloride fluxes display a trend during the reference period, while seasonality in the deposition patterns is quite marked. In juxtaposition to the summer minimum the winter maximum is clearly linked to the higher frequency and intensity of the autumn and winter storms transporting major portions of sea spray from the North Sea (Jensen-Huß 1990; Spranger 1992). The low dry deposition rates of either element characterizing the summer situation of the forest soils is, in particular under the beech and alder stands, due to the scavenging effect of the canopy layers (Branding 1996).

Sodium, considered an inert index element in the framework of canopy interaction modelling by Ulrich (1983, 1991), in fact had negative flux rates in 1993, 1995 and 1996 (Branding 1996), which indicates that it may be involved in various tree-top reactions, when the concentration gradient is reversed, i.e. going from the external surface to the apoplasm of leaves or needles (Fassbender 1977). The temporarily negative canopy difference of chloride may be due to dry deposition of gaseous HCl which can result from heterogeneous gas-phase reactions of NaCl particles with nitric or sulfuric acid or from burning of chlorine-containing materials (Fränzle 1993; Wellbrock 2002).

Potassium fluxes also remained essentially the same during the 1989–1997 period, but again seasonality is distinct, with a clear summer maximum of K in the throughfall solutes attributable to canopy leaching, which corresponds to findings in the submontane beech stands of the Solling Mts (Ellenberg et al. 1986). In accordance with findings by Wetselaar and Farquahr (1980) leaching rates generally increase with leaf age, and in particular with the onset of senescence which brings about a higher permeability of biomembranes, coupled with a rise in the concentrations of organic and inorganic solutes in the apoplasm of the leaf tissue. Furthermore, mechanical damage due to wind, but also other stress conditions such as water shortage or high temperatures, can increase the leaching rate (Tukey and Morgan 1963; Levitt 1980; Lenfers 1994). Comparable stress is imposed on leaves or needles by air pollutants such as ozone (cf. Fig. 8.1) in combination with nitrogenous gases, which leads to more rapid leaf senescence and modification of the chemical composition of the cuticle or both (Guderian 1985; Marschner 1995). Thus, leaching provides an important proportion of the mineral nutrition of the trees which is similar in importance to that from litterfall, as some exemplary figures may show. In 1990 the potassium input onto the beech canopy by dry and wet deposition amounted to 7.8 kg K ha^{-1} year^{-1}, in 1991 it was 6.4 kg K ha^{-1} year^{-1}, which led, owing to leaching, to 15 kg K ha^{-1} year^{-1} or 13 kg K ha^{-1} year^{-1}, respectively, in throughfall, while potassium transport by litterfall amounted to 14 kg K ha^{-1} year^{-1} or 9.4 kg K ha^{-1} year^{-1} (Lenfers 1994).

A comparative evaluation of the magnesium and calcium deposition and canopy throughfall rates exhibits a general decrease in Mg fluxes, while the Ca deposition rates are increased, which however is coupled with lower throughfall rates. This points to efficient re-absorption processes which offer a possibility for plants to supply the sites of demand with mineral nutrients whose re-translocation is limited. As a consequence, average element fluxes by litterfall of the different beech stands (W1) were 5.1 kg Mg ha^{-1} year^{-1} and 31.5 kg Ca ha^{-1} year^{-1} during the 1990–1993

Table 8.1 Deposition rates (kg ha^{-1} year^{-1}) at 80 beech stands of the Level II Programme in Germany (after Gehrmann et al. 2001)

	Na	Cl	K	Mg	Ca	NH$_4$-N	NO$_3$-N	SO$_4$-S
Throughfall								
Average	5.9	12.5	15.4	2.1	8.9	10.2	8.8	14.1
Var. coeff.	56	53	44	50	34	51	43	58
Above canopy deposition								
Average	3.8	6.6	2.0	0.8	4.4	6.1	4.8	6.8
Var. coeff.	50	51	67	53	38	37	25	28

period (Lenfers 1994). Additional information is provided by Table 8.1, which summarizes the results of long-term deposition measurements of the Level II Programme.

For comparison purposes, making allowance for differences in sampling and analytical procedures and differences in reference periods, the corresponding total deposition figures determined in the framework of the Solling Project in a beech stand ~500 m above sea level are mentioned: Na 12.4, Cl 30.1, K 12.1, Mg 3.9 and Ca 21.2 kg ha^{-1} year^{-1} (Mayer and Ulrich 1982). In a montane spruce forest catchment of the Fichtelgebirge (Bavaria) the corresponding figures for the 1988–1996 period are (Matzner et al. 2001): Na 4.8, Cl 3.5, K 2.0, Mg 0.7 and Ca 4.4 kg ha^{-1} year^{-1}.

The long-term average deposition of nitrogen species to the beech stand (W1) amounts to 1911 mol N ha^{-1} year^{-1} or 27 kg N ha^{-1} year^{-1} with a minimum of 1524 mol N ha^{-1} year^{-1} (21.3 kg N ha^{-1} year^{-1}) in 1995 and a maximum of 2385 mol N ha^{-1} year^{-1} (33 kg N ha^{-1} year^{-1}) in 1989. Taking also inputs in form of organic nitrogen compounds into account, the minimum deposition rate amounts to 27 kg N ha^{-1} year^{-1} while the maximum is about 40 kg N ha^{-1} year^{-1} (Wellbrock 2002). The preponderance of ammonia is a result of increased emissions from decomposing animal manure which has been produced not only by cows and other ruminants grazing outdoors, but also by the increased number of cattle, pigs and poultry in highly intensive husbandry (cf. Section 2.2.6).

Important for the deposition process are the preceding heterogeneous gas-phase reactions, which involve suspended particles. Among these, the buffering effect of ammonium on SO$_2$ adsorption and oxidation, including the intermediates SO$_2$NH$_2$– and SO$_3$NH$_2$–, is of particular relevance. In general, however, the fluxes of particulate ammonium compounds exhibit a distinctly lesser degree of variability than those of ammonia, which are much more susceptible to surface characteristics and atmospheric conditions. Thus, a spatially and temporally variable combination of physical and chemical boundary conditions defines the direction of NH$_x$ transport from or to a given surface, i.e. emission or deposition. Integrating over the whole of the year, manured and fertilized plots of the study area, e.g. pastures (A8), are sources of ammonia emission, while natural-like ecosystems such as the beech stands exhibit sink character (Branding 1996), which is in agreement with findings of Farquahr et al. (1980), Sutton (1990), Schjørring (1991) and Erisman (1992). Turning from annual averages to shorter-term exchange processes, however, the picture becomes more complicated. A forest may act as an NH_3 source, for instance, when ammonia concentrations in the atmospheric background are low, while it becomes a sink when the opposite holds (cf. Langford et al. 1990).

Comparing the atmospheric transport of sulfur species with those of the above nitrogen compounds, the faster conversion of NO_x to nitric acid and nitrates than the comparable conversion of SO_2 to sulfuric acid and sulfates and the higher deposition velocity of nitric acid involves that NO_x species contribute more to the acidification of rain nearby the source areas than at large distances. As Table 8.2 shows on the basis of comparative throughfall and stemflow analyses (W1.3, W1.6), the bulk deposition of S species decreased distinctly during the whole observation period, while the concomitant reduction of canopy interaction (net canopy effect) was less pronounced.

The above flux rates implicitly reflect the complicated interplay of deposition and canopy reactions, in particular leaching and uptake processes of the foliage, which at present prove only imperfectly reproducible in net canopy interaction modelling (Gehrmann et al. 2001). In balancing perspective it must be noted that discrepances between cation and anion equivalents may result from the analytically ill-defined role of organic compounds in throughfall and stemflow waters.

Irrespective of analytical bias seasonal sulfur flux dynamics proves strong with a pronounced winter maximum due to higher SO_2 emission rates as a consequence of elevated energy consumption on the one hand and much intensified sulfate transport from the North Sea by storms on the other. In 1991, 1993 and 1996 the net canopy effect was negative, indicating analogously the behaviour of ammonium ions stomatal uptake of sulfur dioxide, which is in agreement with findings by Fowler (1984) and Wesely (1989).

In view of the importance of dead organic matter for the element budgets of forest ecosystems Table 8.3 is presented, based on consecutive sampling campaigns. Particularly interesting are the results of biweekly sampling at eight representative beech sites (W1, W2) during the 1990–1993 period, which contains both two normal years and an exceptional one (1993) with unusually high beech-nut production (cf. Lenfers 1994). The reference period is the sampling year, beginning in May when the trees come into leaf and ending in the following April.

Table 8.2 Throughfall and stemflow chemistry of the beech stands (W1.3, W1.6) during the 1989–1998 period (mol_c ha^{-1} $year^{-1}$) (after Wellbrock 2002, completed for 1998)

Year	H	Na	K	Mg	Ca	Cl	NH_x-N	NO_x-N	SO_4-S	PO_4-P
1989	170	1422	481	419	533	1673	986	600	1447	7
1990	170	2234	496	634	603	2934	1050	593	1521	23
1991	106	1208	380	346	451	2011	1207	708	1388	22
1992	150	1135	558	345	407	2665	1040	478	1270	19
1993	125	1613	348	436	301	1602	1280	661	1332	33
1994	199	1584	386	411	440	1790	1223	668	975	25
1995	71	806	913	372	387	–	716	408	992	
1996	50	767	506	271	465	1006	1094	487	870	
1997	14	891	421	253	437	1277	1018	555	810	
1998	29	1167	653	346	520	1389	960	587	945	
Mean	108	1283	514	383	454	1224	1057	575	1155	22

Table 8.3 Annual aboveground production of dead organic matter and associated element fluxes of beech stands (W1, W2) during the 1988–1998 period (kg ha^{-1} year^{-1}; after Wellbrock 2002). *DOM* Dead organic matter, including leaves, buds, beech-nuts, cupulae and dead wood

Year	DOM	Na	K	Mg	Ca	N	P
1988/89	5897	3	23	5	31	78	9
1989/90	6998	2	14	3	23	53	–
1990/91	5041	2	14	3	23	52	–
1991/92	3279	5	9	5	30	28	3
1992/93[a]	10 644	2	60	12	61	105	15
1993/94	5214	2	20	7	46	60	–
1994/95	4119	–	–	–	–	48	–
1995/96	No sampling						
1996/97	4183	4	20	18	–	–	–
1997/98	5797	–	–	–	–	73	–
Average	5696	3	26	8	36	62	9

[a] Year with exceptional beech-nut production

8.3 Element Cadasters and Nutrient Fluxes in Arenic Umbrisols of Beech Stands and Eutri-Cambic Arenosols of Arable Land

8.3.1 Element Cadasters of Soils

Within the framework of reliably assessing the elemental source or sink functions of soils, a precise characterization of their minerochemical composition is necessary. It involves a comparison of the subhorizon-related total concentrations of physiologically essential and beneficial elements with those of the index element titanium which is not liable to vertical translocation and thus permits to distinguish primary vertical inhomogeneities as a consequence of sedimentation processes (cf. Section 2.1.1) from secondary ones due to weathering (Section 2.2.3).

Table 8.4 summarizes the pertinent element spectra of the Arenic Umbrisol of the beech stand (W1, W2) and field A3 which came into existence high above the present groundwater level (cf. Chapter 11). In view of the importance of the topographic situation for element fluxes they are analysed on a comparative basis in the following Section. As regards the C and N dynamics of these ecosystem types as compared to that of the different alder carr (W6) and wetland facies (A8, A9) which developed in contact with or (relatively) close to the groundwater, the reader is in particular referred to Chapters 4 and 5. The regulatory role of these vegetation units in the framework of medium-scale particulate and solute transport processes is analysed in detail in Chapter 11 under the particular perspective of ecotone dynamics.

The Ti concentrations listed can be subdivided into two groups, i.e. in the upper horizons of either soil they vary between 30 mmol kg^{-1} and 40 mmol kg^{-1}, in the lower parts of the profiles between 15 mmol kg^{-1} and 20 mmol kg^{-1}, which marks the transition from the glacial coversands to the underlying fluvioglacial meltwater sands.

Table 8.4 Total element contents of the Eutri-brunic Arenosol under tillage (A3) and the Arenic Umbrisol under forest (W1) (after Wetzel 1998). For technical reasons the profiles had to be opened in the vicinity of the profiles described in Tables 2.2 and 2.10; therefore the original horizon descriptors were retained: *Forest soil, Field soil*

Horizon	Depth (cm)	Ti	Al	Fe	Mn	Ca	Mg	K	Na
					\($mmol\ kg^{-1}$\)				
Forest soil									
Aeh	5	37	762	165	5	61	42	122	163
RAp	28	35	886	181	7	67	49	139	187
AhBv	38	36	1003	254	7	86	74	135	201
IIBv1	60	29	979	382	6	112	82	114	206
Bv2	91	20	757	190	5	80	48	116	171
rG Bv	110	14	730	127	4	89	46	119	180
Bbsv	130	15	665	165	8	83	41	110	174
IIIBvC	154	15	546	71	4	64	23	103	141
IVBvC	165	16	861	156	6	104	49	150	247
C	400	24	555	99	3	202	48	105	188
Field soil									
Ap	35	36	934	184	11	169	53	146	208
ApBv	47	39	1008	165	8	261	56	158	220
Bv1	66	33	1068	146	5	243	63	182	253
Bv2	84	34	1098	188	5	232	82	179	249
IIBv	112	66	1733	373	6	223	193	236	229
IIIBv	120	41	899	276	6	205	77	127	185
IvrG Bv	147	18	825	182	5	205	62	119	198
BvC	180	15	694	106	4	188	39	116	187
C	400	25	586	91	3	357	46	117	185

Considering the element concentrations at the 400 cm reference level, the differences between the forest and field soil are generally rather small, with the exception of carbonate-bound Ca. With regard to the latter, the forest soil has been liable to stronger decalcification than the field soil (Aue 1993). Weathering and acidification of the forest Arenic Umbrisol on the one hand, and addition of nutrients by manuring and fertilization of the field Eutri-cambic Arenosol on the other have brought about a 40% reduction of K, Ca and Mg in the basal part of the forest soil in comparison with the arable land, going up to nearly 60% in the upper horizons.

8.3.2 Bacterial Populations and Degradation of Soil Organic Matter

Microbial enzymes are specifically adapted to catalyse a host of reactions involved in the degradation of organic matter or environmental chemicals. Therefore, microbial populations are well suited as indicators of site qualities and the intensity of turnover processes (cf. Chapter 3). The latter aspect was a subject-matter to comparative analyses of heterotrophic bacterial populations in 1992 and 1993 in order to more precisely define the influence of forest, pasture and field management practices on the site-specific turnover rates of soil organic matter (Bach 1996). Essential quantitative results are summarized in Tables 8.5 and 8.6; the underlying data sets are most probable numbers (Alexander 1982).

Comparative analyses in different seasons of the reference period showed that the abundances of cellulose decomposers and ammonium or nitrite oxidizers were lower in autumn than in spring, while the populations of the other physiological groups did not change substantially in the course of the year.

On the basis of the above determination of bacterial abundances in situ investigations of denitrification losses and N_2O emissions were carried out from July 1992 to August 1994 on two agricultural soils under crop rotation (A3) and maize monoculture (A2), on the forest soils of the beech (W1) and alder stands (W6) and on wetland (A9) (Mogge 1995). The results indicate that forest and agricultural soils contribute in a comparable way to the emissions, but in either case significantly higher amounts of nitrogen were lost by denitrification than by N_2O emissions. The highest emission rates due to denitrification amounted to $30\,g\;N\;ha^{-1}\;day^{-1}$ or $35\,g$

Table 8.5 Mean logarithmic abundances of physiological bacterial groups in the upper soil horizons of selected Bornhöved sites (A3, A6, A9, W6, W1; after Bach 1996)

Group	Field Ap	Grassland Mah	Wetland Ah	Alder He	Beech RAp
Heterotroph. bacteria	7.09	7.46	7.36	7.4	7.39
Starch decomposers	6.66	6.90	7.20	6.55	7.25
Xylan decomposers	5.71	6.02	6.11	5.60	4.96
Cellulose decomposers	5.57	5.95	6.16	5.45	5.40
Protein decomposers	5.36	5.42	5.58	5.2	5.60
Denitrifiers	4.80	5.54	5.59	3.49	5.49
Ammonium oxidizers	4.64	4.99	5.41	4.87	0
Nitrite oxidizers	3.56	3.32	2.7	2.56	1.20

Table 8.6 Mean logarithmic abundances of physiological bacterial groups in the B horizons of selected Bornhöved sites (A3, A6, W1; after Bach 1996)

	Field	Grassland	Beech
Heterotrophic bacteria	6.52	6.41	6.59
Starch decomposers	6.36	6.02	5.39
Xylan decomposers	5.28	5.07	4.66
Cellulose decomposers	5.24	4.84	3.75
Protein decomposers	4.82	4.99	4.97
Denitrifiers	5.45	5.25	3.70
Ammonium oxidizers	3.72	3.82	1.60
Nitrite oxidizers	2.14	2.04	1.38

N ha^{-1} day^{-1}, respectively, in the fields under crop rotation and the alder forest, while N$_2$O emissions did not exceed 19 g N ha^{-1} day^{-1} on either site.

The high gaseous N emissions of the alder carr are largely dependent on the considerable amounts of available carbon (cf. Section 2.2.3) and the high nitrogen fixation rates of the *Alnus–Frankia* symbiosis, but also exhibit a clear relationship to soil temperature. In the beech forest, in contrast, gaseous N emissions appear limited by low contents of available carbon and nitrate in the RAp horizon. In the fields under crop rotation nitrate is also the most limiting factor, while under maize monoculture soil temperature and C content largely control emissions. On the pastures in slope position the latter boundary conditions are also decisive, while in wetlands the quality of the carbon compounds plays the essential role in the regulation of N emissions.

8.3.3 Annual Course of Element Concentrations in Field Eutri-Brunic Arenosols and Forest Arenic Umbrisols

Material fluxes in the soil and the aeration zone of forest ecosystems reflect the influence of the above minerochemical and microbial situation and the atmospheric deposition and associated canopy, throughfall and stemflow processes and the precipitation-controlled seepage rates. The resultant soil solution represents the aqueous phase of the soil at or below field moisture capacity; it is differentiated from soil water which is the moisture that fills the pores between soil particles at above field moisture capacity. For technical reasons, the soil solution must be operationally defined by the particular methods used to obtain samples. In the present case the methods applied during the 1989–1999 reference period were: (a) continuous direct vacuum extraction in spatially representative high-resolution sampling networks during consecutive biweekly intervals from the litter, and from 5, 12, 50, 150 and 400 cm depths and (b) modelling of seepage rates by means of the WASMOD and VAMOS approaches (Reiche 1991; Bornhöft 1993), calibrated and validated at various spatial scales by water budget estimates (Trepel and Kluge 2004). The enquiries and the underlying methodology (cf. Section 1.3) are described in detail by Aue (1993), Bornhöft (1993), Mansfeldt (1994), Rambow (1996), Reiche (1991), Schimming (1991), Wetzel (1998) and Wellbrock (2002).

These studies show that only the elements Na, K, Ca, Mg, Mn, Al, Fe, Si (the latter four in different speciations) and the cation NH$_4^+$, together with the anions Cl$^-$, SO$_4^{2-}$, HCO$_3^-$ and NO$_3^-$, are the dominant solutes in the soil solutions considered. These species commonly occur in millimoles per litre or lower concentrations, while all the other elements attain only micromoles per litre or lower concentrations. With the exception of PO$_4^{3-}$, therefore, the latter are not made subject to a detailed description in the framework of the present terrestrial flux analyses; a greater number of these, however, will play a role in the analysis of element budgets of the adjacent Lake Belau (cf. Chapter 10).

8.3.3.1   Solute Concentrations of K, Na, Ca, Mg, Mn

In juxtaposition with the solid phase, the soil solution comprises readily available fractions of the total element contents in soils where the individual elements occur as chemical species in solution and in exchangeable, non-exchangeable, structural, or mineral forms. The chemical speciation of elements depends on the architecture of their atoms, while equilibrium and kinetic reactions affect the concentration levels of elements and their availability at any particular time. Generally, mobility and uptake are functions of the total element quantities in soil; more specificially they depend on the concentrations of competitive reactants and sorbents, whereas mobility refers to the transport rate of an element to the point of uptake by plants.

Under conditions of reduced concentrations in the rhizosphere, for instance, roots extract potassium from soil minerals, and this release of non-exchangeable K^+ is enhanced by increasing concentrations of calcium in the ambient soil solution. Owing to "dilution" effects by easily soluble components of fertilizers the non-exchangeable and mineral potassium fractions of the Ap horizon of the Eutri-brunic Arenosol of the field under consideration (A3) comprise only 80% of the total, while it goes up to 90% in the Aeh and RAp horizons of the beech stands (W1; Rambow 1996; Wetzel 1998).

Together with the rapid decrease of potassium in the underlying subhorizons the concentrations of potassium depicted in Fig. 8.3 indicate maximum K^+ recycling out of the litter into the vegetation cover, which corresponds to findings of Ellenberg et al. (1986) in the submontane beech stands of the Solling Mts. The maintenance of high potassium concentrations is a physiological necessity since the element is required in metabolic functions, i.e. most enzyme reactions (cf. Läuchli and Pflüger 1978; Marschner 1995). In the tilled Arenosol both the order of magnitude and the temporal pattern of potassium concentrations in the upper soil horizons differ considerably from those illustrated in Fig. 8.3 as a result of intensive manuring and fertilizer application.

A comparison of the depth-function sodium concentrations in Fig. 8.3 with the deposition, weathering and seepage rates exhibits a continuous ionic transport of the element through the soil without any appreciable storage (cf. Section 10.2.5). Na^+ acts as a growth stimulator, which is caused mainly by its effect on cell expansion and on the water balance of plants (Zehler 1981). Not only can Na^+ replace K^+ in its contribution to the solute potential in the vacuoles and consequently in the generation of turgor and cell expansion; it may even surpass K^+ in this respect since it accumulates preferentially in the vacuoles (Jeschke 1977). Thus, nearly all the K^+ can be substituted by Na^+ in old leaves and made available for specific functions in meristematic and expanding tissues (Wetzel 1998).

The comparative analysis of calcium concentrations of the field and forest soils (cf. Fig. 8.4) exhibits a distinct influence of manuring and fertilizer application; thus, the amount of mineral calcium is higher in the field soils, and also the proportion of Ca in coordinative bond or exchangeable form is considerably enhanced. In this respect an analogous situation exists only in the Aeh and RAp horizons of the acid forest soils (W1, W2), which is indicative of biogenic calcium accumulation,

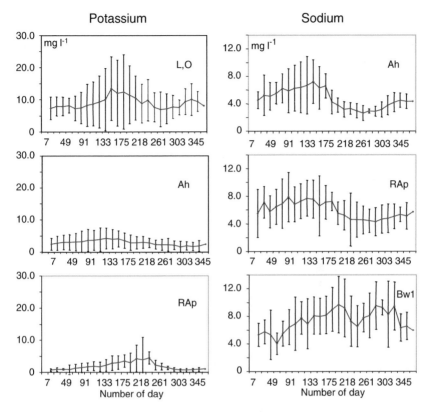

Fig. 8.3 Seasonal fluctuations of potassium and sodium concentrations in the L,O and A horizons of the forest Arenic Umbrisol (W1) during the 1989–1999 period. *Bars* indicate double standard deviation

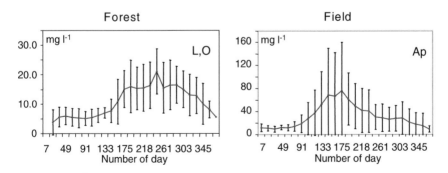

Fig. 8.4 Seasonal fluctuations of calcium concentrations in the L and O horizons of the Arenic Umbrisol under forest (W1) and in the Ap-horizon of the Eutri-brunic Arenosol under tillage (A3) during the 1989–1999 period. *Bars* indicate double standard deviation

while the elevated amounts of easily soluble calcium compounds in the underlying RBgw horizon are due to small-scale heterogeneities of the parent material (Reiche and Dibbern 1996). Comparably elevated concentrations of extractable calcium also occur in the 2Bw horizon of the forest soils with an intercalated loamy layer (cf. Section 8.3.1) and, due to their carbonate content, in the Ap, AB and Bw horizons of the Eutri-brunic Arenosol under tillage (A3; Wetzel 1998).

The progressive decalcification of the gravelly parent material and the subsequent rapid acidification of the rhizosphere of the beech stands observed during the 1990s is indicative of long-term negative changes the beech stands undergo and which are also documented in changes of foliation and other health indicators (cf. Wellbrock 2002; Section 8.5.1). The diagrams of the average seasonal changes in calcium concentration of the soil solution (Fig. 8.4) in fact indicate of a superposition of internal recycling processes and progressive mineral weathering, which leads to a continuous increase of calcium fluxes below the 2Bw horizon or the 50 cm level, respectively (Peters 1990; Schimming 1991; Wetzel 1998).

As a consequence of manure and fertilizer application both the order of magnitude and the fluctuation patterns of calcium concentrations in the upper horizons of the Eutri-brunic Arenosols of the fields (A1, A2, A3) differ considerably from those observed in the above beech stand.

Contrary to calcium, potassium and sodium only a minor portion of magnesium exists in persistent silicate form, the major part is a component of readily weatherable minerals. The exchangeable magnesium stock is enhanced in the upper soil horizons of both soils, which is presumably, as in the case of potassium and calcium, the result of biogenic accumulation processes (cf. Section 8.5.1).

The following representation of the average seasonal changes in magnesium concentration of the soil solution (Fig. 8.5) exhibits, in conjunction with the corresponding seepage rates, efficient biogeochemical recycling mechanisms, which suffice to maintain a meta-stable magnesium supply of the beech stand. A comparison with the seasonal fluctuations of the equilibrium concentration of manganese

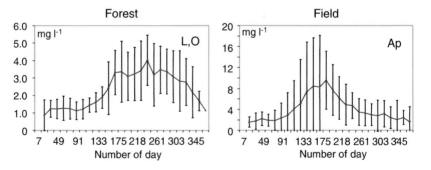

Fig. 8.5 Seasonal fluctuations of magnesium concentrations in the soil solution of the L and O horizons of the Arenic Umbrisol under forest (W1) and in the Ap horizon of the Eutri-brunic Arenosol under tillage (A3) during the 1989–1999 period. *Bars* indicate double standard deviation

indicates, indeed, that the magnesium demand of the trees must be partly satisfied by recourse to, and uptake of, equivalent quantities of manganese.

In analogy to the above human impact on calcium supply also the magnesium concentration in the soil solution of the tilled Eutri-brunic Arenosol (A3) is largely influenced by manuring and fertilizer application, as Fig. 8.5 shows.

In both the field and forest soils 40–50% of the total iron are bound in persistent silicate form without exhibiting major horizon-related differences. Some 72–87% of lithogenic iron compounds are indicative of advanced weathering, the proportion of pedogenic oxides being particularly elevated in the Ah, RAp, Ap and AB horizons. The Aeh horizon of the forest Arenic Umbrisol (W1, W2) is characterized by an 80% proportion of iron bound in the form of Fe(III) and ocassionally Fe(II) chelates, while these iron compounds amount to only 13% in the Ap horizon of the field soil (Fig. 8.6). In both soils, the deeper horizons are equally characterized by a 10% content of pedogenic iron compounds. Owing to the lower content of organic substances in these horizons, however, the majority of soluble iron

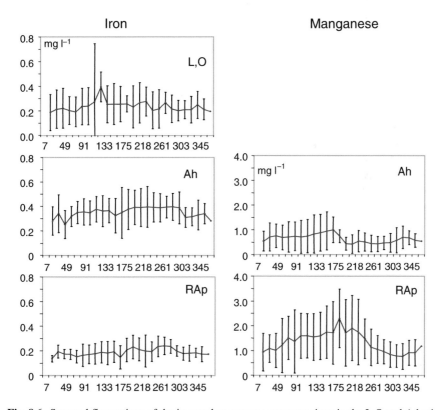

Fig. 8.6 Seasonal fluctuations of the iron and manganese concentrations in the L,O and A horizons of the Arenic Umbrisol under forest (W1) and in the Ap horizon of the Eutri-brunic Arenosol under tillage (A3) during the 1989–1999 period. *Bars* indicate double standard deviation

compounds should be composed of specifically adsorbed iron and iron carbonate. Altogether the horizon-specific differences in iron content of the forest and field soils are due to primary differences in stratification and, considering the Arenic Umbrisol under forest, also due to former gleyification.

Unlike iron only minor amounts of manganese are bound in silicates of the Ap and B horizons of the tilled Arenosol or the Bws horizon of the forest soil, i.e. the major part occurs in a less persistent form and is thus easily bio-available. In the upper horizons of the field soil Mn(IV) predominates by far as a consequence of manuring or enhanced biotic activity, while the Mn(II) portion of the BW1 and BW2 horizons, as a result of the above mentioned relic gleyification, is comparable to that of other horizons. Altogether crystalline manganese oxides decrease from the upper soil horizons toward the lower ones due to the pronounced soil acidification which is, in the upper horizons, accompanied by leaching and downward translocation of acid-soluble amorphous Mn oxides. The Aeh and Ap horizons, but no less the Bw1 and Bw2 horizons of the acid forest soil, are characterized by distinctly elevated amounts of organic Mn complexes, which is especially in the latter case due to specific bonding and in so far analogous to the above fractionation of iron compounds. The occurrence of such higher quantities of exchangeable manganese and the resultant equilibrium concentrations of Mn^{2+} in the soil solution are indicative of biogenic recycling processes which are clearly reflected in Fig. 8.6, in particular when comparing the seasonal changes in manganese concentration of the litter and the following (upper) horizons with those of magnesium as illustrated above in Fig. 8.5. The broad parallelism between the Mn and Mg curves indicates, in the light of physiological interchangeability and the higher manganese demand of green plants than of fungi or animals, a distinctly higher strain on the manganese pool during the vegetation period.

Aluminium concentrations in mineral soil solutions are below $1\,mg\,l^{-1}$ at pH values >5.5 but rise sharply at lower pH. As a consequence, the amount of crystalline alumino-silicates decreases from the uppermost horizon down to the Bw1; subsequently it increases down to the BwC horizon. Conversely, the quantities of amorphous hydroxyaluminium species reach their maximum in the A and B and the following subhorizons down to the Bw (Wetzel 1998; see also Beyer et al. 1999), which results from translocation out of the two uppermost subhorizons. The acid soil reaction of the Ah and RAp horizons is inducive to elevated solution rates of aluminium and correspondingly higher proportions of exchangeable and EDTA-soluble Al. In contrast to the forest soil, in the A horizon of the field Arenosol there is neither residual enrichment of persistent aluminosilicates nor an indication of an appreciable translocation of aluminium on the basis of the depth function of the amorphous inorganic proportion of weatherable Al compounds.

It ensues from the diagrams in Fig. 8.7 and the related seepage rates that Al translocation becomes maximum in the Ah and RAp subhorizons and then decreases continuously down to the BC horizon of the forest soil. In the field Arenosols with higher pH values aluminium leaching from the Ap and Bw horizons is much reduced and becomes negligible in the BwC horizon.

8.3.3.2 Solute Concentrations of NH_4^+, NO_3^-, Cl^- and SO_4^{2-}

In the beech stands, the dominant nitrogen compound percolating into the soil is nitrate. The resultant concentrations in the litter (Fig. 8.7) attain maximum values in summer due to temperature-controlled nitrification processes of the organic matter and of the coincident seasonally elevated NH_4^+ inputs. In the deeper soil horizons both the concentrations and seasonal differences decrease rapidly owing to plant uptake, attaining the minimum below 150cm with only very small intermonthly oscillations. Owing to intense fertilization and manuring the nitrate concentrations in the soil solution of the field Eutri-cambic Arenosol are distinctly higher, and the corresponding flux rates in the upper C horizon and the aeration zone amount to 200mmol m^{-2} year^{-1}.

Considering the seasonal variations of the generally low ammonium concentrations in the soil solutes of both the forest Arenic Umbrisol (with the exception of the litter value) and field Eutri-cambic Arenosol, the rapid decrease with depth is

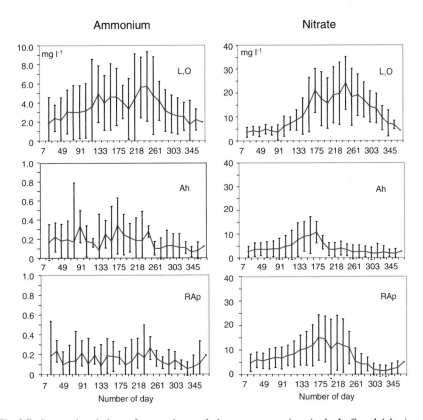

Fig. 8.7 Seasonal variations of ammonium and nitrate concentrations in the L, O and A horizons of the Arenic Umbrisol under forest (W1) during the 1989–1999 period. *Bars* indicate double standard deviation

in either case indicative of intense uptake and oxidation processes. The resultant flow rates are consequently practically negligible (Section 8.3.4). The distinctly higher NH_4^+ concentrations in a series of springs feeding Lake Belau are therefore the result of reduction processes affecting nitrates and dissolved organic carbon compounds percolating down to the basal anoxic domain of the aquifer (Fränzle et al. 1996; Fränzle and Kluge 2003; Chapter 9).

In the Bornhöved soils the proportion of the three sulfur pools is 71% organic S (range: Ap 93%, Bws 59%), 13% adsorbed S (Ap 0%, Bw 25%) and 16% S in solution (A and B 4%, Bw 27%) for the tilled Eutri-brunic Arenosol. The corresponding values for the Arenic Umbrisol under forest are 55% organic S (range: Ah 89%, Bw 21%), 26% adsorbed S (Ah 4%, AB 48%) and 19% S in solution (Ah 7%, Bw 42%; Mansfeldt 1994). During the 1989–1999 period the average monthly sulfate concentrations in the soil solution of the forest Arenic Umbrisol have varied between $4\,mg\,l^{-1}$ and $12\,mg\,l^{-1}$.

In comparison, the whole set of diagrams (Fig. 8.8) does not exhibit a clear seasonality of sulfate concentrations for any of the horizons of the forest soil, which seems to be due to the depth-related different influence of SO_4^{2-} deposition, canopy interaction, throughfall, soil transport, uptake and adsorption–desorption processes. The situation is quite different in the Eutri-brunic Arenosol, where the average maximum concentrations in the upper soil horizons attain values of about $15\,mg\,l^{-1}$ in late summer due to the combined influence of manuring and crop removal. In the lower soil horizons the situation is similar to that in the beech forest.

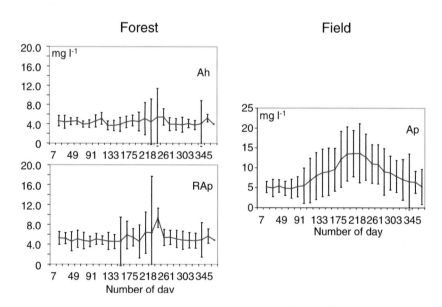

Fig. 8.8 Seasonal variations of sulfate concentrations in the A horizons of the Arenic Umbrisol under forest (W1) and the Eutri-brunic Arenosol under tillage (A3) during the 1989–1999 period. *Bars* indicate double standard deviation

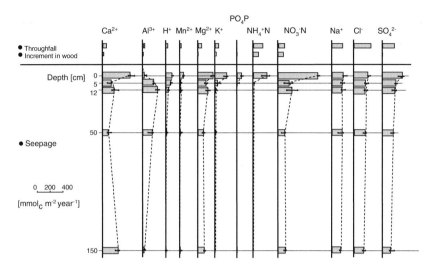

Fig. 8.9 Average annual element fluxes of the beech ecosystem (W1) during the 1989–1999 period

As indicated above, the solute concentrations of chloride ions in the forest soil are closely correlated with those of sodium since both have an almost exclusively marine origin with highest atmospheric deposition rates in winter. The resultant seasonal differences in concentration with April and December maxima of about 20 mg l^{-1} in the litter and about 15 mg l^{-1} in the upper Ah horizon are gradually smoothed out between 50 cm and 150 cm depth in the forest soil (cf. Fig. 8.9). Owing to manuring and fertilization the situation is different in the adjacent fields. In the Ap horizon maximum concentrations of about 40 mg l^{-1} occur in early summer; with increasing depth the lowered maximum is shifted to late summer. Below 150 cm seasonal differences disappear and random fluctuations around a 15 mg l^{-1} concentration level are typical.

8.3.4 Long-Term Element Budgets of Forest Arenic Umbrisols and Eutri-Brunic Arenosols under Tillage

Estimates of the source or sink functions of individual soil horizons, soils or ecosystems must be based on reliable element balances. One elementary problem involved in the procedure is the appropriate definition of the respective system boundaries, another the assessment of the role of the vegetation cover in modifying external inputs due to wet and dry deposition and specific uptake or leaching processes. In practice, the differentiation between the different types of internal and external fluxes is difficult and can be achieved only approximately by recourse to modelling approaches (cf. Ulrich 1983; van der Maas et al. 1991; Wetzel 1998). As a consequence, the present pragmatic budget approach defines the input rates of the beech stand in terms of canopy drip and stemflow and equates the lower boundary of the soil system with the maximum effective root zone. For both the field and forest soil this boundary is fixed at 150 cm; farther below biotic interactions with element fluxes are considered negligible.

The following description of element fluxes through the Arenic Umbrisol of the beech ecosystem and the Eutri-cambic Arenosol of the field catena (Figs. 8.9, 8.10) is based on the comparative analysis of the above monthly concentration values derived from the original biweekly measurements and the corresponding seepage estimates modelled by means of the WASMOD (Reiche 1991) and VAMOS (Bornhöft 1993) approaches. Since each model describes vertical water movement in the sense of matrix flow, i.e. neglecting lateral hydrodynamic dispersion and preferential flow, the multiplication of measured monthly concentration values with the corresponding seepage rates modelled might allege an unrealistically precise temporal resolution of the element fluxes. Therefore, the following budgets are formulated as annual integrals of the monthly horizon-related flux rates, whose seepage components were subject to additional independent validation by means of larger-scale water budget approaches (cf. Chapter 9).

The diagrams in Fig. 8.9 show that the elemental budgets of Na⁺ and Cl⁻ are slightly positive with a sink in the uppermost compartment, indicating the decisive importance of internal element cycling. Maximum flux rates of K^+, Ca^{2+}, Mg^{2+} and Mn^{2+} originate from the litter. The export rates of magnesium and manganese

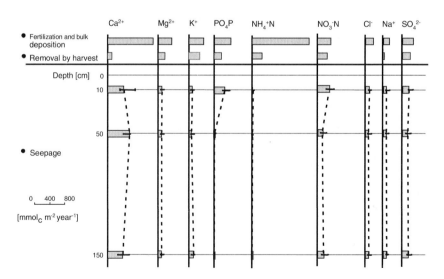

Fig. 8.10 Average annual element fluxes of an agrarian ecosystem (A3) during the 1989–1999 period

below the 50 cm level are only slightly higher than, or practically equal to, the input rates due to throughfall and stem flow, which is indicative of a biogeochemical steady state situation due to efficient recycling processes. In contrast, calcium translocation reflects the superposition of internal recycling and weathering processes, the latter causing the constant increase of Ca^{2+} flux rates below 50 cm depth (Peters 1990; Blume and Schimming 1991; Schimming 1991).

Also the ammonium balance is strongly positive, which means that the high deposition rates of NH_4^+ are apparently not subject to transformation in the litter. However, it ensues from the clearly negative nitrate and sulfate balances of the horizon that a substantial mineralization of dead organic matter takes place here. Thus, the NH_4^+ budget proves to be the result of balanced nitrification and ammonification processes. Owing to nitrification and (or) ammonium uptake by vegetation protons are produced, which leads to an equivalent H^+ flux out of the litter. The major part of these protons is neutralized in the uppermost decimetre of the mineral soil by release of an equivalent amount of Al^{3+} ions; in the underlying soil horizons the aluminium flux decreases to become practically negligible in the BC horizon.

Altogether the ionic balances of the soil solution of the upper part of the forest Arenic Umbrisol (W1) indicate that the positive surplus charge down to 50 cm is likely to be due to fluxes of organic acids (Wetzel 1998). The marked increase in positive charge between the 150 cm and 400 cm levels (not depicted in the diagrams). HCO_3^- is indicative of correspondingly intensified hydrogen carbonate fluxes which compensate for the positive charge, while the HCO_3^{2-} flux rates below the 400 cm level mark the transition to the original minero-chemical composition of the fluvioglacial deposits with average pH values of the soil solution of ~7.5.

The diagrams in Fig. 8.10 show that the Eutri-brunic Arenosol under tillage (A3) is characterized by very high calcium and ammonium inputs due to fertilization and manuring. Ammonium is either taken up by plants or undergoes nitrification, but contrary to the situation in the Arenic Umbrisol of the beech stands the protons released are buffered.

Furthermore it is noteworthy that phosphorus is accumulated in the Ap horizon, while there is depletion in sulfur. Regardless of a remarkable amount of biotic uptake of calcium in the Ap horizon, there is sorption-controlled Ca accumulation in the underlying Bw horizon due to intense fertilization. The resultant calcium fluxes from this reservoir are nearly seven times higher than those of the comparable horizon of the beech stand (Fig. 8.9). Conversely, and irrespective of the considerable amount of plant uptake, potassium exhibits an opposite behaviour with a slight increase of flux rates down to 150 cm, i.e. a positive K balance or source function of the root zone. The export balance of the 400 cm level is characterized by potassium retention, i.e. a sink function.

Finally, in order to demonstrate the extent of interannual variability averaged out in the mean values of the 1989–1999 period in Figs. 8.9 and 8.10, Table 8.7 summarizes exemplary data of element fluxes through the Arenic Umbrisol and the Eutri-brunic Arenosol during the contrasting years 1990, 1995 and 1998.

Table 8.7 Element fluxes through the Arenic Umbrisol under forest (W1) and the Eutri-cambic Arenosol under tillage (A3) in 1990, 1995 and 1998

					Sampling location in soil profile					
					Element					
	Year	Na	K	Mg	Ca	Cl	NH$_4$-N	NO$_3$-N	SO$_4$-S	PO$_4$-P
Forest Arenic Umbrisol										
L, O	1990	47.1	59.0	16.9	80.3	97.9	26.3	74.6	49.0	11.3
5 cm		47.2	40.0	17.0	58.5	105.0	0.9	71.7	36.0	1.1
12 cm		40.6	12.0	14.1	57.5	80.8	1.1	55.4	31.2	0.5
50 cm		31.7	1.3	6.4	29.2	63.5	0.3	21.5	23.3	0.3
150 cm		24.9	1.4	7.3	45.5	46.3	0.2	18.6	16.8	0.2
400 cm		31.4	1.3	5.4	120.6	57.7	0.4	11.2	23.0	0.3
L, O	1995	32.4	57.6	11.0	53.2	64.0	0.2	0.5	115.1	0.0
5 cm		33.0	8.8	9.5	25.8	56.0	1.2	25.1	23.0	0.0
12 cm		26.7	2.0	10.1	27.8	61.7	0.5	22.0	8.4	0.0
50 cm		11.6	0.1	5.8	13.8	43.4	0.2	11.4	13.4	0.0
150 cm	Missing values									
400 cm		38.1	1.5	5.0	141.9	65.9	0.4	6.8	34.5	0.0
L, O	1998	28.4	56.5	11.3	45.2	56.0	8.7	49.4	30.1	8.9
5 cm		28.7	4.3	7.1	10.2	50.8	0.3	5.9	35.5	0.2
12 cm		28.4	5.6	10.4	24.6	55.5	0.3	21.9	25.6	0.1
50 cm		28.5	0.3	5.0	11.6	51.2	0.2	4.7	21.2	0.0
150 cm		31.9	1.4	6.6	29.8	62.8	0.1	0.4	22.4	−2.2
400 cm		24.4	0.9	3.0	85.6	55.4	0.1	3.0	20.0	0.0
Field Eutri-cambic Arenosol										
10 cm	1990	52.1	43.9	18.6	137.9	96.4	2.6	52.1	61.2	18.9
50 cm		31.0	70.0	23.6	138.4	60.1	1.4	72.3	57.2	0.8
150 cm		27.0	81.0	19.6	139.6	51.7	0.6	91.4	48.4	0.7
400 cm	Missing values									
10 cm	1995	6.2	5.3	8.0	66.4	39.4	0.8	18.2	20.2	0.0
50 cm		13.5	12.4	7.0	80.2	18.1	0.3	24.0	14.2	0.0
150 cm		16.1	18.8	5.6	49.5	11.9	0.2	12.4	16.5	0.0
400 cm		21.7	7.4	15.4	321.2	24.7	0.4	40.3	46.5	0.0
10 cm	1998	20.6	39.2	12.7	94.5	28.9	1.6	49.5	33.7	6.7
50 cm		25.0	25.7	9.5	105.4	29.9	0.5	32.7	24.4	1.2
150 cm		20.9	41.1	9.6	77.4	22.2	0.3	37.8	21.3	0.0
400 cm		31.1	26.4	21.9	366.3	54.9	0.7	72.3	40.6	0.0

8.4 Nutrient Fluxes in Alder Stands and Wetlands

8.4.1 Elemental Concentrations in Alder Stands

Owing to the symbiosis with *Frankia*, black alder carrs (cf. Sections 2.2.5, 2.3.3) exhibit an enhanced fixation of atmospheric nitrogen and can therefore function as an important sink of this element on the ecosystem and landscape levels.

For assessment purposes, the above- and belowground biomass of the stands was estimated by harvesting and analyses of tree rings (Lenfers 1994; Eschenbach

1995; Eschenbach et al. 1997); soil microbial biomass was determined by means of substance-induced respiration and chloroform fumigation extraction (Dilly 1994). The measured carbon fluxes and the resultant budget are described in detail in Chapter 4. Therefore, Table 8.8 documents the size of the major carbon pools of the alder stands only for comparison purposes with those of the above beech stands.

The calculated gross primary production of *Alnus glutinosa* was 2280 g C m^{-2} year^{-1} in 1992 the calculated aboveground respiration 135 g C m^{-2} year^{-1}, while the estimated net primary production amounted to 850 g C m^{-2} year^{-1}. The figures suggest a root, rhizosphere and mycorrhizal respiration of 1295 g C m^{-2} year^{-1}. The same value results from a root respiration model (Kutsch et al. 2001b) calibrated with the specific respiration rate of fine roots colonized with ectomycorrhizae as measured by Gansert (1994) for beech and Staack (1996) for alder roots.

Measurements of the symbiotic nitrogenase activity were complemented by the determination of the natural ^{15}N abundances of alder and non-fixing plants in order to obtain reliable estimates of the N fixation rates. The results are in good agreement with the analyses of nodule densities, i.e. in the wet alder stands 70% of the total alder nitrogen originates from fixation, while in the drained facies about 45% is derived from the air. In combination with estimates of the above annual biomass production and with measurements of the annual litterfall, as described below, this permits to deduce that 70–85 kg N ha^{-1} year^{-1} in the wet alder facies and 40–45 kg N ha^{-1} year^{-1} in the drained facies are fixed by the *Alnus–Frankia* symbiosis.

The soils under alder forest are Rheic Histosols (cf. Section 2.3.3) with peat on top of a calcareous gyttja (or Gyttji-limnic Fluvisol). In a small-scale perspective, there are considerable differences with regard to thickness, groundwater level and eutrophication. The sites of the wet alder facies where peat thickness attains 5 m are flooded more than 36 days year^{-1}, while the drained sites near the footslope are liable to subsidence and a considerable amount of oxidative loss of organic matter. Accordingly, both the litter production of the alder stands (Lenfers 1994) and the related microbial communities (Bach 1996) differ considerably, and also the infiltration rates vary between 450 mm year^{-1} and 590 mm year^{-1} (Schleuß et al. 2001). Figure 8.11 summarizes the essential solute characteristics of the Histosols under alder and, for comparison, those of the neighbouring ecosystems.

Table 8.8 Organic carbon pools of the alder (W6) and beech stands (W1) of the Bornhöved Lake District. *DM* Dry matter

Component	Units	Alder stands	Beech stands
Total phytomass	t C ha^{-1}	82	152
Wood biomass	t C ha^{-1}	75	147
Biomass macrofauna	kg DM ha^{-1}	41	24
Biomass mesofauna	kg DM ha^{-1}	9.3	6.7
Biomass testacea	kg DM ha^{-1}	1.9	2.6
Microbial biomass	t C ha^{-1}	0.93	0.93
Soil organic carbon In the top 50 cm	t C ha^{-1}	175	80

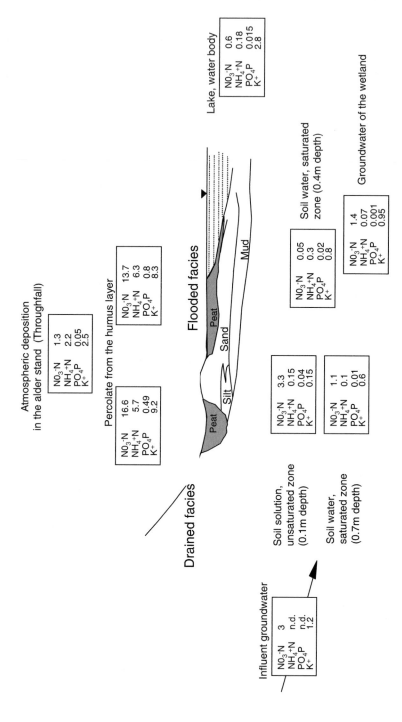

Fig. 8.11 Average solute concentrations (mg l⁻¹) of essential elements in different compartments of the alder stands (W6) and Lake Belau (after Schleuß et al. 2001)

In the Histosols under alder the K, N and P concentrations decrease rapidly below the humus layer; only the potassium concentration in the 0.7 m soil solution of the unflooded alder forest is higher than in the upper parts of the profile, thus reflecting the increasing influence of the groundwater flow from the footslope (cf. Chapter 9).

8.4.1.1 Litterfall in Alder Stands and Related Element Fluxes

In comparison with the above beech stands, the alder stands are characterized by normally higher rates of litter production, resulting in L and O horizons which are liable to rapid microbial degradation. Table 8.9 summarizes the results of a special sampling campaign during the years 1990–1993 in the alder stands (cf. Table 8.3 for the beech stands).

The calculated decomposition of organic matter by soil microbial communities amounted to 523 gC m^{-2} year^{-1}. Thus, all respiratory fluxes sum up to 85.6% of the gross primary production of the alder stand in 1992, with a net ecosystem balance of 328 g C m^{-2} year^{-1} remaining in the system. It should be noted that these figures are subject to marked interannual variability owing to the temperature sensitivity of the processes involved.

The above carbon balances indicate essential differences between the beech and alder ecosystems. *Alnus glutinosa* yields much of its carbon to the symbiosis with *Frankia*, to the mycorrhizae, and to the rhizosphere, whereas *Frankia* provides the tree with large amounts of nitrogen (Dittert 1992); furthermore acid root exudates promote phosphorus availability (Bar-Yosef 1991). This involves, on the one hand, high energy costs for evapotranspiration purposes as a basis of nutrient uptake by means of concentration processes in the wet habitat with low nutrient availability. On the other, the alder trees eutrophy their habitat with litter rich in nitrogen (cf. Table 8.10). As a consequence, the microbial communities in the litter and the topsoil of the alder stands are dominated by fast growing r-strategists (Bach 1996; Atlas and Bartha 1998), which leads to high decomposition rates of the readily biodegradable litter.

A comparison of the annual carbon input rates with the decomposition rates determined by means of litterbags on the one hand and a comparison of both with

Table 8.9 Annual production of dead organic matter and associated element fluxes of alder carrs (W6) during the 1990–1993 period (kg ha^{-1} year^{-1}; after Lenfers 1994). *L* Litter, *OF* other fractions (wood, cones, etc.) of dead organic matter

Year		Dom	Na	K	Mg	Ca	N	P	Mn
1990/1991	L	3479	1.7	6.7	5.1	38.6	68.7	–	0.5
	OF	2248	1.6	4.8	2.3	30.4	35.2	–	0.3
1991/1992	L	3483	4.9	8.5	7.3	54.7	70.2	2.5	2.6
	OF	3115	2.9	6.7	3.9	32.1	42.5	2.8	1.6
1992/1993	L	3135	0.8	5.8	6.3	51.1	72.9	2.5	2.3
	OF	3764	1	8.3	4.8	43.1	70.6	3.5	2.4

Table 8.10 Carbon pools and input rates compared with duration of litter decomposition (after Irmler 1995; Wachendorf 1996). *n.d.* Not determined

Site	Pool (t C ha^{-1} 50cm^{-1})	Input (t C ha^{-1} year^{-1})	Decomposition (years)
Beech forest	57	3.7	8
Field	60	4.5	n.d.
Grassland on slope	72	4.8	n.d.
Mixed forest	77	2.5	7
Wetland	100	3.4	n.d.
Alder (wet facies)	176	10	2
Alder (dry facies)	394	5	4

the carbon pools of the various ecosystem types of the core area on the other are indicative of the different turnover rates. Table 8.10 summarizes the results of pertinent analyses by Irmler (1995) and Wachendorf (1996).

The above input figures include the inputs from the herbaceous layer and the root zone, the latter accounting for (estimated) 20% of the total input in the case of both beech and mixed forests and about 40% in the case of the alder carr. Thus, the annual fine root production of the Bornhöved beech stands compares well with that of the Solling beech forests (Ellenberg et al. 1986). Although the highest carbon inputs are linked to the alder carr, the corresponding pool is distinctly lower than in the drained alder stands because of less compaction of the underlying organic horizons.

8.4.2   Nitrogen Budgets of Wetlands

About 8% of the total study area are peatlands, 20% of which are poorly drained or undrained, while 80% are more or less well drained and subject to agricultural land use with a prevalence of pastures. The latter, a *Ranunculo–Alopecuretum geniculati typicum* in terms of phytosociology (Schrautzer and Wiebe 1993), is the dominant vegetation type on grazed, eutrophic Histosols with fluctuating groundwater levels. The presence of species like *Ranunculus repens*, *Glyceria fluitans* and *Agrostis stolonifera* is indicative of stagnant water during the growing season, caused by soil compaction (the bulk density of 0.58 g cm^{-3} is 5–10 times higher than that of the alder Histosols) and a correspondingly reduced infiltration capacity (Schrautzer and Trepel 1997). As a consequence, grazing can only be extensive and mowing must be reduced, which fosters the occurence of Molinietalia species like *Cirsium palustre*, *Cardamine pratensis* or *Lychnis flos-cuculi* that regenerate only under such low-disturbance conditions.

Analysis of the throughflow and groundwater dynamics of the wetlands (W6, A9) reveals a predominance of infiltration and percolation, while lateral water exchange with Lake Belau is largely controlled by the hydrophysical structure of

Table 8.11 Average nitrogen budgets of three wetland ecosystems (W6, A9) in 1992 and 1993 (kg N ha⁻¹ year⁻¹) after Schleuß et al. 2001). *n.d.* Not determined

	Alder, dry facies	Alder, wet facies	Pasture
Input			
Atmospheric deposition	35	30	20
Nitrogen fixation	45	65	–
Manure			30
Total	80	95	50
Output			
Increment in wood	10	15	–
Harvest	–	–	75
Leaching/litterfall	15	20	–
Humus accumulation	5	–	–
Total	30	35	75
Balance	50	60	–25
Net mineralization	155	30	470
Gaseous losses	20	n.d.	n.d.

the Histosol and the underlying peat layers (for details see Chapter 9). Under these circumstances, the pastures reach the highest annual rates of net nitrogen mineralization in the uppermost 30 cm of all wetland sites studied, as Table 8.11 shows.

In comparison with the Histosols under pasture, where both groundwater level and soil pH provide optimum conditions for mineralization, the lower groundwater level in the near-natural alder carr has practically stopped peat accumulation, while in the drained alder stands soil acidity proves to be the major controlling mechanism of nitrogen mineralization (with regard to the relevant decomposer groups the reader is referred to Table 8.5).

8.5 Stress, Strain and Metastability of Beech, Alder, Pasture and Agro-Ecosystems

Stress is the state of a biotic or abiotic system under the conditions of a "force" applied; *strain* is the response to the stress, i.e. its expression before damage occurs, while *damage* is the result of too high a stress that can no longer be compensated for (cf. Csermely 1998). As long as the strain is completely reversible, it is said to be elastic; beyond this point or threshold, the strain will be only partially reversible, and the irreversible part is called the permanent set or plastic strain. In the present context of a novel application of ecological stoichiometry (Sterner and Elser 2002), elemental imbalances serve first as indicators of plastic strain. They are defined in terms of bioconcentration factors, i.e. the ratio of elemental concentrations in leaves to elemental concentrations in soil solution, litter or mineral soil. Second, element budgets form more complex expressions of stand-specific stress reactions. Finally, as exemplified by the alder carrs, changes in the rooting system and composition of the undercover vegetation are of indicative quality.

8.5.1 Elemental Imbalances as Strain Indicators

In terms of ecological stoichiometry, which focuses on the balance of multiple chemical substances in ecological interactions and processes elemental imbalance is the dissimilarity in nutrient content between two things, such as consumer and food resources or between an autotroph and the inorganic medium. If both had identical stiochiometry, they would be perfectly balanced; the greater they differ, the more their imbalance.

Figures 8.12 and 8.13 compare the elemental composition of different organs of beech and alder trees (in terms of C:N and N:P ratios) with the composition of the soil solutes in the corresponding rooting zones. The primary focus is on leaves, since they (along with secondary roots) are a major form of annual production by most terrestrial plants. Thus understanding the factors regulating the C:N:P stoichiometry of primary production in an ecosystem context includes understanding those factors affecting leaf nutrient content (Aerts and Chapin 2000). Second, leaves are the targets of consumption by many terrestrial herbivores and bear on the dynamics of belowground food webs which are increasingly recognized as key regulators of terrestrial ecosystem dynamics (Strong et al. 1996). Finally, leaves respond more to changes in fertility than do stems and roots (Marschner 1995).

The juxtaposition of Figs. 8.12 and 8.13 shows that the C:N ratios of the beech leaves with their elevated cellulose and lignin contents are much higher (by a factor of >2) than those of the alder leaves while, in contrast, the N:P ratios of alder leaves are distinctly higher (by a factor of >4). It is interesting to note that the N:P ratios of the foliage of the wet and dry alder facies do not differ substantially, nor do those of cones and wood.

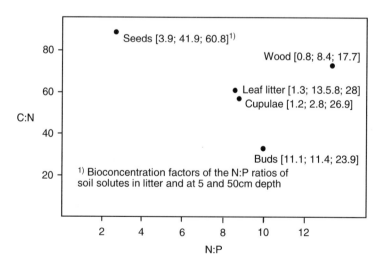

Fig. 8.12 C:N and N:P ratios of leaf litter, buds, beech-nuts, cupulae, and wood of Bornhöved beech stands (W1) during the 1990–1993 period (after data by Lenfers 1994)

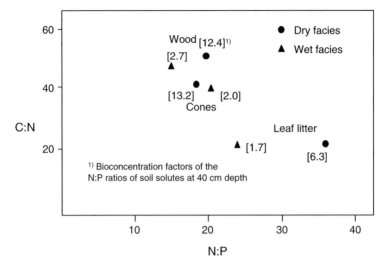

Fig. 8.13 C:N and N:P ratio of leaves, cones, and wood of the dry and wet alder facies (W6) during the 1990–1993 period (after data by Lenfers 1994)

Table 8.12 Bioconcentration factors of beech (W1) and alder (W6) foliage. Bioconcentration factors are defined as the ratio of element concentrations in leaves to element concentrations in soil solution of L, O horizons and at 12, 50 and 150cm depths

Location/depth of soil	N	P	Ca	Mg	K	Na
Beech						
L, O	553	559	703	404	352	264
12 cm	810	3.0×10^4	1083	465	1632	230
50 cm	1324	3.0×10^4	1560	620	7750	179
150 cm	1324	4.5×10^4	722	465	6200	200
Alder (dry facies)						
~50 cm	9304	6.0×10^4	563	563	3300	55
Alder (wet facies)						
~12 cm	2.6×10^4	4.5×10^4	594	594	2625	18

Compared with the N:P ratios of the soil solution the above ratios permit to define bioconcentration factors, which are added in parentheses to the respective plant organs in the above diagrams. In the sense of the introductory definition these factors indicate elemental balance the nearer they come to 1, or imbalance the further they deviate from this value. Thus, they are a measure of stress reactions or plastic strains, since imbalances involve energy-consuming concentration processes in dependence on the respective element concentration in plant organs and soil solution (Fränzle 1994).

Defining elemental imbalances in terms of bioconcentration factors (BCFs) of the elements N, P, Ca, Mg, K and, for comparison, Na provides additional insight into the stress situation, as Table 8.12 shows. BCFs are, in analogy to the above ratios, defined as the ratio of element concentrations in beech and alder leaves to element concentrations in soil solution of L and O horizons and mineral soil at 12, 50 and 150 cm depth.

Focussing first on elemental concentrations in L, O horizons as a reference level for the corresponding concentration values in leaves, the comparative analysis of the above BFCs shows that the elemental imbalance of N and P is one to two orders of magnitude greater for the dry alder facies than for beech, and the difference is even more marked when considering the wet alder facies. The compensatory strain reaction is the development of the *Alnus–Frankia* symbiosis described above.

Except for Mg and Na, the comparison of the depth-related BCFs further indicates a pronounced element recycling between litter and leaves, which was also registered in the Solling Mts. and the Harste beech stands (Göttsche 1972), while the solute pools in deeper soil layers are distinctly less drawn on (cf. Section 8.3.3). The Ca and Na figures exhibit interspecific variation in autotroph stoichiometry, but for either vegetation type the depth functions characterize sodium as a typical "throughflow" element. Finally, the extremely high BCFs of phosphorus from soil solution below the litter horizons underline the role of this element for both beech and alder in the sense of Liebig's law of the minimum. Second in importance in terms of BCFs is potassium as a minimum element.

Summarizing, it may be said that the element-specific differences in concentration processes indicate meta-stability of the beech ecosystem as related to P and K supply; to a lesser degree this applies also to Ca. The situation is similar for P and K in the case of alder, while the Ca supply of the system is better owing to lateral groundwater transport from the adjacent slope. Magnesium supply does not seem to pose particular problems for either ecosystem.

8.5.2   Element Budgets and Strain Reactions of Beech Stands

In the light of the comparative analysis of the annual and interannual variability of macro- and micronutrients and trace element concentrations also the resultant element budgets may be interpreted as complex indicators of the "health" or the strain reactions of the respective ecosystems, i.e. their ability to cope with externally forced change including human impacts like exploitation of the ecosystems or pollution. Analogously, also the strain reactions of the various energy subsystems powering the operation of the ecosystem could be used as indicators of ecosystem resilience or stability, respectively.

Principally, the element budget of a system can be formulated as follows (cf. Schlichting 1975):

Initial element content + material inputs − outputs = pool

The difference of pool and initial content then yields the elemental balance, while the juxtaposition of material inputs and outputs defines element-specific turnover rates. In view of the inherent difficulties to retrospectively define the initial element content of ecosystems with a sufficient amount of accuracy, however, the basic strategy consists in determining changes in pool quantities by means of balancing the input/output relations of the system (Feger 1993). In detail, the analysis of turnover processes involves a differentiation of interior and external element cycles as illustrated in the following Sections on the basis of elemental flux models.

Owing to its essential role as a macronutrient, nitrogen exhibits a particular indicator quality (Beese 1986) for the assessment of ecosystem health. In near-natural forests without anthropogenic load, for instance, nitrogen is largely retained in the system by efficient recycling processes in the plant and soil compartments (Ellenberg 1977). In the case of the Bornhöved beech stands, as depicted in Figs. 8.14 and 8.15, the total atmospheric input of 40 kg N ha^{-1} year^{-1}, as deduced from the EDACS modelling approach (cf. Bleeker et al. 2000) which proved distinctly better suited for estimation purposes than the Ulrich (1983, 1991) and van der Maas et al. (1991) approaches, is coupled with an average storage rate of 12–16 kg N ha^{-1} year^{-1} in the wood fraction.

It ensues from the nitrogen cycle modelled in Fig. 8.14 that the Norg pool of the soil has an increase rate of 2 kg ha^{-1} year^{-1}, while an amount of 79 kg N ha^{-1} year^{-1} is mineralized and 17 kg N ha^{-1} year^{-1} leaves the rooting zone of the ecosystem by seepage (Wetzel 1998; Schimming et al. 2001; Wellbrock 2002). The average mineralization rate during the 1989–1998 reference period estimated from the balance of throughfall and the flux from the L and O horizons attains a value of 62 kg N ha^{-1}

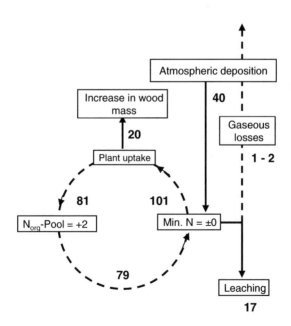

Fig. 8.14 Average nitrogen fluxes (kg ha^{-1} year^{-1}) of the Bornhöved beech stands (W1) during the 1989–1991 period of intensive investigations on internal cycling processes (Schimming et al. 2001). Cycling rates are based on modelling and corroborated by measured seepage rates

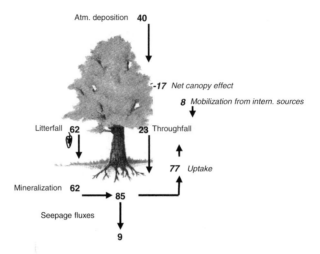

Fig. 8.15 Average measured and balanced (*in italics*) nitrogen fluxes (kg N ha^{-1} year^{-1}) of the Bornhöved beech stands (W1) during the 1988–1997 period

year^{-1}, and the same amount of nitrogen is deposited in the form of litter. Internal cycling rates in both approaches indicate remarkably short residence times for nitrogen in litter. Here, and in the following figures, the term "internal immobilization" comprises middle and long-term storage of elements in the system.

As a consequence of high atmospheric input rates, nitrogen accumulates in the ecosystem, thus increasing the considerable pool of easily available nitrogen compounds in the forest Arenic Umbrisol. Considering the distinct mobility of nitrogen and sorption capacity of the soil, the difference between the net effect of canopy processes and the nitrogen fluxes in the soil is equal to storage in wood. The rates range from 20 kg N ha^{-1} year^{-1} during the 1989–1991 period and have an average of 8.4 kg N ha^{-1} year^{-1} during the whole 1989–1999 period. The ammonia deposited is completely oxidized and leaves the system in the form of nitrate, which contributes to soil acidification owing to the concomitant release of two protons for each ammonia ion transformed.

The temporary increase in pH of the soil solution registered from 1997 to 1999 (Fig. 8.16) seems to result mostly from lower soil leaching rates of nitrate-N which, together with comparatively high ammonium throughfall rates, reduced the acidity.

In comparison with the nitrogen cascade the potassium budget of the beech ecosytems is characterized by highly efficient recycling processes, as Fig. 8.17 shows. This means that considerable K amounts reach the soil by litterfall and leaching from the canopy and are taken up directly from the litter again (cf. Section 8.3.3), whereas neither atmospheric deposition nor translocation processes to the groundwater are of major importance. Thus, recycling processes are essential mechanisms to maintain a non-equilibrium steady state coupled with a (relative) minimum of total entropy production which the metastability of the ecosystem is based upon (Fränzle 1994).

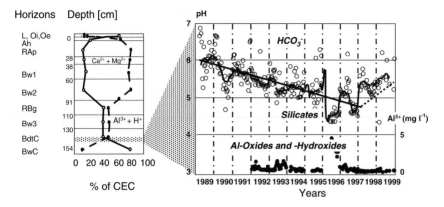

Fig. 8.16 Progressive soil acidification under beech forest (W1) during the 1989–1998 period

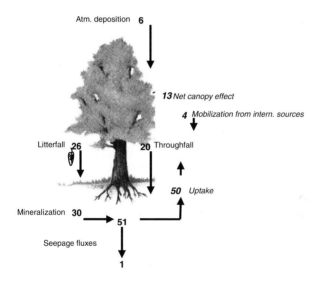

Fig. 8.17 Average measured and balanced (*in italics*) potassium fluxes (kg ha⁻¹ year⁻¹) of the Bornhöved beech stands (W1) during the 1988–1997 period

External magnesium input into the beech forest system is rather low (Section 8.2.2) but sufficient to cope with the internal physiological demand. Thus, the biogeochemical release from primary minerals and from clay minerals leads to a considerable translocation to the groundwater as Fig. 8.18 shows.

The atmospheric input rates of calcium are nearly as low as those of potassium and consequently insufficient to cope with the normal annual demand of the vegetation cover (Fig. 8.19). In addition, considerable Ca quantities are lost from the root zone owing to the pronounced acidification of the Arenic Umbrisol and the seepage-related

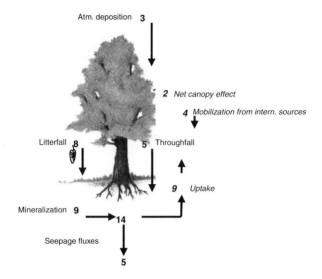

Fig. 8.18 Average measured and balanced (*in italics*) magnesium fluxes (kg ha⁻¹ year⁻¹) of the Bornhöved beech stands (W1) during the 1988–1997 period

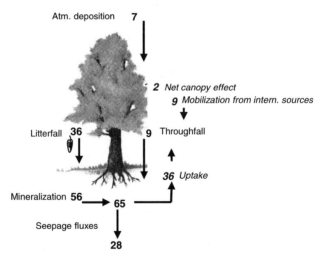

Fig. 8.19 Average measured and balanced (*in italics*) calcium fluxes (kg ha⁻¹ year⁻¹) of the Bornhöved beech stands (W1) during the 1988–1997 period

input of major amounts of carbon dioxide, 0.7% of which occur in the form of carbonic acid. Therefore, the calcium supply of the beech stands is mostly based on internal recycling processes in the ecosystem compartments vegetation, litter and Ah horizon, while the subsequent soil horizons act as Ca sources for the underlying aeration and groundwater domains of the ecosystem.

In comparison with potassium and calcium which exhibit intensive recycling as described above, atmospheric sulfur and sodium and chloride inputs are mostly transported through the biotic and abiotic compartments of the ecosystem into the groundwater; only a minor proportion of sulfur is liable to long-term storage in the wood.

8.5.3   Strain Reactions of Alder Stands and Pastures

Prior to human control of Lake Belau's water table in 1934 and 1936 and the concomitant lowering of the groundwater level, the Histosols of the alder stands originated from reed and sedge peat which were largely prevented from mineralization due to water logging and correspondingly low redox potentials. Thus, nitrogen fixation by the *Alnus–Frankia* symbiosis was essential for the nitrogen supply of the plant communities. Therefore, the drainage of the upper horizons as a consequence of the lowered groundwater table has become a serious stress factor resulting in plastic strain reactions. The top soil dries up during the summer months, which gives rise to elevated mineralization rates of the soil organic matter and consequently a high availability of organically and minerally bound nitrogen. In combination with soil-leaching processes this leads to a severe acidification of the upper soil horizons, which in turn is inducive to less nodulation (Dittert 1992).

Altogether, the most conspicuous integral strain symptoms are changes in the floristic composition which the dystric, drier alder facies has been liable to (cf. Sections 2.2.5 and 2.3.3). Furthermore, they are reflected in marked differences in both carbon pools and input rates of the two alder facies (Table 8.10) and the concomitant differences in soil solute chemistry.

Under near-natural conditions, the wet meadows on fen sites (i.e. the Magnocaricion, Calthion, and the Lolio–Potentillion syntaxa) are very productive systems, comprising functional groups of plants which are well adapted to the different intensities of stress versus competitive conditions occurring within and between such systems (Hills and Murphy 1996). They can achieve such high productivity because water is generally not limiting to growth, though there are variations in water availability (Scholle 1997) and nutrient availabilty may be high (Chapter 9). Therefore, the unflooded pastures of the study area are mostly affected by drainage which leads to an intensified decomposition of organic matter and, as a consequence, an essential change in physical and chemical soil properties (Zeitz 1991). Thus, the bulk density of the upper soil horizons has increased considerably, owing to the declining groundwater level which varies in correspondence with the lake level position; and due to partial mixing with sands a mollic epipedon with an organic matter content of approximately 9% has developed. In combination with disturbances by grazing cattle and the habitual harvesting and fertilization practices these changes are coupled with a marked decrease in biodiversity (Schrautzer and Trepel 1997; Schleuß et al. 2001).

8.5.4   Agricultural Impact

The agro-ecosystems of the study area are normally crop-rotation systems and vary accordingly in the amount of subsidy which they receive from agricultural activities in terms of energy or material inputs (Trümpler 1996; Schmitt 1997). With increasing subsidy productivity rises and so does the intensity of competition experienced by the plants occupying the ecosystems. There is a gradient of increasing production running from the managed grassland on the slopes (A5, A6) below the fields, receiving cow-dung but only low quantities of N-P-K fertilizer (if any) to subsidize the growth of grasses for the animals, according to the various types of agro-ecosystem. These are subject to more or less periodical disturbance inherent in agricultural management practices and receive much higher manure and fertilizer subsidies, with high competition from the crop plants against any other plant species present.

In general the more intensively managed the agro-ecosystems are the more productive they are and the greater is the intensity of competition. Herbicides as a form of indirect subsidy to crop plants actually work by placing weeds under severe toxic stress, while leaving the crop plants (more or less) unaffected. Thus, the diversity of the plant community is extremely low, comprising the crop itself plus a handful of weed species. It is considerably increased, however, if the non-arable parts of the land such as field fringes, hedgerows, etc., are also taken into account.

Analysing the Bornhöved Lake District from the viewpoint of a sustainable development policy, attaching equal importance to the sets of socio-economic and ecological aspects, with their bewildering number of stress and strain situations, requires a comparative analysis of the nutrient-related sink and source functions of the agro-ecosystems on the landscape level on the one hand and a novel combination of planning approaches on the other. While the first subject matter is described in greater detail in Chapters 9, 10 and 11, the latter approaches, as developed in particular by Reiche et al. (1999), Dibbern (2000), Meyer (2000) and Herzog (2002), are taken up in the wider context of Part III of the present book.

8.6   Conclusions

Considering element fluxes through the ecosystems of the study area in the light of ecological stoichiometry exhibits strong relationships between flux rates and plant growth. The high variability in growth processes has a basis in cellular structure and in the ability to flexibly adjust growth rate and cellular allocation to fit the local site conditions, which is in strong contrast to the homoeostasis of C:N:P maintained by animals that obtain their energy and elements together in preformed food items (cf. Section 2.3; Chapters 6 and 7). Plants must obtain their energy and materials from solar radiation and uptake of inorganic forms from soil or water and the atmosphere. In soil metal ions as components of the mineral nutrition of plants interact with organic matter both physically by adsorption and desorption and chemically by ligation (forming complexes) with donor sites available in biomass, and, like in organisms, they display catalytic functions in metalloproteins. Thus, the organotropic bioconcentration of metals in the organs of plants depends on the chemical properties of the metal, its distribution in the different horizons of the pedon and the biochemical structure of the plant organelles or organs affected, which permits to formulate quantitative relationships between the stability of metal complexes and bioconcentration equilibria (Fränzle et al. 2006).

These processes involve a close interdependence of organic matter breakdown by microbes and edaphic metazoans and the balance of numerous elements in the substrate relative to the physiological demands of the microbial communities, which finds a summary expression in the respective C:N and C:P or N:P ratios. Thus, litter decomposition is faster for nutrient-rich detritus as produced by fast-growing plants with high nutrient content (see alder), while slow-growing plants (e.g. beech) produce low-nutrient litter that breaks down slowly, reinforcing slow nutrient-limited plant growth. This means, inter alia, that the stoichiometric yield, i.e. the carbon or biomass per nutrient, is inversely related to ecosystem fertility (cf. also Sections 3.7 and 4.2.3). This relationship, which also follows from generalized thermodynamic considerations (Fränzle 1984), has far-reaching consequences for macro- and micronutrient fluxes through the ecosystems of the study area, which are quite frequently liable to intensive manuring or fertilization, as Chapters 9, 10 and 12 will show in different connections. Furthermore Sections 13.3.5 and 13.3.6 deal with these phenomena from the complementary viewpoints of landscape planning and ecological economics.

Chapter 9
Transport Processes between Lake Belau and its Drainage Basin

Winfrid Kluge and Otto Fränzle

9.1 Introduction

Air and water are the fluid media transporting dissolved and particulate mineral and organic substances between terrestrial and aquatic ecosystems. With respect to the physics of such flow processes, two theoretical approaches are distinguished in general. Using Euler's concept, the multidimensional partial flow transport equations are solved simultaneously for all points in space at a given time. Lagrange's approach describes mainly the temporal change of tracking particles which can shift along flowlines within catchments and may be subject to transformation processes. Although both approaches, from a theoretical point of view, are equivalent, labouring under the apprehension of non-turbulent flow and yielding compatible results (e.g. in groundwater hydraulics), the Lagrange approach provides a detailed description in direct correspondence with the water and material cycles in ecological systems and catchment areas. Between the origin, e.g. the entry into a terrestrial system, and the spatially distinct and temporally retarded effect in the subordinate systems, e.g. the entry into an aquatic system, a direct correlation is established (Kluge et al. 2003).

9.2 A Path Concept as a System-Linking Methodological Platform

The lateral transport processes which cause the hydrological and hydrochemical interactions between ecosystems can be divided, as shown in Fig. 9.1, into the following categories (the cross-section corresponds to the forest catena of the core area):

1. The aerodynamic drift within the boundary layer between the soil and the atmosphere or the canopy surface and the atmosphere and between the water surface and the atmosphere with particular reference to (1a) wind erosion/sedimentation, (1b) emission, (1c) dry and wet deposition, (1d) litter drift;
2. The overland flow, in the sense of Horton's runoff (2a) on slopes which either reach the riparian wetland/floodplain or local closed depressions (2b), the

O. Fränzle et al. (eds.), *Ecosystem Organization of a Complex Landscape.*
Ecological Studies 202.
© Springer-Verlag Berlin Heidelberg 2008

saturation overland flow (2c), due to an infiltration excess and an upwards directed discharge of subsurface water in groundwater-influenced areas, or the overflow of water stored temporarily in overflooded depressions (2d);

3. The interflow within the soil zone where slope-parallel low-permeable layers restrict deeper seepage;
4. The upper subterraneous runoff of oxic, usually younger groundwater which is recharged in the catchment area and mainly emerges at the foot of slopes, in riparian wetlands, and in transition to aquatic systems;
5. The lower subterraneous runoff of anoxic, usually older groundwater which is recharged predominantly in the proximity of the geohydraulic divide and discharged directly into water bodies;
6. The fluctuating exchange processes in riparian wetlands and floodplains during flooding by lakes and running waters;
7. Hydrodynamic processes and displacement of matter in lakes, such as (7a) wind-induced non-periodic currents, (7b) wind-induced periodic waves in the littoral zone, (7c) wave erosion and sedimentation along the shore line, (7d) bioturbation, (7e) diffusion of dissolved matter through the sediment-water contact zone, as well as (7f) a multitude of wind-induced mixing and periodic waves, density-dependent buoyancy processes, and sedimentation and displacement processes in the water body;
8. Active faunal exchange between terrestrial and aquatic ecosystems.

In Fig. 9.1, only the transport processes relevant to Lake Belau are marked. Figure 9.2 summarizes how the vertical and lateral exchange paths between lakes and their surrounding catchment areas are linked; and for the sake of better legibility the number of compartments is deliberately reduced. Waters of different genesis and origin mix in the banks and the water bodies. To reflect the microscale spatial structures of transport processes in sufficient detail, the non-point exchange of matter is divided into two categories, namely diffuse inputs not directly observable and small-point or punctiform such as tile drains, small ditches, springs. From a temporal viewpoint, a distinction between permanent, seasonal, and event-dependent inputs is indicated.

Evaluating the substantial interactions between the catchment and the lake, the following balance terms are of interest: (a) input into the plant-soil or water systems from the atmospheric layer, (b) exchange of matter at the surface, (c) near-surface exchange within the soil zone or littoral sediments, (d) upper groundwater exchange, (e) lower groundwater exchange, (f) exchange of matter between littoral and pelagial.

In this connection the retention coefficients R as defined below are useful for comparisons of the material balance and source or sink functions of individual boxes, compartments, or entire catchments, respectively. They are derived from the equation:

$$R = [(\text{external matter input} - \text{external matter output})/ \\ \text{external matter inputs}] \times 100 \ (\%)$$

In addition to the lateral material flows, the vertical inputs and outputs through the soil surface (deposition, fertilization, harvest) are added to the external exchange fluxes. The internal transformations, e.g. enrichment or mineralization of organic

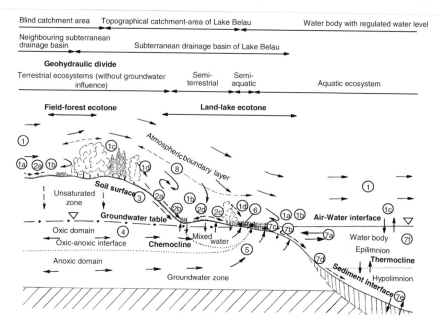

Fig. 9.1 Simplified spatial ecological structure and lateral transport processes between a lake and its surrounding upland in a glacial landscape

matter, wood increase, denitrification of nitrate, sorption or release of phosphorus, are reflected in the difference between external fluxes but do not figure explicitly in the summary formula. R is positive when the inputs exceed the output, for $R = 0\%$ external inputs are equal to external outputs, and $R = 100\%$ indicates a complete retention of material in the balanced compartments. Negative values exist only if the outputs exceed the inputs, due to internal release processes, e.g. mineralization.

9.3 Exchange of Water between Lake Belau and its Catchment

9.3.1 Hydrological Structure of the Catchment

Based on official hydrological maps and a 12.5 m grid elevation model on the one hand, and some 60 geological bore-holes of 2500 m total depth sunk in the framework of the project on the other, the size and structure of both the topographic and the underground catchment areas of Lake Belau were determined. As Fig. 9.3 shows, the topographic catchment has an acreage of 3.2 km², but due to the widespread occurrence of closed depressions (Fig. 9.1) only about 1 km² of it, i.e. the effective catchment area, is liable to direct overland flow.

The latter area comprises 75% of groundwater-far slopes and 25% of groundwater-influenced wetlands. Because of the predominance of sandy soils and the distribution

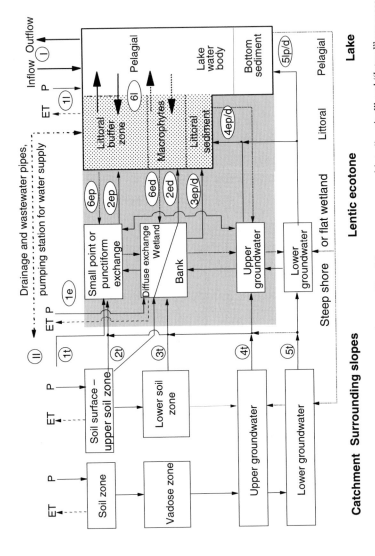

Fig. 9.2 Basic structure of hydrological pathways linking a catchment with a lake system, grouped into "vertical" and "lateral" processes. The *shaded area* represents lentic ecotones; *t* terrestrial area, *e* ecotone, *l*= lake, *p* punctiform input, *d* diffuse input

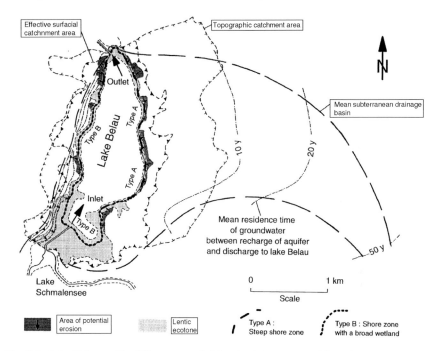

Fig. 9.3 Lake Belau and its different superficial and subterraneous catchment areas

of forests, hedgerows, and pastures contiguous to the lake runoff and interflow occur only under exceptional circumstances, however.

The subterraneous catchment exhibits a marked asymmetry with a 4.2 km² eastern and a substantially smaller (0.4 km²) western subcatchment. Both sensitivity analyses with variable groundwater recharge rates and the evaluation of groundwater level records indicate that the subterraneous catchment is reduced in size by about 50% after periods of extremely little rainfall, e.g. in 1996. This special characteristic of glacial landscapes is due to the following three conditions: (a) the aquifers extend continuously between two water bodies like lakes or water courses, (b) the vertical groundwater recharge exhibits a clear seasonal course, and (c) the levels of neighbouring water bodies differ substantially. Because the mean water level of Lake Belau with 29.3 m a.s.l. lies 0.7 m or 1.4 m above the levels of the Fuhlensee or Schierensee, respectively, the hydraulic divide runs across the upland of the core area (Kluge et al. 2003). Lake Plön is situated approximately 6 km northeast of Lake Belau. With its mean water level at only 22.1 m a.s.l., it induces marked regional geohydraulic gradients (Rumohr 1996).

In the 1989–1998 period, the underground hydraulic divides varied considerably. Extreme conditions occurred during a long-lasting wet period with a maximum catchment area of 5.5 km² in spring 1995 and during a long dry phase in summer 1997 with a resultant minimum acreage of 2.7 km². The southern wetland between the Schmalensee and Lake Belau is drained predominantly by ditches. Because of the nearly identical water levels of both lakes and a supply based predominantly on precipitation, local exchange paths dominate. Within the northern range of Lake Belau both an increased discharge from the drained lowland and a marked drawdown of the groundwater level are indicative of exfiltration of lake water into the aquifer. Figure 9.3 shows that the age of the groundwater which recharges Lake Belau varies between less than one year and >100 years. Using numerical groundwater models and analytical solutions, it could be concluded that the age of the groundwater depends in particular on the thickness and pore volume of the aquifers and on the groundwater recharge rate.

To facilitate the understanding of the geohydrologic conditions of Lake Belau, a W–E cross-section through the point of maximal water depth is given in simplified form in Fig. 9.4. It is to be noted that the aquifer consists of two layers (the upper Weichselian and the lower Saalian fluvioglacial outwash sands), whose mean conductivity is $k_f = 7\times10^{-4}$ m s^{-1} for the upper and $k_f = 2\times10^{-4}$ m s^{-1} for the lower complex. In combination with Fig. 9.3 it shows that the lateral groundwater exchange occurs preferentially in the upper sediment complex of the eastern catchment, where the horizontal tracking flow velocity close to the banks amounts to 0.3–0.7 m day^{-1} in contrast to <0.2 m day^{-1} in the lower complex (Rumohr et al. 1996).

To characterize the hydrological regime of Lake Belau and its surroundings, selected hydrological variables are compiled for the 1989–1998 period in Fig. 9.5. The monthly average values are based on the evaluation of partly measured and partly simulated daily values. Inflow, outflow and lake water levels were registered by high-resolution electrical sensors. Using a correction factor of the precipitation measured at the Ruhwinkel meteorological station (DVWK 1996) the mean value of 795 mm year^{-1} for the 1989–1998 period, as based on rain gauge records, has to be

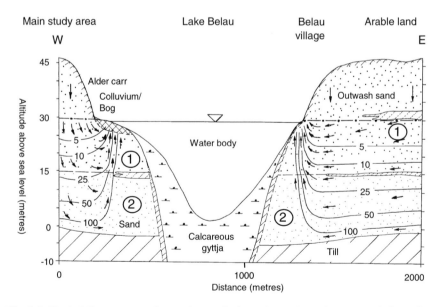

Fig. 9.4 Vertical flow patterns of groundwater discharge into Lake Belau. *Arrows* indicate flow direction; *solid lines* represent the residence time of groundwater in years. Hydraulic conductivity of layer 1: kf = 7×10^{-4} m s^{-1}, and layer 2: kf = 2×10^{-4} m s^{-1}

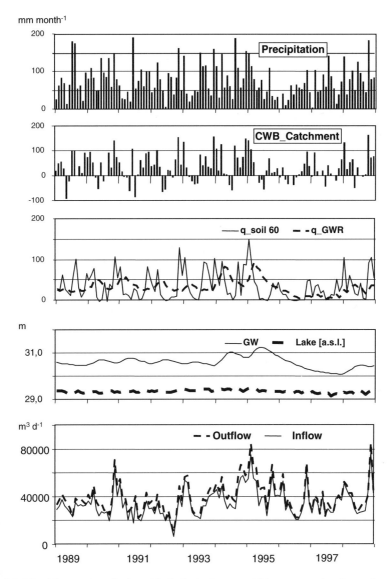

Fig. 9.5 Monthly structure of selected hydrological state variables representing the catchment of Lake Belau from 1989 to 1998. *CWB* Climatic water balance, *q_soil60* mean percolation rate at a soil depth of 60 cm below ground under crop cultivation, *q_GWR* mean groundwater recharge of the catchment, *GW* piezometric groundwater level, *Lake* lake water level

increased by approx. 12%, yielding an effective (balance-true) areal precipitation of 890 mm year^{-1}. Additionally, note that the groundwater recharge persists normally over the whole year in the Bornhöved region except for a low-precipitation period which began in winter 1995/1996 and lasted to the end of 1997 in a less marked form. The specific geohydrological situation of Lake Belau, i.e. increased geohydraulic resistance between the aquifers and the water body, the large volumes of water stored in the aquifers and the lake, leads to a distinctly reduced reflection of the groundwater recharge in the in- and outflow rates of the lake. Thus, in the 1989–1993 period for which the nutrient fluxes are estimated in Sections 9.4.3 and 9.5.3, the hydrological regime is largely equivalent to the long-term average.

9.3.2 Ecohydrological Structure of Lentic Ecotones

Land–lake interfaces or lentic ecotones occupy a special position among the ecotone types (Kluge et al. 2003; Lachavanne and Juge 1997). In contrast to what is designated by the term "littoral", lentic ecotones comprise the entire semiaquatic (lake-influenced) and semiterrestrial (groundwater-influenced) transitional zone between the pelagial and the marginal slopes, as depicted in Figs. 9.1, 9.6–8. Because the ordinary spatial resolution of the GIS data base (elevation model, vegetation and land use, geology) fell short of the inherent requirements of detailed ecohydrological process analyses, microrelief, soils, vegetation and hydrologic characteristics of the land-lake-ecotones were mapped on the basis of a 10-m grid. Figure 9.6 provides a summary of relevant details.

The inflow of groundwater from the larger eastern subterraneous catchment locally leads to basal water-logging and a multitude of small springs between the footslope and the littoral water body. During early spring, effluent discharge clusters can be traced by geothermic underwater surveys. Surface runoff is to be expected especially on the steep eastern slopes. There is no sewerage into Lake Belau.

Lake Belau has two typical shore types. Variant A in Fig. 9.7 is found on the wind-exposed steep eastern bank with an intensive discharge of groundwater inflowing from the catchment (cf. Fig. 9.3). Variant B in Fig. 9.7 is linked to a predominantly flat bank configuration with a wide groundwater-influenced wetland which is typical for the southern basin of Lake Belau.

For farther-reaching analytical purposes a subdivision of these lentic ecotones into the following ecohydrological zones, depicted in Fig. 9.8, appears indicated:

1. The sublittoral shallow water zone which reaches a mean water depth of up to 1.5 m, including the major part of reed stands;
2. The range of fluctuating water level, varying between 29.20 m and 29.45 m a.s.l. under mean annual flood conditions;
3. The rarely flooded adjacent wetland with a mean groundwater level <0.3 m below ground;
4. The non-flooded wetland with a mean groundwater level >0.3 m below ground, where short-term water logging in depressions after heavy rainfalls is possible;
5. The footslopes forming a gentle transition between wetland and slope.

In the sublittoral and within the range of fluctuating water levels the aquatic influence dominates, while the semi-terrestrial character of ecohydrological zones 4 and 5 is in particular reflected in groundwater discharge and surface matter input. Thus, the hydromorphy of the semi-terrestrial zones depends primarily on the groundwater level of the aquifer and the actual lake water level.

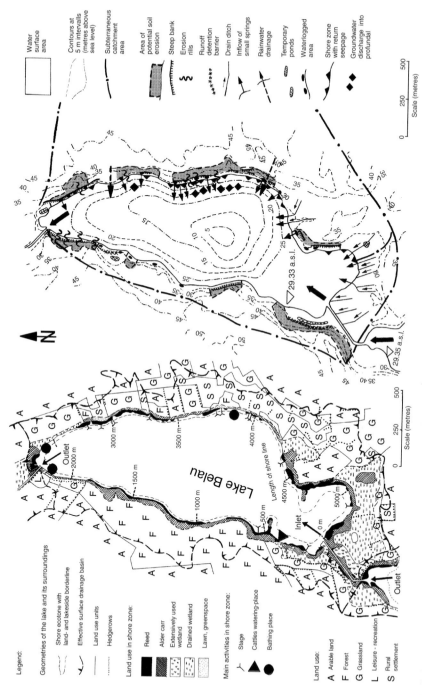

Fig. 9.6 Spatial structure, land use, human activities, and hydrological features of the lentic ecotones of Lake Belau

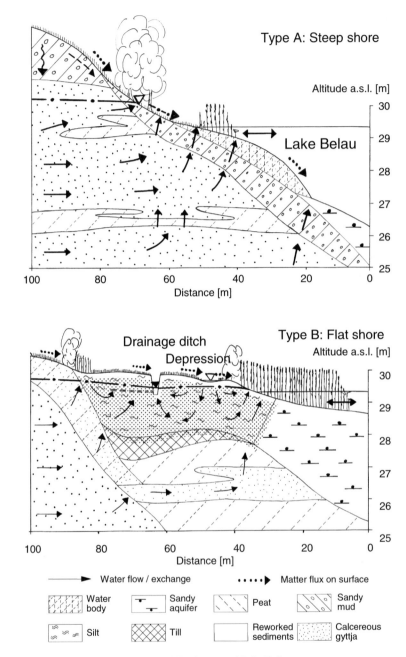

Fig. 9.7 Basic geo- and ecohydrological bank types of Lake Belau

The differentiation of the banks into distinct inflow-sectors provides a novel picture of land–lake-interactions, which link the general geohydrological flow conditions at the catchment scale to the detailed ecohydrological features of the shore

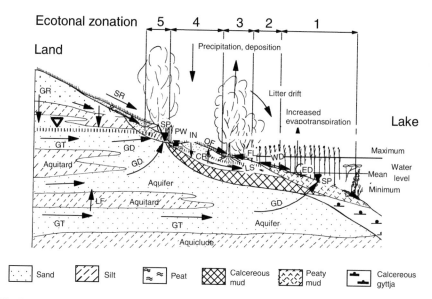

Fig. 9.8 Characteristic microscale features and zones of water exchange between a lake and its adjacent upland. *CR* Capillary rise, *DI* displacement from littoral to profundal, *ED* eddy diffusion, *GD* groundwater discharge, *GR* groundwater recharge, *GT* groundwater transfer, *FL* flooding, *IF* interflow, *IN* infiltration, *LF* leakage flow, *LS* lateral water seepage, *OF* overland flow, *PW* ponding water, *SP* spring (preferential discharge), *SR* surface runoff, *WD* wave displacement

ecotones at the site scale (Fränzle et al. 1996). Thus, not only the diversity of lateral water exchange processes but also the intensity of matter fluxes become manifest. In detail, the comprehensive interpretation of the experimental data in combination with model-based hydrological studies lead to the definition of 12 shore sectors of Lake Belau (cf. Chapter 10), which illustrates the spatial variability along the shore line. In the different sectors the flow pattern of the groundwater which contributes to more than 95% of the water exchange between land and lake was simulated with the three-dimensional groundwater model MODFLOW (McDonald and Haubaugh 1988; Rumohr 1996). For higher spatial resolution the groundwater model FLOTRANS (University of Waterloo, Centre of Groundwater Research, Canada) served to simulate geohydrologic vertical profiles (Piotrowski and Kluge 1994). Furthermore and specifically, the path-box model FEUWA (Kluge et al. 1994; Dall'O et al. 2001) was developed as a prototype of riparian wetland models which provides a deeper insight into the different spatial patterns of both the surface and the subterranean water paths through lentic ecotones.

For undisturbed water exchange in the homogeneous aquifer sands outcropping directly below the lacustrine water body the specific geohydraulic resistance amounts to only $0.03\,\mathrm{d\,m^{-1}}$ (per metre of shore length). Experimental results based on measured gradients of the piezometric water levels at 20 places and simulated lateral water fluxes indicate geohydraulic conductivities, however, which exceed the above value 10 to 100 times. This fact leads to the conclusion that an increase in geohydraulic shore resistance is connected with an increased lateral preferential exchange of water within the ecotone in the form of spatially heterogeneous macrodispersion, effluent seepage water, local springs, and ditches.

In the wetlands southwest of Lake Belau, the lateral water fluxes are influenced by both short-term weather events and the longer-term variation of the groundwater level as follows from a series of simulation runs using the wetland model FEUWA (Dall'O et al. 2001). The example of the alder stands (W6) showed that the evapotranspiration on sunny summer days can lead to daily fluctuations of the groundwater level and to a temporally limited reversal of the water exchange which is normally directed from land to lake. Particularly for bank ecotones a close interaction between the highly variable site hydrology, local vegetation, and the weather regime is characteristic. Within the drained wetlands between the shallow southern part of the lake and the Schmalensee the precipitation surplus is led over by ditches to the mouth of the river Alte Schwentine or into Lake Belau directly.

9.3.3   Water Distribution Matrices of Lentic Ecotones

Lentic ecotones function as hydrologic-scale interfaces (Fränzle and Kluge 1997). The water distribution matrices which are shown for the shore types A and B in Fig. 9.9 visualize the temporally averaged transfers and the mixing of waters differing in both origin and genesis. The arrows in the upper left corner indicate the different water paths recharging as surface and subterraneous runoff from the surrounding uplands and slopes. The inflowing water (horizontal lines in Fig. 9.9) with a 93% groundwater proportion is distributed at the points marked by circles on the water paths (vertical lines) leading to Lake Belau (cf. Trepel and Kluge 2002a, b; Kluge et al. 2003; Trepel and Kluge 2004). Both the western bank and the southern wetlands receive a comparatively low inflow from the catchment, which may be caused by the low extent of the subterraneous catchment in this area (cf. Fig. 9.3). Along the banks the small springs and the effluent seepage water on the steep shore (type A) are connected with the lake by permanent overland flow. In comparison, the inflow from ditches, the runoff from sealed surfaces and the heavy-rain surface runoff play a rather subordinate role for Lake Belau.

9.4   Non-Point Inputs of Nitrogen

9.4.1   Bonding Forms and Concentrations

The determination of concentration values of the waters involved in lateral exchange processes of nitrogen compounds is an essential step for balancing matter fluxes. The following nitrogen fractions are distinguished:

- DIN as the dissolved inorganic nitrogen of the filtered samples with the NO_3-N, NH_4-N, and NO_2-N in smaller concentrations (without dissolved N_2);

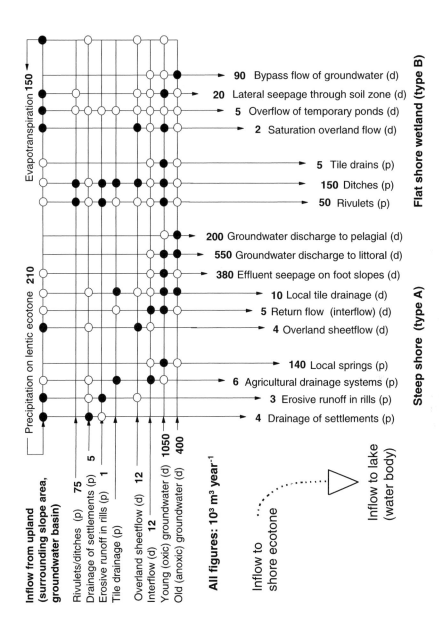

Fig. 9.9 Distributive network of waterways through the lentic ecotones of Lake Belau during the 1989–1993 period. *Black solid circles* Strong interaction, *open circles* weak interaction, *(p)* punctiform source, *(d)* diffusive flux (all units in 10^3 m^3 year^{-1})

- DON as the dissolved organic nitrogen of the filtered samples;
- TDN as the total dissolved nitrogen (TDN = DIN + DON);
- TN(unfil) as the dissolved and suspended total nitrogen of the unfiltered samples [TN(unfil) = TDN + SN; SN meaning suspended nitrogen compounds];
- TPN as total particulate nitrogen of solid samples (soil substrate, loose sediments, litter, particulate biomass, manure and mineral fertilizer);
- TN as total dissolved, suspended and particulate nitrogen (TN = TDN + SN + TPN);
- The gaseous nitrogen in the atmosphere, (N_2, NH_3, NH_4^+, nitrous compounds).

The separation of the dissolved from the suspended fractions of aqueous samples is based on filters with a mesh size of $0.45\,\mu m$.

The results of a complex analysis of pertinent data which were collected in the framework of different subprojects, are presented in Table 9.1.

The concentrations of dissolved nitrogen in the seepage water of the core area attained mean values of $21\,mg$ TN(unfil) l^{-1} under tillage and $9\,mg$ TN(unfil) l^{-1} under beech (Aue 1993; Schimming et al. 2001; Wetzel 1998). Normally DON can reach proportions of 25–40% of the total nitrogen TDN in soil water. In contrast to the soil zone, the content of dissolved organic nitrogen decreased to less than $0.3\,mg$ DON l^{-1} in the aquifer (Scheytt 1994). The values summarized in Table 9.1 represent the site scale within which soils and land use exhibit only relatively small differences. To evaluate the mean nitrate nitrogen contents and their variability at the catchment scale data simulated by WASMOD-STOMOD (Reiche 1996) were used. For the approximately 750 quasi-homogeneous elementary units of the underlying GIS, the mean seepage water concentration of nitrate is approx. $15\,mg$ NO_3-N l^{-1} (Meyer 2000), whereas the spatial variation coefficient CV_s amounts to approx. 50% and varies within broad limits. The variability depends in particular on the land use pattern and the soil characteristics.

To analyse the vertical variability of hydrochemical conditions in the aquifer, 11 multi-level wells were sunk. The on-site measurements of the relevant hydrochemical parameters (oxygen, pH, redox potential, temperature) and the ion concentrations inform about the vertical hydrochemical structure of the aquifer (Scheytt 1994). The diagrams in Fig. 9.10 are indicative of a hydrochemical interface (chemocline) which varies in depth between $8\,m$ and $20\,m$ below the groundwater surface. In contrast to the anoxic domain where nitrate is practically no longer detectable, the depth functions of nutrient concentrations within the oxic domain can differ considerably in neighbouring wells (i.e. at a distance $<50\,m$; Rumohr et al. 1996). The frequently underestimated macrodispersion in the heterogeneous aquifer leads to a superposition of waters of different retention time and origin. Lateral flow takes place preferentially in the more permeable sands. The small nitrate concentrations in the multi-level well M11 (see Fig. 9.10), which clearly differ from those of the other locations on footslopes near Lake Belau, may on the one hand result from reduced input rates under forest. On the other they may be due to increased denitrification in the wetland supported by a groundwater level approx. 0.5–$1.0\,m$ below ground and a high supply of dissolved organic carbon. A significant long-term change in depth of the chemocline, which would be linked to a partial consumption of dissolved oxygen and carbon, could not be observed during the 1990–2001 period.

Table 9.1 Nitrogen concentration of representative water samples: mean values and coefficients of variation. *n* Number of evaluated samples, *m* number of sampling points, *b.g.* belowground, CV_s mean spatial coefficient of variation, CV_t mean temporal coefficient of variation, *n.d.* not determined, *n.r.* not relevant

	n	m	Mean NO$_3$-N	Mean NH$_4$-N	Mean TN (unfil)	CV$_s$ (TN)	CV$_t$ (TN)	Period
				(mg l^{-1})		(%)		
Catchment of Lake Belau								
Soil water, 1.4 m b.g. (arable land)	116	2	15.5	0.15	20.9		70	1989–1993
Soil water, 1.5 m b.g. (beech forest)	115	2	5.6	0.09	9.1		65	1989–1993
Soil water leaching (WASMOD: catchment)	260	80	15.2	n.d.	n.d.	50	50	1989–1993
Groundwater (GW)								
Oxic groundwater	710	41	16.6	0.35	17.2	60	45	1993–2001
Anoxic groundwater	550	37	<0.05	0.8	0.85	200	125	1993–2001
Land-lake ecotone								
Near-surface GW (unflooded alder wetland)	144	6	1.1	0.1	2.3	200	35	1990–1992
Near-surface GW (flooded alder wetland)	108	4	0.06	0.4	0.75	40	30	1990–1992
Small springs at the foot of slopes	53	28	16.0	0.09	16.2	40	35	1993–1994
Effluent seepage of water at the shore zone	50	40	4.75	0.24	5.2	90	50	1993–1994
Wetland ditches	34	13	0.08	1.3	3.0[a]	90	150[a]	1993–1994
Lake Belau								
Bulk deposition (arable land)	312	1	0.67	0.99	1.66	n.d.	30	1989–1996
Inlet of Schwentine River	272	1	0.63	0.21	2.15	n.r.	50	1989–1993
Outlet of Lake Belau	272	1	0.49	0.22	1.86	n.r.	60	1989–1993

[a] Estimated in comparison with results from other study areas.

In the alder stands of the forest catena (W6) a set of some 20 piezometers yielded water samples from different depths during the 1990–1992 period. The analyses recorded in Table 9.1 confirm the existence of lateral gradients which are particularly remarkable in the transition of the non-flooded to the temporarily flooded alder stands (Fränzle and Kluge 1997). These gradients are not caused by lateral exchange processes but rather occur as localized hydromorphic phenomena in the soil zone depending on microrelief and groundwater level. In the upper groundwater domain, the nitrate concentrations clearly decrease towards the bank from approx. 5 mg l^{-1} to <0.1 mg l^{-1}, while the mean ammonium concentrations increase only minimally from approx. 0.1 mg l^{-1} to 0.4 mg l^{-1}. With increasing depth, contrasting effects occur,

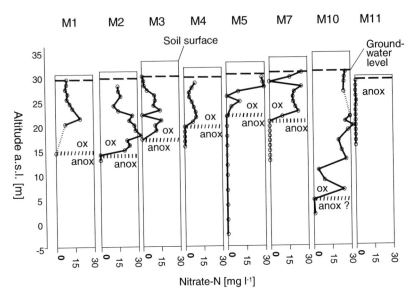

Fig. 9.10 Nitrate concentration of groundwater in different depths recorded in multi-level wells situated on the basal slopes near Lake Belau. Sampling periods: *M1* 1993; *M2, M3, M4, M5, M7, M11* 2001; *M10* 1993. *ox* Oxic regime, *anox* anoxic regime

which means that increased nitrogen inputs are associated with a preferential bypass underflow through the wetland between aquifer and lake (Schleuß et al. 2001).

Figure 9.11 informs about the hydrochemical variability of the inflowing water around Lake Belau. The highest concentrations are found in local springs part of which are due to interflow, while others yield young groundwater. Because of small discharge rates effects of land use on the slopes on soil water and young groundwater cannot be precluded. Another characteristic is that the ammonium concentration in the drainage ditches is clearly higher than the nitrate concentrations (Jelinek 1995). Small nitrate contents in the anoxic groundwater of the water-saturated wetlands and the drainage ditches, which cannot always be correlated with increased ammonium contents, suggest a potentially high retention effect of the lentic ecotones, mostly due to denitrification processes (Haycock et al. 1997).

In the uplands around Lake Belau the transition between oxic and anoxic groundwater corresponds to a residence time of approximately 15–20 years. A simple balance model comprising analytical solutions and numerical groundwater modelling approaches shows that only 25% of the waters are older than 28 years (Kluge et al. 2003; Rumohr 1996).

The values of bulk deposition were used to estimate the contribution of atmospheric deposition to the total nitrogen input into the catchment of Lake Belau (Branding 1996; Wellbrock 2002). With a mean total concentration of approx. 1.7 mg TN(unfil) l⁻¹ the values of the bulk deposition are within the range of the lake water, whose mean concentration amounts to 1.9 mg TN(unfil) l⁻¹ (Naujokat 1997; Schernewski 1999). In Lake Belau the inflowing water has a TN concentration

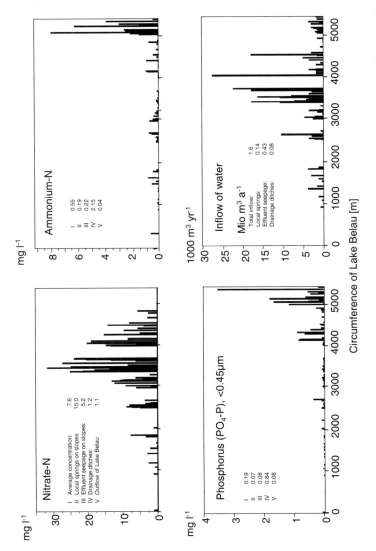

Fig. 9.11 Spatial variability of nitrogen and phosphorus concentrations in effluent groundwater and ditches emerging as small-point inflows to Lake Belau. Sampling date: 16 February 1994; number of water samples: 84

of 2.15 mg l^{-1}, the outflow 1.86 mg TN l^{-1}. Thus, the nitrogen concentrations observed in Lake Belau attain only some 10–15% of those measured in the lower soil zone where deep seepage recharges the terrestrial water cycle.

9.4.2 Path-Based Balance of Nitrogen Inputs from Contiguous Uplands

For 11 hydrologically representative inflow sectors (cf. Chapter 10) the respective water fluxes (Fränzle et al. 1996) were multiplied by the corresponding nitrogen concentrations. Relating the results to shore types A und B (cf. Figs. 9.3, 9.7) yields the nitrogen fluxes listed in Fig. 9.12 in terms of total nitrogen (TN) for the 1989–1993 period. A geohydraulic divide situated at some 2–4 km distance from the lake, land use practices and the specific configuration of shore type A cause the high nitrogen input of ca. 18 t TN year^{-1} from the eastern upland to the banks which is 96% of the total fluxes. The anoxic groundwater path (5) and the other paths contribute less than 2% each. Along the paths which characterize type B, only 10% of the nitrogen surplus of the whole catchment reaches the wetland. South and east of Lake Belau, the oxic groundwater and small ditches dominate with proportions of 75% and 15%, respectively.

9.4.3 Influence of Lentic Ecotones on Non-Point Inputs of Nitrogen

The geohydrologic exchange type and the intensity of land use along the lentic ecotones (see Figs. 9.6, 9.7) determine their buffering effect between the catchment and the water body of the lake. Between the slope and the wetland more and more oxic groundwater flows, mainly on the surface, to the eastern banks (cf. type A in Fig. 9.7). It supplies the local springs on the slope (path 12 in Fig. 9.12) and the effluent seepage water (path 13), providing more than 40% of the nitrogen from the catchment as permanent surface runoff to the littoral zone. Presumably denitrification processes on the banks account for 3.1 t N year^{-1} or 15% of the total nitrogen input by oxic groundwater from the contiguous uplands. However, denitrification affects only seepage water on the slopes but not the spring water.

In the southern and western wetlands the lateral exchange of nitrogen is comparatively small and only about 1 t TN year^{-1} out of the 1.9 TN year^{-1} coming from the adjacent areas, reach the littoral zone. Thus, only approx. 5% of the entire nitrogen input comes from the type B domain. Runoff from built-up areas, overland runoff on slopes, saturation overland flow, and the overflow from depressions of the wetland as a result of precipitation events (paths 1, 2, 8–10 in Fig. 9.12), contribute in each case less than 1% to the total input into the littoral zone. The input due to interflow and tile drains on slopes (paths 3, 11) accounts for nearly 3% of the whole nitrogen input, exhibiting a clear seasonal variation with maxima from early spring to early summer.

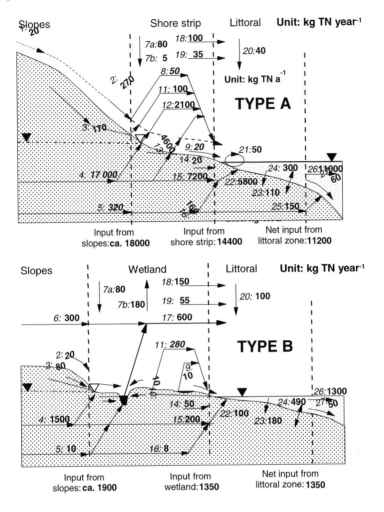

Fig. 9.12 Mean annual fluxes of nitrogen (1989–1993) through the typical ecohydrological bank types. *Type A* Bank with intensive recharge from the subterraneous subcatchment. *Type B* Wetland with low recharge from the subcatchment. *1* Drainage of built-up areas, *2* surface runoff, *3* interflow and tile drainage on the neighbouring slopes, *4* oxic groundwater,: anoxic groundwater, *6* inflow of rivulets or ditches, *7a* atmospheric deposition onto shore zone, *7b* fertilization minus harvesting, *8* surface throughflow, *9* saturation overland flow, *10* overflow of ponding water, *11* direct inflow from tile drains, *12* local springs recharged with oxic groundwater, *13* effluent seepage water, *14* lateral seepage through wetland soils, *15* discharge of oxic groundwater, *16* discharge of anoxic groundwater, *17* inflow of rivulets and drainage ditches, *18* organic input by cattle and waterfowl, *19* atmospheric input of litter, *20* atmospheric deposition onto littoral zone, *21* bank erosion (undercutting), *22* input by groundwater inflow and interstitial water exchange, *23* net uptake of nutrients by reed, *24* net balance of organic matter in littoral sediment, *25* discharge of anoxic groundwater, *26* advective net input into pelagial, *27* net contribution of detritus to pelagial matter fluxes (kg total N year⁻¹), referring to the total length of A and B bank types, i.e. 2750 m in either case

There is a close correspondence between the ecohydrological function and the geohydrological structure of the lentic ecotones as can be demonstrated on the basis of the above shore types A and B. The data compiled in Table 9.2 inform about the dependence of lateral fluxes on the internal balance terms that are typical of lentic wetlands. While in type A the lateral matter transfer is most important, in type B the quantity of nitrogen which finally reaches the lake is determined by drainage intensity, land use, nutrient cycles of the biomass and the microbially catalyzed internal transformations, in particular mineralization and denitrification.

Using a balance approach, the net nitrogen discharge from the littoral into the pelagial zone was calculated on the basis of path-based inputs, the internal nitrogen transformation processes and storage. The underlying (simplifying) assumption is that the nitrogen pools remain essentially unchanged, except for the organic matter in the sediment. The nitrogen necessary for primary production in the littoral zone of Lake Belau is just 1.5 t N year^{-1}, much less than the 15.7 t N year^{-1} crossing the banks. The net input of the littoral into the pelagial zone amounts to approx. 12.5 t TN year^{-1}. The figures in Fig. 9.12 (paths 23, 24) imply that: (a) 1.4 ha of reed belong to type A and 2.5 ha to type B, (b) only 10% of the nitrogen demand of the reed is taken up from the water by the adventitious roots, (c) approx. 300 kg TN year^{-1} (i.e. approx. 75 kg TN ha^{-1} year^{-1}) of the dead rhizomes and the detritus remain as a stable organic pool in the sediment, (d) nitrogen fixation does not exceed 10 kg TN ha^{-1} year^{-1} in the seaward border of the alder stands, and (e) the denitrification of the oxic groundwater, emerging in the sediment, only takes place in the sediment but not in the water body. During the preferential passage of the oxic groundwater through the sediment 1.7 t TN year^{-1} are denitrified which is nearly 25% of the total input of 7.5 t TN year^{-1}.

9.4.4   Inter-Scale Balances of Lateral Fluxes

Nitrate has the character of an ideal tracer in oxic groundwater. Its flowlines represent the spatial reference between the agro-ecosystems as the sources of excess nitrogen in the catchment and the lake as the influx system. The budgets that depend on mass conservation were used to examine the compatibility of the measured and simulated data at different scales. By up-scaling, the "punctiform" data were integrated in such a way that the geohydrological and hydrochemical structures (storage pools, boundary zones, and ecotones shown in Fig. 9.2) are transferred to a group of compartments as depicted in Fig. 9.13. The exchange rates of nitrogen between the boxes, the residence time T of the water within the boxes and the retention coefficients R (cf. Section 9.2) of nitrogen are mean values over the 1989–1993 period.

It ensues from the differences between average inputs and outputs or the source or sink terms of the respective boxes, compiled in Table 9.2, that the TN(unfil) pool in the budget period is not subject to a trend. A negative sign of the internal budgets

Table 9.2 Exchange of nitrogen and phosphorus through the ecohydrological shore types A and B of Lake Belau. All data are mean values of the 1989–1993 period. *TN* Total nitrogen, *TP* total phosphorus, *n.r.* not relevant

Ecohydrological shore types	Unit	Nitrogen		Phosphorus	
		Type A	Type B	Type A	Type B
Area of lentic ecotone					
Wetland	ha	3.9	18.6	3.9	18.6
Littoral	ha	2.5	5	2.5	5
Pelagial Lake Belau	ha	105			
Total Nutrient fluxes					
Input from upland (sum)	kg year^{-1}	17 500	1950	75	10.3
Surface transport of TN, TP, resp.	kg year^{-1}	350	90	13	6
Percentage of dissolved fraction	%		40	25	25
Subterraneous flux of TN, TP, resp.	kg year^{-1}	17 000	1800	60	4.5
Percentage of dissolved fraction	%		99	65	65
Shore wetland balances (sum)	kg year^{-1}	−2800	1600	10	38
Total atmospheric deposition	kg year^{-1}	80	460	2	15
N$_2$ fixation	kg year^{-1}	50	220	n.r.	n.r.
Fertilization	kg year^{-1}	58	750	7	40
Accumulation in soil (net sum)	kg year^{-1}	8	2700	5.5	25
Harvest	kg year^{-1}	−40	−930	−2.5	−60
Gaseous losses (denitrification)	kg year^{-1}	−3000	−1600	n.r.	n.r.
Lateral input to littoral (sum)	kg year^{-1}	14 500	1400	87	59
Surface transport of TN,TP, resp.	kg year^{-1}	7100	1130	47	49
Percentage of dissolved fraction[a]	%		80	65	50
Subterraneous flux of TN, TP, resp.	kg year^{-1}	7300	260	40	10
Percentage of dissolved fraction[a]	%		97	80	75
Lateral input to pelagial (sum)	kg year^{-1}	11200	1350	155	52
Aquatic transport of TN, TP, resp.	kg year^{-1}	11060	1350	155	52
Percentage of dissolved fraction	%		80	70	65
Subterraneous flux of TN, TP, resp.	kg year^{-1}	150	10	2	0
Retention coefficients (*R*) of lentic ecotones					
Total of lentic ecotones	%	35	45	−90	−25
Wetland	%	20	40	−10	−40
Littoral zone (sediment included)	%	25	0	−75	15
Littoral zone (water body only)	%	15	20	0	5

[a] Percentage of dissolved fraction TDN or TDP (in fractions of TN or TP <0.45 μm).

indicates N-sinks and corresponds to an increase in stable particulate or solid organic matter (peat formation or wood increase, sedimentation and long-term immobilization of organic matter). It furthermore refers to the transformation (in particular denitrification and volatilization) of dissolved nitrogen into gaseous form, which makes those fractions unavailable as nutrients. In contrast, a positive sign (indicating an N-source) corresponds to mineralization of organic matter.

For the evaluation of the in- and outputs of the 4.4 km^2 catchment (excluding the lentic ecotones) the following figures were taken as a basis: mineral fertilization

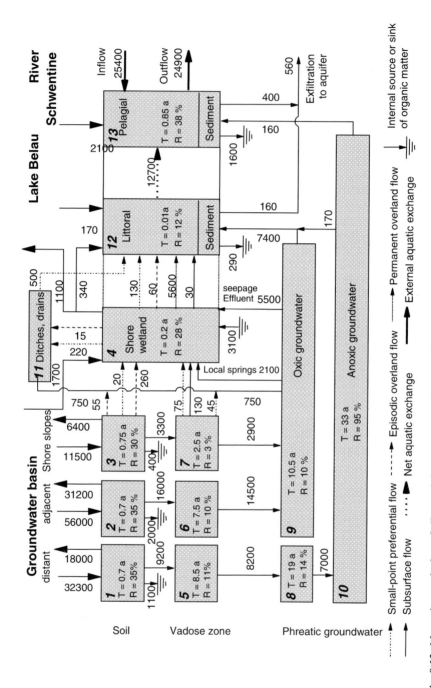

Fig. 9.13 Mean nitrogen budget of all essential compartments of the Lake Belau catchment and the water body during the 1989–1993 period. Fluxes in kg N year^{-1}. T Mean residence time of water in the respective compartments in years, R nitrogen retention coefficient

182 kg TPN ha^{-1} year^{-1}, deposition 20 kg TN(unfil) ha^{-1} year^{-1}, and harvesting 127 kg TPN ha^{-1} year^{-1} with a return rate of 27 kg TN ha^{-1} year^{-1} as organic fertilizers (Dibbern 2000). The deposition, fertilization and N fixation (only in the wetlands) are combined in Fig. 9.13 to obtain the total vertical inputs at the surface. The vertical discharge of the soil compartments reflects only the net crop effect. Leaving out of account some minor lateral transport processes on slopes, this means that approx. 58 kg TDN ha^{-1} year^{-1} of the 65 kg TN(unfil) ha^{-1} year^{-1}, constituting the mean budget surplus of the soil zone, reach the groundwater. Denitrification in the vadose zone and in the oxic groundwater retains only 5.5 t TDN year^{-1} or 19% of the TDN freight, due to exponential degradation with a half-life period of 50 years (Wendland and Kunkel 1999). But then the seepage through the chemocline between the oxic and anoxic groundwater is combined with an almost complete denitrification equivalent to a reduction of the load around 6.7 t TDN year^{-1}. With approximately 500 kg TDN year^{-1} the anoxic groundwater contributes only relatively little to the total discharge.

The net sink effect of the lentic ecotones depends both on the amount of organic matter accumulated and on the respective denitrification rates. For the entire wetlands near the shore and the littoral zone, including sediment, balance deficits of 2.4 t TN year^{-1} and 1.2 t TN year^{-1}, respectively, are determined. While the mean concentration in the groundwater flowing from the wetland into the littoral zone amounts to 8.5 mg TN l^{-1}, the concentrations in the littoral and pelagial zones deviate only for a short time and only slightly from each other (Schernewski 1999). This is an indication of a permanent intensive water exchange between the littoral and pelagial zones. The remaining 12.6 t TN year^{-1} which flow from the littoral to the pelagic zone represent the total diffuse discharge from the catchment of the lake, including the export of particulate nitrogen in form of detritus which is not likely to exceed 120 kg TPN year^{-1}, however.

The mean residence times T of water in Fig. 9.13 are those of dissolved nitrogen in the respective compartment. These values (encompassing the whole range from the surface to the deeper aquifer and the lake) vary between <0.02 years in the drainage ditches and the littoral zone and approximately 30 years in the anoxic groundwater. With the exception of part of the drained wetland, all compartments exhibit a net sink effect which is particularly pronounced in the chemocline between oxic and anoxic groundwater and in the pelagial zone of the lake. The retention coefficients in Fig. 9.13 illustrate the influence which the individual compartments exert on the exchange of nutrients between land and lake. With retention coefficients of approximately 40% for the subterraneous catchment this value amounts to 28% for the near-shore wetland. For the littoral zone, nitrogen retention is small, as indicated by an R value of approximately 10% (cf. Chapter 10).

The retention capacity of the lentic ecotones of Lake Belau is small in comparison with that of the majority of lotic lowland ecotones in Schleswig-Holstein (Kluge et al. 2003b). Thus, from the 4.6 km^2 catchment approx. 26 kg N ha^{-1} year^{-1} reach the pelagial. The lake apparently functions as a nitrogen-sink, where approx. 4.5 t TN year^{-1} are deposited and ca. 14 t TN year^{-1} undergo denitrification (Schernewski 1999), whence follows the above retention coefficient.

There is a relationship between the origin of the waters and the type of nitrogen input into the lake. 86% of the water carrying the diffuse inputs into Lake Belau are groundwater, 8% are precipitation onto the wetland, 4% come from small drainage ditches, 1% from surface runoff and only <1% from interflow (cf. Fig. 9.9). Correspondingly, also the external N input from the surrounding upland to the riparian wetland via the groundwater dominates with 78% (see Fig. 9.13, box 4). Surface runoff or interflow including drainage systems contribute 1.4% or 2.6%, respectively, to the total input. The wetlands contribute 24% to the total input by mineralization of organic matter and only 1.8% due to deposition. The intricate structure of the flowpaths in the wetlands entails that the input modes into the littoral zone (see Fig. 9.13, box 12) differ in the following way: about 42% or 5800 kg TN year^{-1} are due to overland flow, about 55% or 7400 kg TN year^{-1} are the effluent seepage of groundwater through the sediment, <0.3% reach the wetland rim of the littoral as surface-parallel seepage. Surface inflow can be attributed to permanent small-point inputs (ditches and small springs with about 35%) and to diffuse inflow of seepage water on the foot of the slope (about 65%). The diffuse input, dependent on precipitation events (rain discharge of built-up areas, saturation overland flow, depression overflow of wetland areas), contributes only <2% to the total input on the surface.

With the transfer of nitrogen through the bank ecotones the bonding forms of nitrogen compounds also change. The data compiled in Table 9.3 show that the particulate fraction only occurs on the surface while the solute fractions occur mainly in the underground. With the input from the catchment, the dissolved

Table 9.3 Synopsis of the nitrogen budget of all essential compartments of the drainage basin of Lake Belau during the 1989–1993 period

	Total volume of water (10³ m³)	Mass of TN(unfil) fluid (kg N)	Mass of TPN solid (10³ kg N)	Mean conc. of TN(unfil) (mg l⁻¹)	Internal balance solid[a] (kg N year⁻¹)	Internal balance gaseous[a]	Net input TN[b]	Net output TN[b]
1[c]	1400	8	2800	25	−1100	−4000	14 500	−9200
2[c]	2500	13	5000	25	−2000	−7100	25 000	−16 000
3[c]	500	3	1500	22	−400	−1500	5100	−3600
4[c]	230	1.1	2300	4	+3100	−5500	8500	−62 000
5	24 000	75	100	18.5	−	−1000	9200	−8100
6	37 000	110	150	18.5	−	−1600	16 000	−14 000
7	2500	9	20	17	−	−100	3300	−3100
8	28 000	160	<0.01	17.2	−	−1150	8200	−7000
9	61 000	340	<0.01	16.8	−	−1700	17 000	−16 000
10	69 000	15	<0.01	0.8	−	−6700	7000	−320
11	5	0.005	0.05	3	−	−500	900	−500
12[d]	45	0.1	2.5	1.5	290	−1500	14 000	−13 000
13[d]	10 200	26	5000	1.4	−1600	−13 500	40 000	−25 000

[a] Sign is negative when acting as a sink.
[b] Total nitrogen (dissolved, suspended, and particulate fractions) without gaseous exchange.
[c] Thickness is 1 m.
[d] Inclusive of a 0.3-m sediment layer.

inorganic nitrogen (DIN) amounts to 96% which can be broken up into 95% nitrate, <4% ammonium and <2% nitrite. The remaining percentage comprises the organic fractions of dissolved, suspended and particulate nitrogen. After the transfer through the wetlands, the dissolved inorganic nitrogen still determines the exchange in the transitional zone to the littoral (DIN attains about 92%), although there is a certain increase in suspended and particulate components. The situation in the pelagial exhibits a great deal of broad adjustment to the hydrochemistry of the littoral zone. According to Schernewski (1999), the following proportions exist: dissolved inorganic nitrogen DIN 51% (nitrate 25%, ammonium 25%, nitrite 1%), dissolved organic nitrogen 36%, suspended and particulate organic nitrogen 13%. The amount of particulate organic matter (litter, detritus) varies along the banks (Lenfers 1994). Altogether, relief, structure of vegetation, land use, and wind exposure are major determinants of the role of lentic ecotones as sources or sinks for organic matter.

9.5   Non-Point Inputs of Phosphorus

9.5.1   Bonding Forms and Concentrations

Considering the transition of the terrestrial to the aquatic systems, the influence of the nitrogen diminishes in favour of the phosphorus as a production-limiting nutrient. Due to fertilization and sorption surplus phosphorus is usually available in the soil zone of modern high-yield agrarian ecosystems. In forest ecosystems, an internal nutrient cycle predominates, where phosphorus is re-used by mineralization of litter (cf. Chapter 8). The long-term release of mineral phosphorus which decreases with progressive soil development is comparatively small (Schernewski and Wetzel 1997). In contrast to terrestrial, and in particular agrarian ecosystems, in aquatic ecosystems a concentration of >0.03 mg total-Pl^{-1} (Dokulil et al. 2001; OECD 1982) can cause eutrophication.

Following pedological and limnological practice (Haygarth and Sharpley 2000; Hupfer 2001; Johnes and Hodgkinson 1998; Schernewski 1999; Schlichting et al. 2002), the following phosphorus fractions in water and soil samples are distinguished, depending on the preparation and laboratory methods and reflecting the multiplicity of the physical, geochemical or biochemical bonding forms:

- RP (<0.45) water-soluble reactive phosphorus after filtering at 0.45 µm;
- TP(<0.45) water-soluble total phosphorus after filtering at 0.45 µm and acid oxidative digestion;
- TP(unfil) total water-soluble and suspended phosphorus in the unfiltered water sample after total digestion;
- TPP total particulate phosphorus of solid samples (soil substrate, sediment, dust particles, litter, particulate biomass);
- TP total phosphorus of the water-soluble, suspended, and particulate fractions.

The term RP(<0.45) corresponds to water-soluble reactive phosphorus (SRP) which is sometimes incorrectly described as orthophosphate in the literature. Only with certain restrictions can the difference between TP(<0.45) and RP(<0.45) be regarded as water-soluble organic phosphorus since it includes weakly bound polyphosphates and a large number of organic bonding forms. The distinction of the water-soluble reactive fraction from the non-reactive one refers more to the analytical methods applied than to the wide spectrum of chemical and biochemical components included. While for TP(unfil) it is always the water that forms the dominant phase, the term TTP is used to designate the phosphorus content of all samples in which the solid phase dominates. In order to achieve additional information on type and intensity of the bonding forms in solid samples, sequential analytical methods are increasingly used (Schlichting et al. 2002) or release rates are determined by batch experiments. In any case there is a strong dependence of the results on the respective analytical procedure, which makes the comparison of results from different sources difficult.

Thus, a scale-related analysis of the interactions between different ecosystems or their compartments is difficult for various reasons:

1. The bonding forms participating in spatial exchange processes appear in highly variable intensities in different system components.
2. The internal transformations depend on a set of physical and biochemical boundary conditions which differ in time scales.
3. Knowledge about the total concentrations is not sufficient to explain the bonding forms.
4. The factual bioavailability is only partly known.
5. The analytical methods correspond but insufficiently to the ecological relevance of chemical species (Haygarth and Sharpley 2000).

Owing to these methodological problems inter-systemic phosphorus budgets like the following one are still the exception rather than the rule.

For comparison purposes Table 9.4 provides a compilation of the different terms of the P budget at the catchment scale. It illustrates that phosphorus is transported in solute [TP(<0.45)], suspended [TP(unfil) – TP(<0.45)], and particulate form (TPP). The sources of external inputs into Lake Belau are: (a) emissions due to industries, traffic, and wind erosion from arable land, (b) the catchments of the upstream lakes including their water bodies, and (c) the direct surface and subterraneous catchment of Lake Belau.

The concentration of reactive phosphorus RP(<0.45) in surface runoff, oxic groundwater, effluent seepage water on slopes, and in the drainage ditches exceeds the mean concentrations in the waterbody of Lake Belau. The small springs and the emerging seepage water on the footslope have distinctly higher concentrations than the oxic groundwater, indicating local enrichment presumably due to fertilization and grazing in the riparian zone. The highest RP(<0.45) values are observed in drainage ditches that discharge into the southern lake basin. In all water samples taken from surface runoff and small springs, the proportion of total water-soluble phosphorus TP(<0.45) exceeds the amount of reactive phosphorus RP(<0.45) by approximately 40%. The total phosphorus concentration TP(unfil) in the unfiltered samples exceeds the amounts of the solute fractions.

Southwest of Lake Belau the wet deposition on arable land (A3) was recorded by wet-only rain gauges. The RP(<0.45) concentrations attain values of $0.04\,\text{mg l}^{-1}$

Table 9.4 Phosphorus concentrations of representative samples. *n* Number of samples, *m* number of sampling points, *n.d.* not determined, *n.r.* not relevant

	n	m	Mean RP (<0.45)	Mean TP (<0.45)	Mean TP (unfil)	Mean TPP (soil) (mg P kg⁻¹)	Period
				(mg l^{-1})			
Catchment of Lake Belau							
Rill erosion	–	–	–	–	–	500	
Wind deflation	–	–	–	–	–	800	
Superficial sheet flow from slope	–	–	0.055	0.08	–	500	
Event discharge from urban areas	–	–	0.3	0.5	–	1000	
Rivulets	65	1	0.022	0.056	0.068	n.r.	1992–1993[b]
Groundwater (GW)							
Oxic groundwater	52	52	0.045	0.063	0.1	n.r.	1991
Anoxic groundwater	39	39	0.015	0.025	0.05	n.r.	1991
Land–lake ecotone							
Near-surface GW (unflooded alder wetland)	144	6	0.010	n.d.	0.015	n.r.	1990–1992
Near-surface GW (flooded alder wetland)	108	4	0.016	n.d.	0.02	n.r.	1990–1992
Small springs at the foot of slope	53	28	0.072	n.d.	0.1	n.r.	1993–1994
Effluent seepage water at the shore zone	50	40	0.075	n.d.	0.1	n.r.	1993–1994
Saturation runoff from wetland			0.075	–	0.12	n.r.	
Depression overflow from wetland			0.1	–	0.2	n.r.	
Wetland ditches	34	13	0.35	n.d.	0.5	n.r.	1993–1994
Lake Belau							
Bulk deposition (arable land)	312	1	0.05[c]	n.d.	0.55[a]	n.r.	1992–1995
Sediments of Lake Belau	20	–	–	–	–	1500[d]	1992
Inlet of Schwentine River	272	1	0.027	0.051	0.087	n.r.	1989–1993
Outlet of Lake Belau	272	1	0.035	0.050	0.071	n.r.	1989–1993

[a] Estimated in comparison with results from other study areas.
[b] Schmalenfelder Au after Naujokat (1997).
[c] Wellbrock (2002).
[d] Zeiler (1996).

which result from washout processes of, and solution from, aerosols or floating dust particles. The mean concentration of total phosphorus TP(unfil) in the bulk samples had the same value of 0.055 mg P l^{-1} as recorded at the Eutin meteorological station approximately 15 km away during the same period of observation (LAWAKÜ 1995). It must be taken into account, however, that chemical reactions in or pollution of the samples, can introduce an undefinable element of bias. With regard to the P-content of the air-borne dust and wind-eroded soil material, there are hardly any precise values available. According to Matschullat and Kritzer (1997) the P-concentration of air-borne dust of the Saxonian Erzgebirge varies between 370 mg and 890 mg TPP kg^{-1}. Schernewski and Wetzel (1997) give a mean value of 800 mg TPP kg^{-1} for the upper soil zone of the fields in the vicinity of Lake Belau which forms the basis for estimating the eolian input path. It must be noted that the error of the measured mean values may vary between 10% and 35%, while the error of the estimated values may exceed 50%. To what extent these errors eventually contribute to the uncertainty of the calculated non-point nutrient input clearly depends, however, on the relative importance of the individual paths in relation to the total phosphorus input.

9.5.2 Atmospheric Input of Phosphorus into Lake Belau

Air-borne phosphorus originates on the one side from burning fossil fuels and on the other from wind erosion of fallow land during winter times. Assuming that the bulk deposition comes near the wet deposition on the water-surface, this leads to a mean P-input of about 55 kg TP year^{-1} or 0.5 kg TP ha^{-1} year^{-1}, respectively, into Lake Belau, given a mean (corrected) precipitation of 890 mm year^{-1}. The mean dust input, measured by means of Bergerhoff samplers (VDI 1996) on the arable land (A3.3) southwest of Lake Belau provides for 0.07 g m^{-2} day^{-1}. Extrapolating this value to the whole lake surface yields a mean dust input of about 29 t year^{-1} or 250 kg ha^{-1} year^{-1}, which corresponds to a mean input of 23 kg TPP year^{-1} or 0.2 kg TPP ha^{-1} year^{-1}. Especially during winter storms there can be an intensified transport of dust particles from arable land under sparse vegetation cover, with grain sizes varying predominantly between 0.01 mm and 0.3 mm.

9.5.3 Path-Related Phosphorus Inputs into Riparian Ecotones

Phosphorus inputs into aquatic systems due to soil erosion have been documented in detail by numerous experimental programmes and modelling approaches (Bork and Schröder 1996; Hasler 1975; Roth 1999; Schmidt 2000). The process-based model EROSION 2D (Schmidt 1991) was used to simulate soil erosion during the 1989–1994 period for several rainfall intensities. The model takes into account both the relief features of slopes (cf. Fig. 9.6) and the deposition of correlative sediments on the adjacent wetland (Jelinek 2000). The following results were deduced from various simulation runs with stepwise increased precipitation intensities and variable slope configurations:

- Number and intensity of heavy rainfalls as well as the nature of the vegetation cover determine the amount of sediment input into Lake Belau. Erosion processes are linked to rainfall intensities >10 mm h^{-1}. Thus, there was no sediment yield in 1993, while in 1994 it attained the maximum value of approx. 4.5 t year^{-1} or 2.3 kg TP year^{-1}, respectively. There is no soil erosion under forest cover.
- The near-natural wetlands which still exist in a considerable part of the riparian zone impede the transport of sediments into Lake Belau, except (probably) on the steep slopes in Fig. 9.6. Mapping the relevant hydrological and erosional features at the local scale shows that effluent groundwater seepage and grazing animals increase the sediment yield.
- For the 1989–1994 period the simulated mean sediment input from the adjacent slopes into the banks amounted to 6.7 t year^{-1}, but only approx. 1.2 t year^{-1} actually reached the lake, which corresponds to a phosphorus input of 1 kg TPP year^{-1}.

Furthermore the results show that soil erosion will particularly contribute to the eutrophication of Lake Belau, if in late winter or early spring heavy showers fall on frozen ground or fertilized soils. The inputs into the wetland in Fig. 9.14 were obtained by multiplying the path-related water flows (cf. Fig. 9.9) which are based on both a simple modelling approach and field data, with the measured concentrations (cf. data in Table 9.4).

In recent literature there is a growing awareness that inputs of water-soluble phosphorus via the groundwater path are still underestimated. The reason is the continuously increased fertilizer pool particularly in the Ap horizon of the field soils while the corresponding concentrations of the solute fraction in groundwater are slightly above the solubility minimum (Driescher and Gelbrecht 1990; Hannappel and Voigt 1997).

9.5.4 Influence of Lentic Ecotones on Phosphorus Transfer

The average exchange rates of phosphorus through the lentic ecotones of Lake Belau are summarized in Fig. 9.14, and additional information on balance terms is provided by Table 9.2. For type A (which encompasses a narrow, relatively steep shore strip) local springs and effluent groundwater contribute 33 kg TP year^{-1} or 38% to the total P input into the littoral zone. The input after heavy rainfall with approx. 1.5 kg TP year^{-1} is small compared with the direct discharge of groundwater into the littoral zone with approx. 40 kg TP year^{-1}. The input of phosphorus is liable to irregular fluctuations due to land use, activities of waterfowl, and the release of particulate phosphorus by wave erosion. The corresponding values average 11 kg TP year^{-1}, 1.5 kg TP year^{-1}, or 30 kg TP year^{-1}, respectively, for the wind-exposed steep

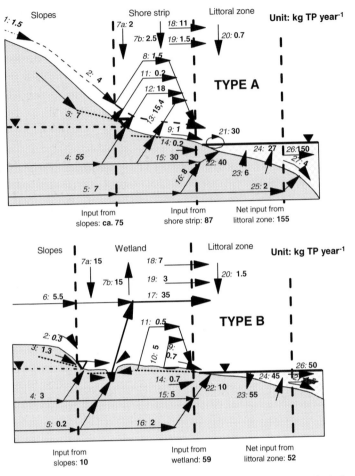

Fig. 9.14 Mean annual exchange rates of phosphorus through the basic ecohydrological bank types A and B (for legend see Fig. 9.12)

sections. Special attention has to be paid to the drainage ditches within the wider wetland areas of shore type B, which transfer approximately 35 kg TP year⁻¹ to the littoral zone. According to long-term experience, the P concentrations in small watercourses and ditches are subject to relatively high fluctuations (Jelinek 1995; Kieckbusch 2003). This implies that during longer stagnation periods P concentrations in the oxygen-deprived ditches can rise very much due to resolution processes bringing about maximum values of 8 mg TP(unfil) l⁻¹. For this reason only concentration measurements in flowing ditches yield unbiased data. Altogether, in the predominantly drained wetland of shore type B the lateral fluxes from the adjacent land and the inputs into the water body are largely uncoupled. Especially the event-dependent surface runoff points to this phenomenon, involving a saturation overland flow with 0.7 kg TP year^{-1} and the overflow of ponding water with 5 kg TP year^{-1}.

Estimates of the nutrient exchange through the hydrodynamically open interface between littoral and pelagial is based upon a balance approach similar to the above for nitrogen (cf. Section 9.4). The input of phosphorus into the pelagic water column is deduced from the mass balance of the whole littoral system including the important pools (water body, macrophytes, phytoplankton, periphyton, zooplankton and other consumers, detritus, mineral and organic sediments) and all of the external vertical and lateral fluxes of the solute and particulate P fractions across the ecotones. Therefore, the deduction of the relevant values in Fig. 9.14 is based on the following physiologically based assumptions: The annual phosphorus demand of *Phragmites australis* amounts to approx. 100 kg P year^{-1} (2.5 g P m^{-2}). From this amount, about 25% is needed for the biomass increase of the rhizomes which attain an average age of 5 years (Schieferstein 1997). About 30 kg TP year^{-1} are released from dead rhizomes, part of the detritus, and organic matter in the sediment. 10% of the phosphorus demand of *Phragmites australis* is met by the adventitious roots from the water body, 50% by P release from the organic substance and 40% are extracted from phosphorus minerals in the sediment. Approximately 90% or 25 kg TP year^{-1} of particulate P, which are mostly due to undercutting of the banks, are redeposited in the littoral zone. The high spatial variability of the macrophytic vegetation (Schieferstein 1997), the heterogeneity and frequent displacement of the littoral sediments (Jelinek 1995), differences in wind exposure and varying C/P proportions (Schieferstein 1997) may lead to somewhat biased estimates. In the light of these uncertainties, an average net export rate of approx. 200 kg P year^{-1} from the littoral into the pelagial water column appears reasonable.

9.6 Comparative Evaluation of Nitrogen and Phosphorus Fluxes

The complex sink or source functions of lentic ecotones with regard to the lateral exchange of nutrients comprise hydrodynamic, hydrochemical, and ecological aspects. Since the nitrogen input from the surrounding catchment predominates in the form of dissolved nitrate, the denitrification aspect is of interest if anoxic conditions are found in the throughflow areas of the ecotones. Anoxic domains represent barriers for nitrate or sinks acting almost without temporal lag. The gaseous nitrogen compounds which appear as a product of the microbially catalyzed reduction of nitrate (cf. Section 8.3.2) are no longer available in the system.

In contrast to nitrogen, with regard to phosphorus increased attention must be paid to the displacement and sedimentation processes of suspended and particulate fractions along the soil surface and within the littoral water body, furthermore to the diffusion of water-soluble phosphorus through the sediment–water interface as well as to sorption and desorption processes in hydromorphic soils or interstitial water. Thus, wetlands act as buffer zones. Adsorbed and organically bound phosphorus which has been enriched over many years in the ecotones can be released later, however. Balancing phosphorus flows through the ecotones causes problems because of the high spatio-temporal variability of the flow patterns and due to the complexity of the hydrochemical processes involved. This is exemplified by the small threshold concentrations of a multitude of organic and inorganic compounds and processes which operate at different scales. The ecohydrological aspect, finally, which involves the influence of vegetation and land use on the site budgets, is of particular importance for designing efficient management concepts (cf. Chapter 13). It has to be taken into account, however, that a high retention effect which lentic ecotones may have with regard to nitrate-nitrogen, must not necessarily apply to phosphorus, too. In fact, in an anoxic hydrochemical environment where nitrate undergoes denitrification, a simultaneous danger of resolution of bound phosphorus exists (Kieckbusch 2003).

Concise information about the essential nutrient fluxes into the water body of Lake Belau is presented in Fig. 9.15. The totality of lateral matter fluxes which cross the lentic ecotones at or below the surface, the mean site budgets and the corresponding retention coefficients (cf. Section 9.2) are compiled in Table 9.2. These coefficients characterize the temporal averages of the sink and source effects of the shore systems. For nitrogen, approx. 90% of which are exchanged across steep banks (type A), the ecotone acts as a little-damped transfer system in the above-ground exchange and as a transfer and sink system in the subterraneous exchange processes. Because of the positive site budgets the near-shore wetland assumes a largely neutral position with regard to nitrogen and the external inputs correspond to the lateral. Denitrification losses of the lateral N input from the surrounding upland correspond to the sum of internal site balances.

With regard to phosphorus, limited release into the drained wetlands and under-cutting of the the wind-exposed steep banks (type A) are responsible for the input into the littoral. Thus, while ecotones react as a sink for nitrogen, for phosphorus a moderate source effect results. The increase amounts to 2–10 times the input depending on the ecohydrological structure of both the wetland and the littoral.

It should be noted that the figures of Table 9.2 represent only a spatial and temporal average. Especially with regard to phosphorus the temporal variability results from the superposition of a set of influences, ranging from single events (surface runoff due to heavy rain, undercutting by waves, fertilization, harvesting), over seasonal fluctuations (medium-term water and energy balances, biomass production, mineralization of organic matter, etc.) to long-term trends (groundwater recharge, changes of land use, plant successions in restored wetlands, compaction and subsidence of peatland, disturbance or partial loss of the reed belt; Pöpperl et al. 2001).

The discharge of subterraneous water onto the surface, the accumulation of organic matter in the flat lentic ecotones, and the increase of preferential flow in the drained peatland result in changes of the granulometry and chemistry of the substances involved in the lateral exchange processes from land to lake. Contrary to surplus nitrogen in form of dissolved inorganic compounds, resulting from increased organic solute and suspension loads which stem from the catchment, the suspended and particulate phosphorus fractions remain largely the same with the transition from land to lake.

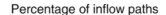

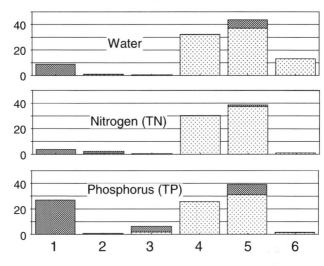

Legend:

1 Rivulets / Drainage ditches
2 Tile drainage systems
3 Episodic surface runoff (urban areas, Horton-runoff on slopes, saturation runoff in wetlands, overflow of temporary ponds in wetlands)
4 Permanent overland flow (small springs and effluent seepage on slopes)
5 Direct discharge of subterraneous water into littoral
6 Direct discharge of groundwater into pelagial

Shore type:	Area of drainage basin [km^{-2}]	Mean annual groundwater [Recharge [mm m^{-2} year^{-1}]	Mean output of nitrogen [kg ha^{-1}year^{-1}]	Mean output of phosphorus [kg ha^{-1}year^{-1}]
Type A Steep shore	4.2	365	26	0.17
Type B Flat shore with wetland	0.4	330	20	1.2
Total drainage basin	4.6	360	25.7	0.28

Fig. 9.15 Comparison of the main input paths of water, nitrogen, and phosphorus into the water body of Lake Belau during the 1989–1993 period

9.7 Conclusions

In the glacially and periglacially sculptured Bornhöved Lake District, the subterraneous catchments represent reference systems where the interactions between the terrestrial and aquatic ecosystems take place in dependence on spatially and temporally variable hydrological and hydrochemical boundary conditions. The lentic ecotones, which can be subdivided into the predominantly groundwater-influenced semi-terrestrial wetland type and the lake-influenced littoral type, act as transfer and buffer systems controlling the non-point nutrient input into Lake Belau. Two shore types are distinguished: a steep one with intensive influx from the adjacent catchment area (type A) and a wetland type with lesser supply from the catchment and a multitude of various sink and sources functions (type B). These types are representative of the major part of the geo- and ecohydrological bank features of lakes in Northern Germany. The identification of the water-borne transport processes between Lake Belau and its catchment permits to more precisely characterize the impact of lentic ecotones on the non-point inputs of nitrogen and phosphorus into lakes. The marked differences between nitrogen and phosphorus with regard to chemical bond, concentrations, mobility, and transformation processes point to difficulties in the formulation of comprehensive descriptions and appropriate models of both fate and behaviour of these essential nutrients.

A detailed geohydrological knowledge of the different water flow paths, linking the land areas to the water bodies in the framework of the terrestrial water cycle, is fundamental to the analysis of the multitude of non-point inputs. Regarding nitrogen, the nutrient load of the lake (like that of the other water bodies of the Schwentine system) depends on the land use pattern in the catchment, the denitrification in the anoxic domain of the aquifer and in the lentic ecotones. Whereas the narrow shore strip on the wind-exposed eastern banks functions predominantly as a nitrogen transfer system, a superposition of internal and external influences occurs in the major part of the wetlands. To a considerable extent the non-point input of phosphorus depends on the ecohydrological features of the bank ecotones. In comparison with nitrogen, the evaluation of phosphorus fluxes is characterized by considerable uncertainties. Natural-like ecotones would reduce nitrogen and phosphorus inputs from the agricultural area of the catchment area by about 40% (cf. Chapter 13).

The whole set of data covers a broad range of scales from the quasi-homogeneous elementary areal units of the GIS up to the 4.6 km^2 catchment or from hours to more than decades, respectively. Nested budget verification ensures the compatibility of the different nutrient flows in the catchment. It must be noted, however, that the study area exhibits special conditions for the quantification of lateral interactions between lakes and their catchments because Lake Belau represents the type of through-flow lakes. This means that the input from the upstream catchments by the River Schwentine exceeds the input from the immediate vicinity by a factor of 3 to 6. The validation of the direct input to Lake Belau is furthermore complicated by the fluctuating divides of the subterranean catchment.

Chapter 10
Lake Belau

Otto Fränzle and Gerald Schernewski

10.1 Introduction

The Bornhöved Lake District comprises six lakes, in two broadly parallel alignments due to Weichselian pleniglacial meltwater and late-glacial dead ice dynamics (cf. Section 2.2.1): Bornhöveder See, Schmalensee, Belauer See in the southeast, and Fuhlensee, Schierensee, Stolper See in the northwest (Fig. 1.1; Section 2.2.4). Hydrographically speaking, they form parts of the Schwentine system which includes a major proportion of ditches in the southern part of the drainage basin. Lake Bornhöved has two outlets, the western one feeding a stream flowing past the Fuhlensee and through the Schierensee to finally empty into Lake Stolpe, while the eastern one forms the Schwentine River connecting the adjacent Schmalensee with Lakes Belau and Stolpe.

The selection of the Bornhöved Lake District as a representative north German landscape or ecosystem complex in terms of ecological setting (cf. Chapter 1) was complemented by a basically comparable procedure to precisely determine the suitability of its lakes for comparative limnological research purposes with a focus on land–water interactions (cf. Chapter 9). To this end appropriate selection criteria in the framework of multivariate statistical procedures were: morphometry of the water bodies, annual temperature cycles, concentrations of total and dissolved carbon (TOC, DOC), total dissolved nitrogen (TDN) and concentrations of the dissolved inorganic nitrogen compounds NH_4^+, NO_2^-, NO_3^-, total dissolved phosphorus (TDP) and finally concentration of dissolved inorganic PO_4^{3-} and conductivity. These data are available for a total set of 65 lakes from the long-term lake monitoring programme of Schleswig–Holstein.

From this comprehensive data set the trophic state indicators TOC, TDN and TDP were specifically analysed by means of biplot techniques (Fränzle and Killisch 1994) and classified on the basis of average linkage clustering procedures (Schernewski and Schulz 1999; Schulz 1996). By means of subsequent T-value analyses representative lakes were selected from each cluster, and among these Lake Belau proved to have the highest degree of hydrochemical representativeness in terms of TOC, TDN and TPD changes during the 1983–1993 period.

O. Fränzle et al. (eds.), *Ecosystem Organization of a Complex Landscape.*
Ecological Studies 202.
© Springer-Verlag Berlin Heidelberg 2008

10.2 Hydrographic Structure of the Lake Belau Drainage Basin

The hydrological setting of Lake Belau is summarized in Fig. 10.1. The morphometric characteristics of the lake are depicted in Fig. 10.2 on the basis of detailed sediment echograph soundings. The relevant hydraulic and geohydrological features of the catchment are described in detail in Chapter 9. The map clearly shows the superposition of two morphogenetic processes. The first is dead ice dynamics as described in Section 2.2.1, the second is the formation of a subaqueous delta in the southernmost part of the lake which is channeled by the Schwentine River. Turbidites are constantly discharged down these channels, and there is a miniature 'abyssal plain'. Slumps occur from time to time, just as in the central parts of the basin where the postglacial sediments attain a maximum thickness of about 30 m.

Both the inset map of Figs. 10.1 and 10.3 exhibit a good deal of broad adjustment of the different types of depositional and residual sediments and the total organic carbon pattern to the prevailing wind-driven currents in the water body and to the intensity of wave action (cf. Figs. 10.7, 10.8). Thus, lag deposits, i.e. coarsegrained residues left behind after finer particles have been transported away, occur at the east coast, where currents attain maximum velocity and lateral undercutting is relatively pronounced. In contrast, weaker currents along the wind-sheltered west coast correlate with sandy sediments, while in the shallow southern bay and the pelagic parts of the lake three variants of calcaric gyttja prevail. The spatial distribution of total organic carbon (TOC) basically reproduces this pattern. The lowest concentrations are found along the east coast and around the small peninsula near the west coast, where currents in the littoral water column experience a local intensification. The highest TOC values are recorded in the central pelagic parts of the lake and, in particular, in the southern bay as related to the above delta formation (Stark 1993).

10.3 Energetic Setting of Lake Belau and its Drainage Basin

10.3.1 Short-Wave Net Radiation

The water body of a lake is in constant exchange of energy and matter with its environment whose energetic structure is basically determined by the solar energy cascade and its different transformations including import from, and export of, latent or sensible heat and wind energy from or to neighbouring ecosystems.

Figure 10.4 summarizes the distribution of the short-wave net radiation in the reference year 1990 as related to local slope characteristics and the land use pattern of the 5.84 km² environs of Lake Belau. The data are based on measurements of the individual balance terms at representative locations of the study area and subsequent

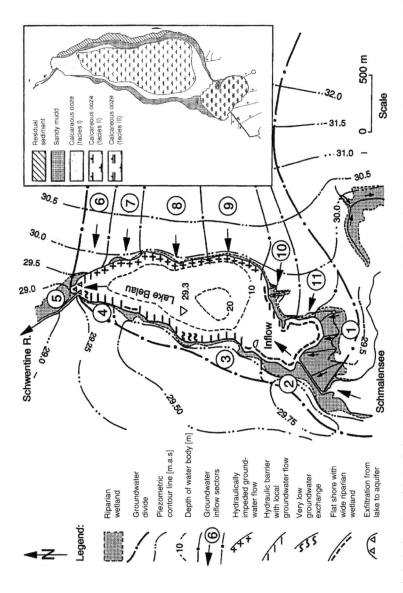

Fig. 10.1 Hydrological setting of Lake Belau with groundwater flow sectors and different hydraulic types of lentic ecotones. *Inset:* facies of bottom sediments

Fig. 10.2 Bathymetric map of Lake Belau (after Müller 1981)

Lake Belau

500 m

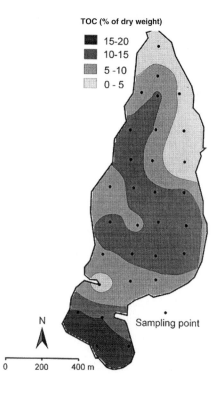

TOC (% of dry weight)

15-20
10-15
5 -10
0 - 5

Fig. 10.3 Organic carbon content (% dry weight) of the uppermost (0–1 cm) sediment layer of Lake Belau (after Stark 1993). *Dots* indicate sampling sites in the pelagic ooze; sampling localities in the littoral are not shown

N

Sampling point

0   200   400 m

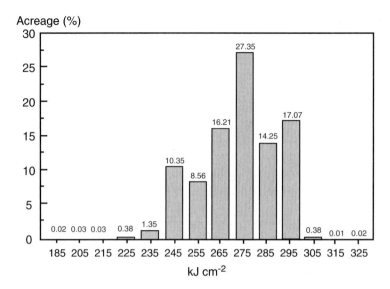

Fig. 10.4 Spatial differentiation of short-wave net radiation in the environs of Lake Belau in 1990 (after Venebrügge 1996)

extrapolation by means of slope-related radiation models and a geographic information system (Venebrügge 1996).

The spatial integral of these 14 classes yields a total of 18.1 PJ year^{-1} for the above reference area, which has the deliberate quality of a minimum estimate, since the reference year was characterized by a distinct reduction of the potential global radiation on horizontal surfaces from 687 kJ cm^{-2} year^{-1} to 310 kJ cm^{-2} year^{-1}. Table 10.1 at http://www.ecology.uni-kiel.de/bornhoeved-report provides in an exemplary manner for 1990 an overview of the seasonal variability of the net radiation of the drainage basin, defining the monthly maximum and minimum values as a function of exposure and slope.

10.3.2 *Water Temperature, Wind and Stratification*

The general energy balance at the pelagic lake surface, which forms the basis for developing models of the water temperature and the seasonal circulation patterns, can be summarized in the following equation:

$$GR + AR + BR + E + C + HF = 0 \qquad (10.1)$$

where GR is net global radiation, AR and BR are long-wave atmospheric and back radiation, respectively, E is evaporation, C is convection and HF the total heat flux. All fluxes into the lake are defined as positive and are measured in kJ m^{-2} day^{-1},

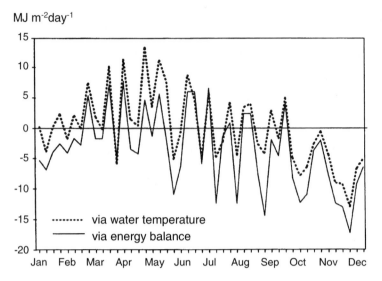

Fig. 10.5 Energy flux across the surface of Lake Belau in 1989 with March, May and June values above the long-term average (after Schernewski et al. 1994)

with the exception of the term HF which is considered negative. Without developing in detail the formulae underlying these terms (cf. Hurley Octavio et al. 1977; Schernewski et al. 1994) the energy flux across the surface of Lake Belau is represented in exemplary form in Fig. 10.5. The comparison with values derived from measured differences in pelagic water temperatures (Herbst and Kappen 1999) shows that the radiative and thermal energy inputs can thus be modelled with a sufficient amount of accuracy.

Considering the littoral water column by way of comparison, however, allowance must be made for strong evaporative cooling processes in the reed belt (cf. Section 10.4.3) which can easily lower the temperature of the water by 3 °C (Schieferstein 1997).

A further important element in the energy balance of the lake, in particular its circulation patterns and the stability of thermal stratification is wind. In the Bornhöved area its speed may attain maximum values of $>15 \, \text{m s}^{-1}$ at 1 m above lake level, but in the case of Lake Belau with its predominantly meridional orientation the resultant average effective fetch is low for the prevailing wind direction (WSW).

The kinetic energy of the wind depends predominantly on wind speed and can be estimated more precisely by means of the following formula (Kerger 1992):

$$E_{kin} = (\rho_{air} C_{10})^{3/2} (\rho_{water})^{-1/2} u_{10}^{3} A \, \Delta t \tag{10.2}$$

where ρ_{air} and ρ_{water} are the air and water densities, u_{10} is the wind speed at 10 m height (m s^{-1}), Δt the time step in seconds, A the cross-sectional area in m^2, and C_{10} the respective wind shear coefficient which is 0.9 for $u_{10} < 5 \, \text{m s}^{-1}$ or else 0.5.

The influence of wind on water movement in general and the stratification of the water body in particular is complicated. Wind pushes the surface waters and generates a current whose flow rate depends on both wind speed and fetch length; as the current reaches the shore it bends downwards and travels in the opposite direction. This causes an initial shift of the epilimnion downwind and of the hypolimnion upwind and a corresponding gradient on the lake surface.

The combined influence of solar heating, advective transport of sensible heat and wind shear on the development and autumnal disappearance of the thermocline above the point of maximum depth is summarized in Fig. 10.6.

Figure 10.6 clearly shows that the vertical shift and the related change in thickness of the thermocline are unsteady processes, coupled with short-term wind events such as one around the middle of July which lowered the 2 °C m^{-1} metalimnetic layer by about 3 m or the heavy storm of 28 August 1989 which displaced the whole metalimnion within a few hours by more than 5 m downwards.

The vertical temperature distribution of the water body can be modelled as a function of the above radiative and advective energy inputs following the Ryan and Harleman (1971) approach in combination with the wind mixing algorithm developed by Stefan and Ford (1975). The accuracy of simulation is such that even details like an occasionally doubled thermocline are represented (Schernewski 1999). This permits the application of physical criteria (cf. Idso 1973; Cole 1975) to estimate the stability of thermal stratification. It can, for instance, be defined in terms of the amount of water necessary to be shifted in order to bring about complete mixing. This parameter has the appreciable advantage of intuitive clearness, but is not suited for comparisons between different lakes, since it relates to the total

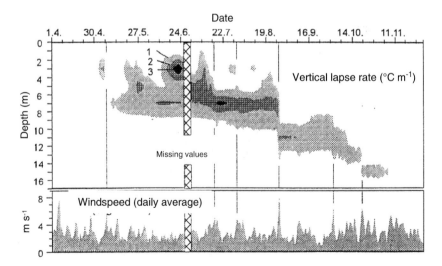

Fig. 10.6 Thermal stratification of holomictic–dimictic Lake Belau in 1989 (after Schernewski 1992). The configuration of the thermocline is indicated by means of the vertical lapse rate [°C m^{-1}]

volume of a water body considered. In the case of Lake Belau, and referring to the comparatively warm year 1989, the parameter fluctuated between zero in April and November, reaching a maximum of about 9.4×10^6 kg in July (Schernewski 1992).

Considering the whole 1988–1999 period from the point of view of stratification, an irregular alternation of dimictic and warm monomictic states can be registered. Examples of the latter occurred during the warm winters 1988/1989 and 1989/1990, when the surface temperature of the water body fell below 4 °C only for very short intervals. Coincident relatively strong winds induced complete circulation which lasted from mid-November until April, characterized by marked short-term fluctuations of the eddy diffusion coefficient (K_z). During summer stratification, however, the coefficient exhibited a remarkably regular vertical profile with monotonously decreasing values from the thermocline ($0.5 \, \text{cm}^2 \, \text{s}^{-1}$) and a hypolimnetic minimum ($0.03 \, \text{cm}^2 \, \text{s}^{-1}$) near the bottom (Schernewski 1992).

The thermocline may start oscillating during storms, especially when winds arise periodically and when the oscillation period of the thermocline resonates with the wind period. Thus the thermocline may ultimately oscillate over a vertical distance of several metres and water from the hypolimnion may 'escape' into the epilimnion causing a partial destratification. The most remarkable of these internal waves (or seiches) were due to the above storm event of 28 August 1989; they had a periodicity of 5 h, an amplitude of >2 m and a damping of approximately 50% of their height per cycle. The maximum intensity was observed in the vicinity of the thermocline, but almost the entire water body was influenced by compensation currents (Beinhauer et al. 1991). Smaller internal seiches with an amplitude of several decimetres and periodicities of 3–5 h occur during the whole summer, especially when the thermocline is located near the surface; owing to distinctly lesser damping they are detectable for several consecutive days.

Another wind-induced phenomenon is turbulence. In general highly variable in size and duration, the eddies may nevertheless form relatively stable medium-scale configurations in the water body as the following figures illustrate. They are based on a two-dimensional vertically averaged flow model (Podsetchine and Schernewski 1999) with a Manning roughness coefficient of $0.015 \, \text{m}^{-1/3}$ s, a horizontal diffusion coefficient of $0.01 \, \text{m}^2 \, \text{s}^{-1}$, and a Coriolis parameter of $1.176 \times 10^{-4} \, \text{s}^{-1}$. An integration period of 2.5 h was sufficient to obtain steady-state solutions. The wind field was kept constant during computations.

The comparison of measured and computed depth-averaged currents exhibits a good agreement in both the central basin and the shallow southern bay. Taking spatially variable Manning coefficients into account permits a quite reliable depiction of even the small-scale effects of the reed belt (cf. Section 10.4.3) on the currents, while the corresponding influence of bottom roughness appears negligible. Wind shelter effects for wind speeds >2 m s^{-1} stand out quite clearly, and the same applies to the influence of variable wind speeds. Thus a spatially variable wind field induces a general one-cell flow with a strong reverse jet along the western shoreline (Fig. 10.7c), which is in good agreement with measurements, while a (hypothetical) spatially homogeneous wind field would lead to a (model-predicted) two-cell circulation system (Podsetchine and Schernewski 1999).

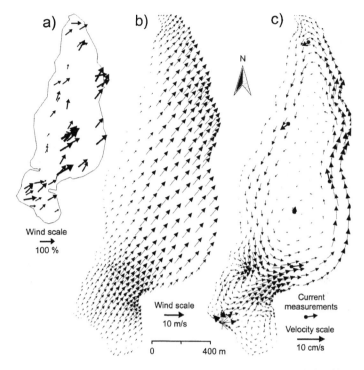

Fig. 10.7 Wind-driven currents in Lake Belau induced by south-westerly winds with speeds >4 m s⁻¹ at 1 m height above lake level. **a** Compilation of data from 7 April, 27 July, 16 September 1997, and 8 August 1998. For easier comparison of the individual data sets and a better illustration of shelter effects relative wind speeds are given. **b** Computation of the spatially heterogeneous wind field (max. 6 m s⁻¹) underlying the simulated flow field. **c** Computed depth-integrated flow field of 16 September 1997 and control measurements at five sites (*arrows*). *Arrow shafts* are proportional to speed

The situation is much more complicated under the influence of a westerly wind field. Figure 10.8, based on wind data from 69 points on 24 February 1998, shows that under these conditions with a typically pronounced leeward shelter effect cyclonic eddies with vertical axes and highly variable dimensions result, whose influence on the development of phytoplankton patchiness will be discussed in Section 10.4.3.

The above figures illustrate that changes in wind direction, velocity and squalliness induce comparatively rapid circulatory adjustments of the epilimnetic water body. In the shallow southern bay of Lake Belau, however, pattern formation is not only due to wind but also to the River Schwentine flowing in with an average discharge rate of about 0.4 m³ s⁻¹ (Fig. 10.9). Due to the configuration of the coastline winds of all directions cause eddies which vary in size, intensity and somewhat in location. Nevertheless two areas appear particularly favourable, namely the eastern part of the bay and an area some 150 m north of the mouth of the Schwentine River,

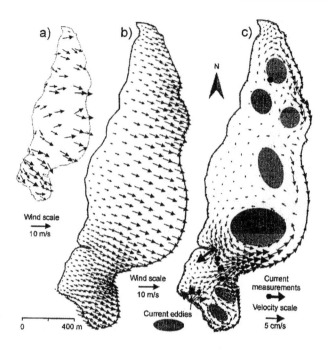

Fig. 10.8 Wind-driven cyclonic circulation pattern of Lake Belau due to westerly winds. **a** Wind measurements as of 24 February 1998. **b** Computation of the wind field (280° direction, maximum speed 5 m s⁻¹). **c** Depth-integrated flow field computed for a comparable weather situation on 6 June 1998 and related control measurements at six sites. *Arrow shafts* are proportional to speed; *shaded areas* indicate eddies

where eddies exhibit such a degree of local stability largely irrespective of wind direction that algal growth can adapt to the situation (Schernewski 1992; Schernewski et al. 2005).

10.4 Dissolved and Particulate Nutrients and Trace Elements in Water and Sediments

The hydrochemical setting of Lake Belau reflects the asymmetric structure of its drainage basin on the one hand and the different land use patterns developed on either side of the lake on the other. Owing to the predominance of forests nutrient input by surface and groundwater is distinctly less on the western banks than in the east, where settlements and agriculture provide for a considerable amount of fertilization, manuring and sewage. Figure 10.10 gives an exemplary, although momentary, overview of the inflow/outflow relationships of the lake with its environment.

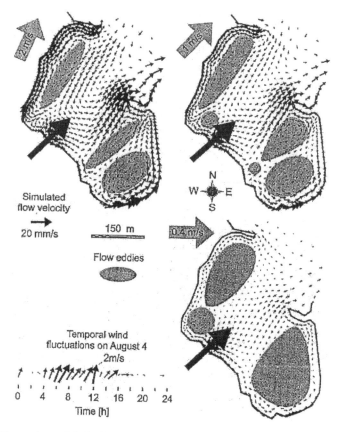

Fig. 10.9 Fluctuations of the basic wind field on 4 August 1993 and associated eddy formation in the southern part of Lake Belau. Computations show the flow fields at 12 a.m.(wind speed 2 m s⁻¹ from SSW), 3 p.m. (1 m s⁻¹, SW) and 6 p.m. (0.4 m s⁻¹, W). *Thick black arrows* mark the River Schwentine

The above situation should be matched with the complex hydrological exchange processes in the Belau catchment area (Fig. 10.11) and the long-term averages summarized in Fig. 9.15 or Tables 9.1 and 9.2, respecively.

Under the seasonally varying influence of lateral macro- and micronutrient inputs and seasonal temperature and wind situations Lake Belau is, as a typical eutrophic hardwater system, characterized by a high productivity of its successive phytoplankton communities. They start with a diatom bloom in spring, after the May clearwater stage, followed by a second bloom of Chlorophyceae, Dinophyceae and Cyanophyceae in June and July and a final bloom of Dinophyceae and Cyanophyceae in autumn (cf. Section 10.4.3). In combination with the pronounced summer stratification resulting from the above morphometric and energetic characteristics, the development and subsequent decomposition of organic matter give rise

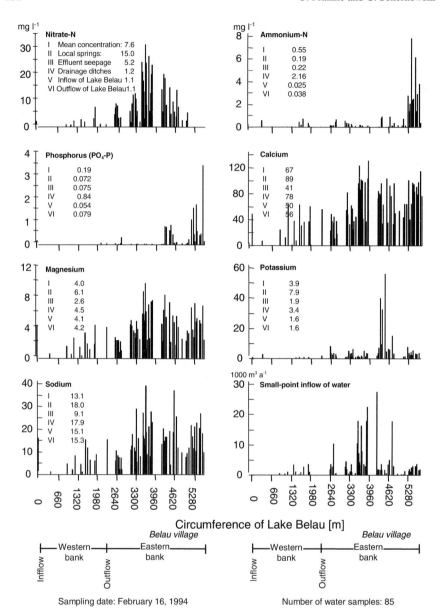

Fig. 10.10 Average concentrations of selected macronutrients in Lake Belau, in and outflowing watercourses and seepage water, determined on 16 February 1994

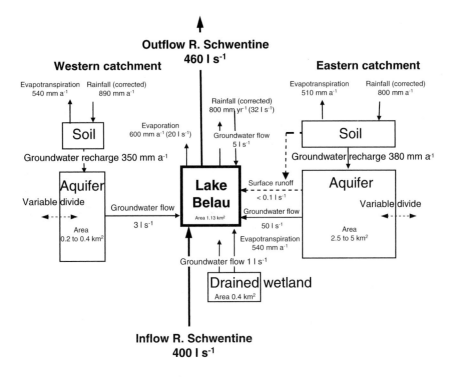

Fig. 10.11 Water balance of the Lake Belau catchment area during the 1989–1998 period (after Fränzle and Kluge 2003)

to strong seasonal redox dynamics which governs the biogeochemical cycling of the nutrients C, N, P and the trace elements Mn, Fe, Cu, Zn, Mo, Cd, Sb, Ba, rare earth elements (REE), Pb and U, besides biological processes in the photic zone and at the sediment/water interface (cf. Section 10.4.2). Figure 10.12 summarizes the essential processes on the basis of comprehensive measurements in the 1993–1994 period, including the most important redox reactions of electron acceptors.

10.4.1 Macronutrient and Carbon Fluxes

Contrary to the situation in terrestrial ecosystems only about 10% of the total carbon are bound in organisms or dead organic matter of the lake and approximately 10% are dissolved organic carbon, while 80% form inorganic compounds. Total organic carbon (TOC) can be subdivided into particulate organic carbon (POC) and dissolved organic carbon (DOC). The average TOC concentration of about 10 mg l^{-1} comprises a proportion of about 86% DOC and 14% POC; either component can be of autochthonous or allochthonous origin.

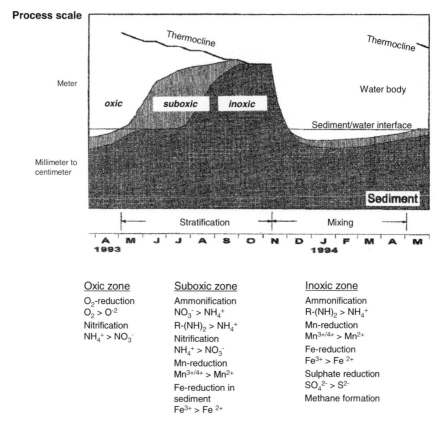

Fig. 10.12 Schematic representation of the seasonal redox dynamics in the water body and bottom sediments of Lake Belau and pertinent reactions of selected electron acceptors (after Zeiler 1996, modified)

Owing to the exchange of CO_2 between water body and atmosphere, carbon balances are difficult and unpromising. Further, there is no point in limiting the analysis to the organic components, because of the multitude of transformation processes between organic and inorganic carbon species in the framework of photosynthetic and decomposition processes. Nevertheless one important process, namely the formation of methane and its subsequent oxidation to carbon dioxide, is worth mentioning. CH_4 results from the decomposition of organic matter which starts already in the hypolimnion but is mostly linked to the bottom sediments. Depending on the redox conditions and the composition of the organic compounds deposited, gases such as CO_2, CH_4, H_2S, N_2O and N_2 are formed which diffuse into the supernatant water according to concentration and solubility. The comparatively high percentage of methane in the gas mixture, varying between 40% and 95% according to findings by Heyer (1990) in a variety of lakes, is due to its low solubility. As a consequence, CH_4 is enriched in the upper sediment layers in the form of bubbles which eventually escape, carrying away substantial amounts of the ambient water-saturated sediments whose shear strength is exceedingly low, which leads to the formation of numerous funnel-shaped depressions. Diving campaigns have shown that parts of the lake bottom can attain the character of a shell-pitted area (Garbe-Schönberg and Zeiler, personal communication), while echo-sounding provides more precise information on the spatial configuration of the uppermost bubble-enriched sediment layer owing to the associated false-bottom effect.

Exemplary measurements of methane fluxes above the deepest point of Lake Belau yielded extremely variable values, both temporally and spatially, and depending on the weather situation. In August 1993, for instance, the mean rates amounted to about 3 mmol CH_4 m^{-2} h^{-1}. Contrary to the literature (Csermak et al. 1992; Schmidt and Conrad 1993), where a distinctly higher methane release is attributed to littoral in comparison with pelagic sediments, measurements by Rembges and Rusch (personal communication) in the reed formations along the western shoreline yielded only flux rates of 93 µmol CH_4 m^{-2} h^{-1}. The reason may be either an intensified microbial methane oxidation in this biotope or a gas transport through the reed blades or a combination of both processes.

Probably the observation made by Rudd and Hamilton (1978) in a Canadian Shield lake that the oxidation of methane does not provide an essential carbon dioxide source for primary producers also holds for Lake Belau. Only during summer stratification such processes could attain a certain importance in the epilimnion, but owing to the above 'eruptive' character of methane release from the bottom sediments the time-span for oxidation in the uppermost part of the water column is rather short. Assuming an average summer release rate of 2 mmol CH_4 m^{-2} h^{-1} with an oxidized proportion of 5%, a daily CO_2 influx of 2.4 mmol into the epilimnion would result. With regard to the whole lake this would be equivalent to a total methane production of >1000 m^3 day^{-1} and, taking the other gases into consideration, a total estimate of about 1500 m^3 day^{-1} appears reasonable (Schernewski et al. 1994).

The following graph (Fig. 10.13) provides a qualitative synopsis of the above reactions and fluxes of carbon species as related to calcite precipitation and co-precipitation of phosphate. The shaded areas indicate partial models which have been subject to computer simulation based on a combination of the physical–chemical approaches by Rossknecht (1977), Lindsay (1979) and Morel and Hering (1993).

Taking the year 1991 as an example, Lake Belau is saturated with CO_2 during turnover in March, which prevents calcite precipitation. In April, with the beginning of the diatom bloom, the gas concentration near the surface decreases to values of 2 µmol l^{-1}, which leads to a shift in the equilibrium and resultant calcite formation with maximum values of about 250 µmol l^{-1}. The onset of stratification in May brings about a warm epilimnion with low CO_2 but high calcite concentrations and a cool hypolimnion with increasing amounts of settling calcite and organic material. The subsequent decomposition of the latter leads to an increase in CO_2, which is coupled with an enhanced tendency to calcite dissolution. During summer calcite formation extends over a superficial 7 m water column while the underlying pelagic water is calcite-aggressive.

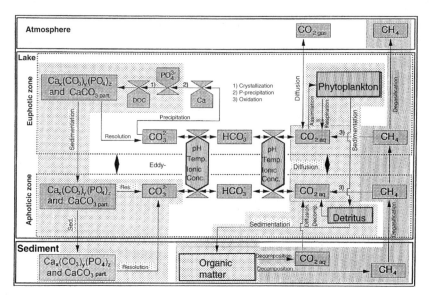

Fig. 10.13 Conceptual model of the carbon cycle in the Lake Belau hardwater system (after Lenz 1992; Schernewski et al. 1994)

10.4.1.1 Nitrogen Fluxes and Balance Estimates

A comprehensive nitrogen balance for the 1989–1998 period is summarized in Fig. 10.14, which shows that the three major terms account for almost 99% of the total nitrogen input into the lake's water body; the rest is inferior to the statistical noise of the main components, as shown by an analysis of the nitrate and ammonium loads of the sectoral groundwater flows into the lake (Fränzle et al. 1996; cf. Fig. 9.15).

Nitrogen species transported by seepage to the groundwater as a result of (partly excessive) fertilization and manuring (cf. Chapters 8, 9) normally reach the oxic domain of the aquifer with a few years' delay. Here the O_2 concentrations range from 1 mg l^{-1} to 12 mg l^{-1} and the redoxpotential accordingly varies between +200 mV and +350 mV, while nitrate concentrations attain maximum values of 125 mg l^{-1}. In the underlying redoxcline the oxygen concentrations decrease within 4 m to <0.5 mg O_2 l^{-1}, which leads to a reduction of the nitrate species and formation of N_2O, N_2 and $[NH_4]^+$ ions; the sulfate content reaches values of about 80 mg l^{-1}, and iron and manganese appear in solution. In the anoxic zone below the redoxcline the redox potential is <100 mV. As a consequence sulfate concentrations drop to 60 mg l^{-1}, while iron concentrations range between 0.5 mg l^{-1} and 2.2 mg l^{-1}; the corresponding figures for manganese vary from 0.2 mg l^{-1} to 0.35 mg l^{-1} (Rumohr et al. 1996). While the gaseous nitrogen species escape, the ammonium is transported with the anoxic groundwater into the lake, where it undergoes uptake or oxidation, depending on the depth of release (cf. Fig. 9.12) and the stability of stratification.

Nitrogen budget (g N m^{-2} a^{-1})

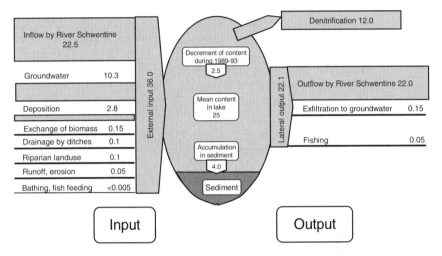

Fig. 10.14 Average annual nitrogen budget of Lake Belau for the 1989–1998 period. Values and units (g N m^{-2} year^{-1}) of the input and output terms of the system are defined in correspondence with the lake surface (113 ha) as a reference unit

Phosphorus budget (g N m^{-2} a^{-1})

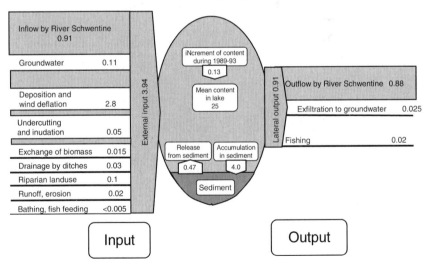

Fig. 10.15 Average annual phosphorus budget of Lake Belau (g P m^{-2} year^{-1}) for the 1989–1998 period. Values and units as in Fig. 10.14

Among the output components of the nitrogen balance the discharge-coupled export by the Schwentine River plays the major role, while sedimentation is of lesser importance, which is also indicated by the comparatively low average nitrogen content of about 1% N found in the bottom sediments. It should be noted, however, that sedimentary nitrogen compounds exhibit a distinct spatial structure with a 1.9% N concentration in the shallow southern basin of the lake with prevalent fluvial sedimentation, and only a 0.6% N content in the shallow littoral of the main basin which is subject to moderate erosion caused by breakers and the orbital water movement of waves.

The importance of denitrification processes is not only reflected in the above nitrogen balance but also in high supersaturation rates of N_2 in the water. On 30 July 1991, for instance, concentrations increased from 1.2 mmol N_2 l^{-1} at the surface to 1.7 mmol N_2 l^{-1} at the bottom at 25 m depth (Witzel, personal communication). Since the maximum solubility of the gas in relation to actual pressure and temperature would have been 0.8 mmol N_2 l^{-1}, this is equivalent to a 100% supersaturation in the major part of the water column. In later years even distinctly higher values were recorded. In light of the limnetic nitrogen balance with an average annual denitrification rate of 98 μmol N m^{-2} h^{-1} relating to the lake/atmosphere interface (Schernewski 1999) it is worth comparing the relative importance of the process in the metalimnetic water column with nitrogen release from the sediment. At the sediment/water interface denitrification occurs during the whole year, while it plays an essential role in the water body only during shorter periods. Clearly, summer is a phase of maximum microbial denitrification, when very high concentrations of denitrifying bacteria occur in the metalimnion, causing maximum rates of about 90 μmol N m^{-2} h^{-1}. Considering the whole year, however, it ensues from the above annual rate that denitrification at the bottom of the lake is likely to play a far greater role. Altogether the above figures substantiate, in comparison with data reported by Seitzinger (1988) which vary between 3% and 62%, that nitrogen retention in Lake Belau is comparatively high and the lake plays a considerable role as a purifying unit in the Bornhöved lake system (cf. Section 9.4 and Table 10.2 at http://www. ecology.uni-kiel.de/bornhoeved-report).

10.4.1.2   Phosphorus Fluxes and Balance Estimates

In comparison with the determination of nitrogen balances and speciation processes reliable estimates of phosphorus species in the water body of Lake Belau and its tributary ground and surface water flows (cf. Chapter 9) prove much more difficult for several reasons, among which the above-mentioned co-precipitation with calcite and redox-controlled reactions with iron are of particular importance. Owing to precipitation and adsorption on trivalent iron compounds and calcite the P concentrations in seepage and groundwater are normally low; average concentrations in the drainage area are about 20 μg l^{-1} dissolved phosphorus (TDP). Total phosphorus (TP) concentrations of about 70 μg l^{-1}, however, are found in springs and drain water due to solution and internal nano-erosion processes in soil (Jelinek 1995).

In contrast to nitrogen, phosphorus input by diffuse sources is small in comparison with the load of the Schwentine River as Fig. 10.15 shows. One reason is the relatively small catchment, the other the high number and distribution pattern of hedgerows which reduce surface runoff and soil erosion (Schernewski and Wetzel 1997; Meyer 2000).

In analogy to nitrogen, the phosphorus load of the Schwentine River also constitutes the major input term of the phosphorus balance while, in contrast to findings by Patrick et al. (1973) and Reddy and Rao (1983, 1999), phosphorus release from temporarily inundated lentic ecotones and bordering wetlands does not play a significant role because the annual lake level fluctuations are normally <20 cm (Jelinek 1995). More important is phosphorus release from the seasonally anoxic bottom sediments of the lake which accumulate with an average net deposition rate of 3 mm year^{-1}, while the maximum value amounts to >20 mm year^{-1} in the central parts of the basin (cf. Fig. 10.1). As soon as their redox potential falls below 200 mV, a series of processes sets in among which the reduction of nitrate, nitrite, ferric iron and sulfate are most important. The release of soluble reactive phosphate from ferric (hydr)oxides and calcium carbonate amounts to about 20 kmol per total littoral and pelagic sediment/water interface per year. Thus, while 0.5 g Pm^{-2} year^{-1} are sediment-bound under present-day circumstances, somewhat less is released from the sediments into the supernatant water column (Zeiler 1996). Nevertheless the P concentration increases from the sediment surface down to a depth of 50 cm as a consequence of high fluvial phosphorus input until the late 1970s, when the upstream Bornhöved sewage plant with a P elimination rate of >80% became operative (Naujokat 1996). Due to the comparatively high release rates from this sedimentary reservoir the net P retention of Lake Belau is less than 10 %. As substantiated in Section 9.4.3, a reasonable estimate of the net phosphorus flow from the littoral to the pelagic water column amounts to 200 kg P year^{-1}.

Deposition of particulate phosphorus compounds is of far less importance for the balance, but seems to exceed the inputs due to groundwater, interflow and tube drainage. In contrast to the average situation in Germany, however, where soil erosion is considered the major external source of phosphorus in lakes (Schwertmann et al. 1987), in the study area the predominantly low relief energy, hedgerows and the comparatively high permeability of soils largely reduce the overland flow rates with a critical drag necessary for erosion and particulate phosphorus transport. Thus, in general the average potential soil erosion exceeds the value of 1 t ha^{-1} year^{-1} only locally; for the total drainage basin of the lake system this means an annual sediment input of 53.3 t into the receiving waters (Meyer 2000).

With regard to the output terms the export of phosphorus compounds by the Schwentine River falls short of the inflow load; the difference measured is a reflection of the retention capacity of the system. The sedimentation of phosphorus species, among the output terms the second in importance, and biotic uptake are dependent on a set of interrelated biotic, chemical and physical processes. Modelling precipitation and re-solution reactions under oxic and anoxic conditions permits a more precise determination of the influence of Ca and Fe species in this connection (cf. Figs. 10.12, 10.13). In the anoxic water column hydroxyl apatite and iron sulfides are precipitated;

when oxidizing conditions develop a massive calcite precipitation sets in, coupled with the formation of particulate Ca-phosphate and Fe(OH)$_3$. Although the solubility product of FePO$_4$ is not exceeded, the redox cycle of the iron species exerts an influence on P sedimentation, since in addition to precipitation adsorption processes also control the phase transfer of P species. In particular iron (hydr)oxides adsorb reactive P species on their mineral surfaces and incorporate phosphorus into oxic sediments once the stratification has broken down (Baccini 1985). In dependence on local pH conditions the phosphorus compounds are partly dissolved in the hypolimnion and partly accumulated in the sediment; only a comparatively small proportion reaches the epilimnion in the same year again (Lenz 1992).

The above horizontal and vertical mixing processes can bring about resuspension, provided their kinetic energy is higher than the cohesion of the particles settled; the consequence may be spurious measurements in sediment traps. In Lake Belau the phenomenon is well documented during the circulation period, when the particulate flow rates in the 18 m and 25 m traps prove to be generally higher than in the 6 m traps. Estimates on the basis of comparative trace element analyses (Zeiler 1996; cf. Section 10.4.2) yield an approximate value of 30% of resuspended particulate material for the 25 m traps; proportionally higher amounts are to be expected in the littoral because of wave action.

10.4.2   Micronutrients and Trace Elements

Iron and manganese play a major role in the biogeochemical cycles of many essential and trace elements (Burdige 1993). Fe and Mn (hydr)oxides are important both as high-surface adsorbents for trace elements and as oxidants in the framework of microbial degradation processes at the sediment/water interface under suboxic boundary conditions (Berner 1981). Under anoxic conditions, prevailing in interstitial or hypolimnetic water, the concentration of these heavy metals is largely governed by the sulfide concentration resulting from the reduction of sulfates (Fig. 10.12, box). When the redox potential of the system falls below 200 mV, bivalent iron and manganese ions appear in solution; under oxic conditions the metals form particulate MnIII/IV and FeIII phases. In addition to the redox potential of the aquatic system inorganic ligands and organic complex-forming compounds, e.g. fulvic acid, govern the manganese and iron speciation also under oxic conditions (Salomons and Förstner 1984; Davison 1993). In dependence on speciation Mn and Fe function as essential micronutrients, e.g. for phytoplankton communities, which explains their potentially limiting influence on photosynthesis in hardwater lakes (Wetzel 1972). A mechanistic model describes the dynamics of the micronutrients and trace elements in five phases (Zeiler 1996; cf. Fig. 10.12 for the seasonal redox patterns).

In phase I, i.e. normally during the spring circulation period in March and April, the nutrient-rich water and the increase in temperature set off a bloom of phytoplankton production, with a prevalence of diatoms inducing high fluxes of particulate trace elements. The concentration of stable iron compounds in the isothermal water body seems to be considerably influenced by complexation reactions with phytoplankton exudates, as can be deduced from the synchronous increase of dissolved (<0.4 μm) iron species (>0.1 μmol l⁻¹) and excretions (Barkmann et al. 1994), while allochthonous iron input due to inflow and atmospheric deposition remains largely constant. Furthermore it ensues from findings by Matsunga et al. (1982) and Nishio and Ishida (1990) that FeIII chelates have a positive effect on algal growth, and consequently the phenomenon observed can be interpreted as an autochthonous mechanism of iron supply for the diatom communities.

High particulate flux rates of manganese are closely correlated with fluxes of particulate organic carbon. In the euphotic zone Mn is adsorbed on algae or taken up by the phytoplankton and transported with the dead organic matter to the bottom. At the sediment/water interface the readily biodegradable components are microbially decomposed, which leads to a marked oxygen deficiency. The latter in turn causes the release of MnII ions into the pore water as a result of microbial oxygen abstraction from the MnIII/IV (hydr)oxides in the framework of suboxic metabolic pathways.

Like manganese and iron, copper, zinc and lead also seem to be stabilized as organometallic complexes (chelates) in the euphotic zone by exudates of phytoplankton and excretions of zooplankton or dissolved humic substances for metabolic or detoxification reasons (cf. Baccini and Suter 1979; Zhou et al. 1989; Morel and Hering 1993; Balistrieri et al. 1994; Williamson and Parnall 1994). Complexation essentially reduces the concentration of free aquo-ions which could otherwise exhibit highly toxic effects on aquatic organisms; this applies in particular to Cu and Zn. The latter remains longer in the epilimnion than Cu and Pb, because it can form stable complexes only with aged exudates which appear after the breakdown of a phytoplankton bloom (Barkmann et al. 1994). In contrast to these heavy metal species dissolved cadmium and barium are not enriched during the spring bloom, since dissolved organic complexes are of minor importance for Cd, while Ba as an alkaline earth metal has a generally low tendency to complexation (Sigg and Stumm 1991; Morel and Hering 1993). Cd is rather adsorbed on biodetritus (Noriki et al. 1985) and Ba on organic matter or particulate Mn and Fe (hydr)oxides (Sugiyama et al. 1992). Contrary to findings by Bruland (1980), Murray (1987), Balistrieri et al. (1992) and de Baar et al. (1994) in marine and limnetic ecosystems, there is no positive correlation between dissolved Cd and phosphate in Lake Belau; therefore an active metabolic role of the element can be ruled out.

At the beginning of summer stagnation, i.e. in phase II, anaerobic micro-organisms decompose fresh algal detritus at the sediment surface and consequently the redoxcline migrates from the sediment up into the hypolimnion. Extensive Mn and Fe recycling takes place, associated with the microbially controlled release of adsorbed or incorporated Cu, Zn, Pb and Ba species from the recently settled diatom detritus. In the oxic epi- and hypolimnion these remobilized trace elements are then adsorbed onto newly formed particulate MnIII/IV and FeIII (hydr)oxide phases (Balistrieri et al. 1994) which are partly transported by the above horizontal mixing processes into the pelagic waters and partly re-sedimentated in the lower part of the hypolimnetic water column. Here they may undergo renewed reductive solution with release of the trace elements and, under locally sub-oxic conditions, repeated re-cycling. In detail there are considerable differences in the behaviour of the trace elements involved. Zn is presumably concentrated in the sub-oxic hypolimnion in form of dissolved or colloidal organometallic complexes, since sufficient quantities of organic degradation products (amino acids, peptides, carboxylates, etc.) are available as potential ligands in this early phase of sediment diagenesis (Zeiler 1996). Two mechanisms play a role in the fate of Pb. Higher aqueous concentrations require essentially reduced fluxes of remobilized Fe and Mn (hydr)oxides from the sediment preventing adsorption to the particulate phase; moreover the subsequent degradation of algal tissue is necessary, to which Pb is strongly bonded. Ba is primarily adsorbed on reactive MnIII/IV (hydr)oxides and accumulated in the solid phase. Once this carrier undergoes reduction under sub-oxic conditions, Ba is desorbed and enriched in the interstitial and lowermost hypolimnetic water. The extremely high particulate Ba fluxes in the medium and lower hypolimnetic water column indicate the importance of newly formed MnIII/IV (hydr)oxides as scavenger minerals for Ba.

During phases III and IV, i.e. in summer and autumn, the concentrations of dissolved Zn, Mo, Sb and U decrease in the epilimnion. Fe and the rare earth elements La, Ce, Pr, Nd, Sm, Eu, Gd, Tb, Dy, Ho, Er, Tm, Yb and Lu are remobilized from the bottom sediments under extremely anoxic conditions, because allochthonous FeIII (hydr)oxides are microbially reduced during the anaerobic degradation of dead organic matter. As a consequence an enrichment of Fe and rare earth elements is observed in the sulfidic hypolimnion (Zeiler 1996). Since thermodynamic models predict iron sulfides under these conditions, dissolved Fe should exist in the form of colloids (1–100 nm) or humic complexes. Irrespective of considerable amounts of copper released from the sediment, the concentration of dissolved Cu species in the stratified water column keeps decreasing, since the hypolimnetic sulfur cycle gains in importance.

Also the concentrations of dissolved Zn species decrease in the water, while simultaneously a significant increase in particulate element fluxes can be observed in the 25 m sediment trap, consisting of ZnS° complexes adsorbed onto, or co-precipitated with, FeS (cf. Jean and Bancroft 1986; Arakaki and Morse 1993). This highly efficient transport mechanism comes to an abrupt end in late summer, when the major part of Zn, mobilized under sub-oxic redox conditions at the beginning of stratification, has been removed from the liquid phase.

The concentration of dissolved Pb attains a first maximum in the sub- and anoxic hypolimnion. According to thermodynamic models it should form PbS° complexes, but probably crystal growth is inhibited by complexation of organic ligands with Pb sulfides (Uhler and Helz 1984). A second concentration maximum of dissolved Pb occurs at the end of the stagnation phase below the thermocline, when this trace element desorbs from settling MnIII/IV (hydr)oxides as a carrier phase under the existing reductive conditions.

Mo and Sb are likely to co-precipitate with iron sulfide. Dissolved hexavalent U is reduced and particulate UO_2 accumulates in the sediment. At the end of the summer stratification rare earth elements and Pb diffuse into the sub-littoral sediment porewaters. In contrast to these elements Cd has particulate fluxes which co-vary with the sedimentation of the summer and autumnal phytoplankton blooms.

After the autumnal breakdown of stratification, i.e. during phase V, dissolved oxygen is mixed into the hypolimnion and the FeII and MnII species are oxidized and precipitated as (hydr)oxides. Trace elements co-precipitate with, or adsorb onto, these endogenous particles, whose net flux is temporally overlapped by the above advection and resuspension processes, which may account to >30% of the total particle flux rate (Zeiler 1996). On the whole the concentrations of the particulate trace element fractions remain largely constant during the autumnal circulation, only Cu and Zn exhibit distinctly higher concentrations in the lower sediment traps, which is indicative of proportionally higher flux rates in comparison with those of Cd, Ba and Pb. Basically the difference observed seems to be due to rapid release of Cu and Zn from newly deposited biodetritus and a subsequent re-sedimentation in lower parts of the water column. Table 10.3 at http:www.ecology.uni-kiel.de/bornhoeved-report summarizes the annual flux rates of the above and some further elements for the 1993–1994 period.

The steady state of the Lake Belau ecosystem is under the influence of both external fluctuations and internal feedback mechanisms, i.e. re-solution processes from the bottom sediments. Significant release rates were determined for ammonia, dissolved reactive phosphorus, Mn and Ba, which can become important for a series of biotic cycles. In addition to ammonia and phosphorus manganese also has the character of an essential nutrient, controlling the trophic state and energetics of the system. The same applies to micronutrients or bioactive trace elements such as Fe, Zn, Mo, Sb or U. Functional interrelationships between abiotic and biotic system components are reflected in the occurrence of dissolved organometallic complexes in the euphotic zone of the lake, where trace elements like Fe, Zn or Pb seem to be stabilized in the aqueous phase by exudates of various phytoplankton communities, which makes them bioavailable on the one hand and mitigates possible toxic effects on the other. In the pelagic water column, for instance, a close connection with the 'microbial loop' (Section 10.5.6) may be assumed for zinc, whose organic complexes can be utilized by bacteria together with dissolved organic carbon. Furthermore, the concentration of lanthanoids, normally highly toxic to aquatic plants, is drastically reduced by phosphate-controlled eutrophication processes (Elschenbroich and Salzer 1988). Finally, in general most micronutrients are predominantly constituents of enzyme molecules and are thus essential only in small amounts. In contrast, the macronutrients are either constituents of complex organic compounds, such as proteins and nucleic acids, or act as osmotica.

10.5   Biocoenoses

10.5.1   Reed Belts

Along 3.7 km of the banks of Lake Belau reed belts (*Scirpo–Phragmiteta*) have developed which may locally attain a width of more than 20 m, in exceptional cases even 50 m, spreading into a water depth of about 1 m, but omitting stony and strictly anaerobic substrates (cf. Chapter 4). Thus, they cover about 3.9 ha which, together with the adjacent semiterrestrial stands with an acreage of 1.3 ha, add up to 4.6% of the total lake area. A comparative aerial-photo interpretation of the years 1959, 1988 and 1992 revealed a retreat of the stands by 4.4%, caused by a variety of disturbances. Of prime importance among these are mechanical stress by waves, floating wood and ice, trampling by drinking animals and bathers and the construction of landing-stages (Schieferstein 1997). Chemical stress is due to changes in the sedimentary character and hydrochemistry of the littoral zone (cf. Rodewald-Rudescu 1974; Jelinek 1995), including a decrease in oxygen concentration, caused by nutrients and sewage. Biotic stressors comprise insects, foraging waterfowl, grazing cattle and fishery (putting out of weir-baskets).

In agreement with previous studies (Rodewald-Rudescu 1974; Kühl 1989) an increase in productivity with temperature is generally found in the Lake Belau reed belt, too. Thus, for instance, productivity increased generally from 1.75 kg m^{-2} (dry weight) to 1.80 kg m^{-2} (dry weight) in adjustment to cumulative temperatures (i.e. sum of daily mean air temperatures >5 °C during the months May to October) which amounted to 2095 °C in 1993 and 2590 °C in 1994; but the increase was quite different at the 12 plots analysed, although the phenological phases proved to be almost identical. Primary productivity varies between 0.3 kg m^{-2} (dry weight) and 2.6 kg m^{-2} (dry weight), while the ratio between above-ground and below-ground production ranges from 1:1 to 1:4. On average the reed belt produces a biomass of 17 t ha^{-1} year^{-1} (dry weight), i.e. about 7 t C ha^{-1} year^{-1} above ground and 10 t C ha^{-1} year^{-1} below ground; thus, the reed stands range among the most productive systems of the study area (cf. Chapter 2).

Analyses of nutrient uptake (Chapter 9) by Lenfers (1994), Gessner et al. (1996) and Schieferstein (1997) indicate that the reed belt contributes 37.8 t C year^{-1} to the total production of Lake Belau in the period under discussion; it comprises, inter alia, some 8.5 t C year^{-1} of reed leaves and 3 t C year^{-1} of alder litter. Nevertheless the littoral cannot generally be considered a carbon sink in view of its structural heterogeneity and the above exchange processes with the pelagic water column. On the one hand detritus can be transported into the littoral by epilimnetic currents and settle there, which makes the littoral a nutrient sink where even peat formation may occur (Ostendorp 1992, 1995; Schieferstein 1997). On the other hand wind or the nocturnal temperature gradient between the littoral and pelagic water bodies can induce a current out of the littoral, which then operates as a source (Schröder 1975; Meissner and Ostendorp 1988). Thus, it ensues from these findings that the controversial discussion of the (predominant) source or sink function of the littoral zone (cf., e.g., Tóth 1972; Westlake 1980; Wetzel 1983) seems to have been influenced either by premature generalizations or an insufficient number of geostatistically valid sampling plots analysed.

In comparison with the above primary production of the littoral zone the productivity of the pelagic phytoplankton amounts to 130 t C year^{-1} or 77.5% of the lake total. The resultant productivity ratio of 0.29 manifests the extraordinary importance of the littoral biota for the carbon balance of Lake Belau, which can be illustrated by comparative figures from the Schöhsee, situated some 15 km east of Belau and comparable in morphometry and trophic state, but with only 1.3% of littoral area. The lesser acreage leads to a total production of only 6.2 t C year^{-1}, comprising 2.3 t macrophytic and 3.9 t epiphytic carbon (Esteves 1979; Ho 1979).

10.5.2   Aufwuchs-Associated Nematodes and Oligochaetes

The 'aufwuchs' (or periphyton) on reed stems is composed of algae, in particular Bacillariophyceae, Chlorophyta and Cyanobacteria, and associated bacterial and fungal communities (cf. Mazumder et al. 1989; Lock 1993) which form a complex three-dimensional structure (Meulemans and Roos 1985) housing a multitude of animals. In the Lake Belau reed belt the assemblages of aufwuchs-associated nematodes and oligochaetes comprise 27 taxa of Nematoda, 16 taxa of Oligochaeta and one genus of Aphanoneura. A comparatively large proportion of these, however, namely six out of the Nematoda and five out of the Oligochaeta, are not really members of the aufwuchs-associated assemblage, but represent sediment-dwelling species which only accidentally appear on the reed stems (Löhlein 1998).

During the 1992–1994 reference period the assemblages of both nematodes and oligochaetes were dominated by only three genera. Among the nematodes, the Chromadoridae *Punctodora ratzeburgensis* and *Chromadorina viridis* made up 81.6% of all individuals found, while the Plectidae genus *Plectus* with five species accounted for further 5.1%. Analogously the assemblage of oligochaetes was dominated by the Naididae species *Chaetogaster diastrophus*, *Nais* spp and *Stylaria lacustris*, which added up to 93.5%. For further taxonomic specifications the reader is referred to Löhlein (1998).

Abundance and biomass transect studies showed that both were 10–20 times higher in the outer parts of the reed belt than near the banks. In each assemblage peak abundances of the dominant species occurred in May, reaching up to 4.5 oligochaetes per square centimetre of reed blade surface, and more than 200 nematodes per square centimetre, respectively. While *Punctodora ratzeburgensis* and *Chromadorina viridis* simultaneously reached their maximum population densities, a seasonality was apparent in the oligochaete communities. *Chaetogaster diastrophus* was the first to reach its maximum in April, followed by *Nais* spp in May and *Stylaria lacustris* in June or July. After the maxima all populations declined within a few weeks. As regards the nematodes, there are both spatial and seasonal shifts in the proportion of the different trophic guilds. Proceeding from the banks to the outer margin of the reed belt, the proportion of fungivorous and bacterivorous taxa declines, while that of algivorous taxa increases. At all sites, algivorous nematodes reach the highest proportions in the assemblage during spring and winter.

Population breakdown is due to a shift in limiting factors. In early spring populations flourish owing to an abundance of food provided by the bloom of aufwuchs algae, but are most probably limited by water temperatures. Later in spring and early summer, declining algal biomass causes a shortage of food and the worm populations break down. This shift in nutritional boundary conditions renders estimates of the secondary production of the nematode and oligochaete populations difficult. In 1993, for instance, it varied from $142\,\mu g$ C cm^{-2} to $670\,\mu g$ C cm^{-2} reed blade surface, made up to more than 90% by oligochaetes. Furthermore enclosure–exclosure experiments showed that potential predators like *Rutilus rutilus* of the age class 0+ did not suppress the spring abundance of oligochaetes, while competitors (i.e. snails of the genus *Bithynia*) impaired their mass development in the aufwuchs for 3–4 weeks (Löhlein 1998).

10.5.3   Molluscs

The distribution of the benthic gastropoda and bivalvia fauna in Lake Belau, comprising 34 species from 12 families (Asshoff et al. 1991), reflects the substrate and vegetation patterns of the littoral on the one hand and the seasonal hydrochemical variations of the water column on the other. Thus, the total mollusc biomass (including shells) of the littoral amounts to some 5000 kg dry weight with an estimated annual production of about 13 000 kg, while the summer oxygen-free profundal remains free from molluscs (Asshoff 1990). The biomass of the dominant species at seven localities with different substrate conditions is depicted in Fig. 10.16.

With an average abundance of 11 400 individuals m^{-2}, the snail *Potamopyrgus jenkinsi* contributed more than 90% to the total mollusc abundance and more than 30% in terms of biomass. A comparison of the above figure with Fig. 10.1 indicates a habitat preference for the north-eastern and north-western littoral, where sandy and stony substrates and some erosion prevail. *Theodoxus fluviatilis* exhibits comparable distribution patterns and attains a maximum abundance of up to 500 individuals m^{-2} on reed-free sites in the northern part of the lake, where the banks are liable to undercutting. In areas with distinctly less water movement, where thicker detritus layers can accumulate, *Bithynia leachi* or *B. tentaculata* dominate and account for nearly 90% of the total mollusc biomass. Small mussels, e.g. *Dreissena polymorpha*, occur in varying abundances with maxima around 200 individuals m^{-2} at all sites investigated. Favourable habitats are in front of the peninsula at the west coast and along the north-eastern coastline, where the TOC concentrations in the uppermost sediment layer are <5% dry weight. Large mussels like *Unio tumidus* are rare in Lake Belau and appear to be limited to exposed sites with relatively high current speeds, where they may attain a maximum density of 5 individuals m^{-2}. For a comparative analysis of the habitat preferences of all of the 34 species studied in relation to differences in euryoecious or stenoecious character the reader is referred to Asshoff et al. (1991).

10.5.4   Chironomids

Comprehensive analyses of the benthic fauna of Lake Belau and the upper Schwentine system in the first half of the 1990s (Otto 1991; Dienemann 1997) provided evidence of 43 Chironomidae genera of these sub-families: Chironominae

Dry ´weight (g m⁻²)

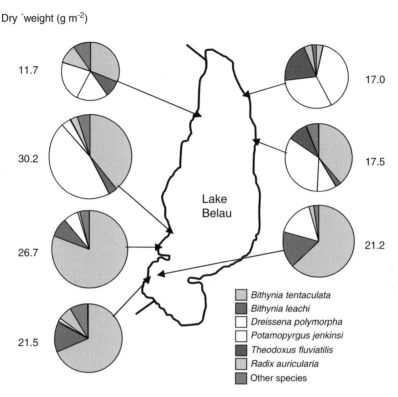

Fig. 10.16 Total shell-free biomass (dry weight, g m⁻²) of dominant mollusc species at seven littoral sites of Lake Belau in 1989 (after Asshoff 1990)

(25 genera or 58.1% of the total chironomid fauna), Orthocladiinae (11 genera), Tanypodinae (five genera) and Diamesinae (two genera). In an ecological respect it is noteworthy that only the following nine species were found at all of the benthic sites of the lake, namely *Chironomus plumosus*, *Cladotanytarsus* sp., *Dicrotendipes pulsus*, *Glyptotendipes pallens*, *Microtendipes pedellus*, *Parachironomus varus*, *Polypedilum nubeculosum*, *Tanytarsus* sp. and *Procladius* sp. The interannual variability of the populations is such that only *Glyptotendipes pallens*, *Polypedilum nubeculosum* and *Procladius* sp. occurred at each site in two consecutive years, while the other species were either lacking in one year or at one or several sites. For a detailed survey of all of the 62 species analysed with regard to habitat preferences and specific ecological characteristics the reader is referred to Dienemann (1997).

10.5.5   Pelagic Phytoplankton and Benthic Algal Assemblages

Systematic inquiries into the planktonic populations of the pelagic water column of Lake Belau yielded a total of 88 species, representing six families in the following specification: 39 Chlorophyceae, 21 Bacillariophyceae, 17 Cyanophyceae, four Dinophyceae, four Chrysophyceae, three Cryptophyceae. Considering the phytoplankton associations in terms of biomass, however, only 14 Chlorophyceae and ten Bacillariophyceae species proved to be of importance. Thus, and in conjunction with the Dino-, Chryso- and Cryptophyceae, a total of 43 species appear to be essential for both composition and productivity of the assemblages (Landmesser 1993; Barkmann 1998).

Comparative quantitative analyses of the populations during the 1989–1994 period revealed a high degree of both seasonal and interannual variability. Under conditions of complete circulation in spring, and in several years also in winter, Bacillariophyceae were dominant, but with considerably variable biomass figures. Following the silicate-controlled diatom bloom, the Cryptophyceae populations attain their biomass maximum in April and May and may then determine the phytoplankton composition, as the years 1991 and 1992 showed. In many years an intensified development of Chlorophyceae follows in May and June; they may make up a high proportion of the total number of phytoplankton species without, however, predominating in biomass. In July and August the Dinophyceae dominate, in particular the big species *Ceratium hirundinella* and *C. furcoides*, whose length exceeds 50 µm. In some years, e.g. in 1990 and 1993, they contributed up to 95% of the total phytoplankton biomass (Landmesser 1993; Schernewski 1999). Normally, however, the Dinophyceae maximum is associated with major proportions of Cyanophyceae species which attain a relative maximum of importance among the phytoplankton taxa in late summer and autumn. Only in exceptional cases like in April and May 1994 may Chrysophyceae also exhibit noteworthy abundances.

A comparison of phytoplankton distribution with the location of wind-driven currents exhibits a great deal of broad adjustment. Thus, provided the wind speed is <3 m s^{-1}, concentration maxima coincide more or less with the interior parts of cyclonic eddies, the minima with the location of fastest currents, as Fig. 10.17 illustrates.

Under these conditions not only is the horizontal flow component inside the cyclonic eddies very low, but also the corresponding vertical turbulence is small enough to allow the quite motile flagellates of the genus *Ceratium*, which can perform daily migrations of several metres (Nauwerek 1963; Galvez et al. 1988; Jones 1993), to develop distinct vertical density patterns according to light gradients. For wind speeds >3 m s^{-1} the resultant turbulent mixing of the water body prohibits crowding of flagellates at the depth of their light optimum (Lampert and Sommer 1993).

Characteristic differences also exist in both the horizontal and vertical distribution of algal populations. Planktonic diatoms have their maximum productivity during the circulation period or at the beginning of stratification and are, as a consequence, distributed relatively uniformly in the whole pelagic water body, except for short-term fluctuations in density due to the above ephemeral (cyclonic) circulation patterns. Dominant in the majority of surface samples were the species *Stephanodiscus parvus*, *S. minutulus*, *S. neoastraea/rotula*, *S. hantzschii* and *Cyclostephanos dubius*. In contrast to these forms, the non-planktonic diatom flora

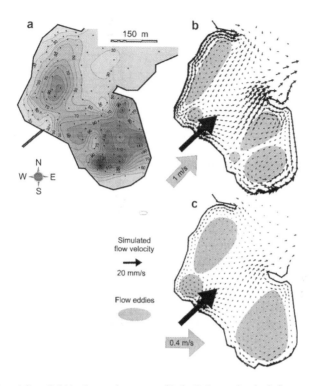

Fig. 10.17 Local flow field in the southern part of Lake Belau under the influence of weak, fluctuating winds on 4 August 1993 and associated chlorophyll-a concentrations (mg m⁻³), largely due to a *Ceratium* bloom. Computations on the basis of 103 fluorimeter measurements at a depth of 0.3 m in the afternoon between 3 p.m. and 6 p.m.

which comprised no less than 77 species at 30 sampling sites analysed in the second half of the 1990s (Håkansson et al. 1998) revealed a much higher spatial diversity. Fragilarioid forms, especially *Staurosira construens* and *Pseudostaurosira brevistriata* were dominant only in shallow or marginal parts of the lake. Superimposed on this dispersal pattern of life-form categories is the influence of the local nutrient sources described above. Thus, while otherwise rare or absent, species like *Amphora pediculus*, *Navicula scutelloides*, *N. cari*, *N. radiosa* and *N. gothlandica* increase considerably in abundance in those parts of the lake which receive the highest input of nutrients (cf. Fig. 10.10). Apart from this relatively large-scale pattern, marginal sites often have site-specific diatom assemblages which are believed to reflect the direct or indirect influence of erosional processes along the shore and inflow from the catchment (cf. Figs. 10.1, 10.8).

In contrast to diatoms the other taxa are limited to the epilimnetic water column. During the exemplary years 1989 and 1990 Chrysophyceae, Cryptophyceae and Chlorophyceae appeared only in late spring and summer, i.e. under conditions of stable stratification. The situation changes in late summer and autumn, when

Dinophyceae and Cyanophyceae are most productive and the thermocline sinks to greater depths, which brings about an increase in both the quantity of epilimnetic water and turbulence.

Following OECD (1982) and Takamura and Nojiri (1994), the chlorophyll-a (chl) concentration is used as an integral indicator of both biomass and productivity of the algal assemblages. However, a comparison of the correlation functions biomass (fresh weight, mg m^{-3}) = 131.6 chl$^{1.1}$ (amount of chlorophyll-a, fresh weight, mg m^{-3}; r^2 = 0.88, n = 24) and productivity (fresh weight, mg m^{-2} h^{-1}) = 288 – 1127 log chl (r^2 = 0.62, n = 51) and the comparatively low correlation coefficient of the latter point to inherent limitations of the method. A comparison with pertinent literature data shows (Landmesser 1993) that the annual phytoplankton productivity figure of 126 g C m^{-2} is comparatively low in relation to biomass and chlorophyll concentration. The corresponding photosynthetic efficiency attains the normal value of 0.35%, which means in terms of the solar energy cascade that only a very small proportion of the incident radiation is stored in the biomass.

An analysis of the biomass and productivity data with regard to dominant species shows that in 1989 more than 40% of the annual biomass and more than 25% of the annual productivity were due to the large diatom *Stephanodiscus neoastraea*. By way of contrast in 1990 the summer Dinophyceae species *Ceratium hirundinella* and *C. furcoides* made up about 50% of the annual biomass and productivity. The examples illustrate the rule that a few algal species with interannually varying abundances dominate the phytoplankton biomass and productivity, while nearly 90% of the species inventory are quantitatively negligible (cf. Meffert and Wulff 1987; Håkansson et al. 1998).

The corresponding potential growth rates of the algal populations, i.e. their change in size, fluctuated around an average value of 0.14 day^{-1} in 1989 and 1990 with an absolute maximum of 0.4 day^{-1} in May and July 1990. The actual growth rates, however, reached only an absolute maximum of 0.18 day^{-1}, which means in other words that the populations needed more than five days for doubling. Halving, in contrast, can take place within two days under the impact of grazing zooplankton (Landmesser 1993). Thus, in comparison with many other lakes both the growth and loss rates of the Lake Belau phytoplankton assemblages appear relatively low, which may be due to high mixing rates.

10.5.6   Zooplankton and the Microbial Loop

The protozooplankton as the smallest fraction of the zooplankton comprises, in addition to Amoebina and Heliozoa, mostly ciliates and heterotrophic nanoflagellates whose size ranges from 2 μm to 20 μm (nanoplankton), while bigger ciliates, heterotrophic flagellates and Rotatoria form the microzooplankton (20–200 μm). The average size of crustaceans normally exceeds 100 μm. The protozooplankton proportion of the total zooplankton biomass varies from 5% to 40% and is characterized by a marked dominance of heterotrophic nanoflagellates; only in April do

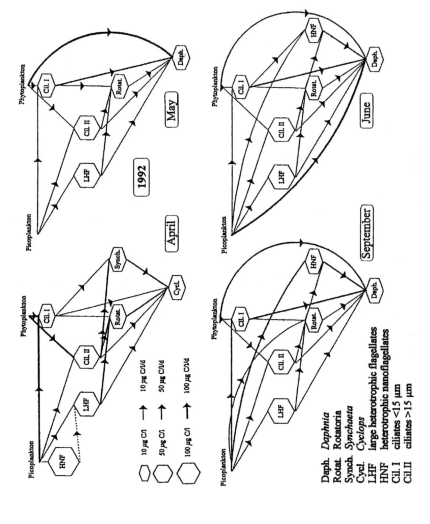

Fig. 10.18 Food web structures of phyto and zooplankton communities of Lake Belau during April, May, June, and September 1992 (after Zimmermann 1994)

ciliates make up 60% and in August the big heterotrophic flagellates may even reach a 70% proportion (Zimmermann 1994). In analogy to the development of phytoplankton communities the different zooplankton components are also quite variable in their seasonal and interannual development. Thus, for instance, among the 30 ciliate species found in Lake Belau the spring species *Tintinnidium fluviatile* made up 31% and *Frontonia leucas* 48% of the mean annual biomass (Barkmann et al. 1994).

The detailed analysis of biotic interactions and related material fluxes during four periods of the reference year 1992 and the graphic documentation illustrate that the traditional food chain concept with herbivorous zooplankton as primary consumers feeding on phytoplankton and carnivorous zooplankton as secondary consumers hardly applies (cf. Fig. 10.18). Only in May and June does *Daphnia* as an efficient filterer bring about a direct material flux from the phytoplankton to the zooplankton subsystem while, particularly in winter and spring, complex foodweb structures with complicated microbial fluxes develop.

In the microbial loop bacteria play an essential role by using the excreted products of primary producers, which enables them to build up major populations that, in turn, fall prey to protozooplankton communities on which bigger zooplankton species finally feed. Excretion is maximum at the end of the diatom bloom when nutrient competition in the algal communities is highest. Thus, ciliates have an optimum supply of feeder algae (<5 μm) and bacteria in April and May, which explains their concomitant maximum productivity. Also Rotatoria, represented by 27 species with a dominance of *Keratella*, profit from the algal bloom and exert a strong grazing pressure on ciliates. In addition they feed, like the crustacean fraction of zooplankton, on protozooplankton (Zimmermann 1994).

The crustaceans, comprising seven Copepoda and nine Cladocera species, make up the major biomass fraction of the zooplankton. They attain their biomass maximum in May, i.e. immediately after the bloom of the <30 μm phytoplankton fraction. Experiments showed that filtering crustaceans of the size class >100 μm are by far superior in grazing capacity to Rotatoria of the 55–100 μm size class (Fleckner, personal communication). Thus, the crustacean zooplankton, and in particular *Daphnia*, contribute most efficiently not only to the spring decline of algal biomass but also to the collapse of the protozooplankton communities, which initiates the clear water phase in May.

10.5.7   Fish

The fish fauna of Lake Belau consists of 19 species with a dominance of bream (*Abramis brama*), roach (*Rutilus rutilus*) and perch (*Perca fluviatilis*) which make up more than 70% of the total fish biomass of about 15 t. For fishing purposes considerable quantities of eel (*Anguilla anguilla*), pike (*Esox lucius*) and pike-perch (*Stizostedion lucioperca*) have been introduced. The other species which account for about 10% are: *Gobio gobio, Blicca bjorkna, Alburnus alburnus, Scardinius erythrophthalmus, Acerina cernua, Cobitis taenia, Leuciscus cephalus, L. idus, Tinca tinca, Carassius carassius, Cyprinus carpio, Coregonus albula* (Pfeiffer 2000). In terms of individuals the figures mean that the total perch population consists of about 70 000 fish, while the bream population has about 20 000 individuals and the total number of pike-perch amounts to about 300 only. The main prey of bream is the large pelagic zooplankton, i.e. the crustaceans and organisms living on the sediment surface like Chironomidae and Chaoboridae. The composition of prey changes during the course of the year and is dependent on the growth and age of the fish. Young roach, bream and perch share the dietary preferences of the adult fish, but prefer the sheltering littoral for food intake.

The two dominant species bream and roach were subject to detailed estimates of condition and the related bioenergetic parameters consumption, defecation, excretion and respiration in order to develop individual-based models. These permitted a simulation of the spatial organization of a population as a consequence of the individual activity patterns of fish and their temporal variability (Breckling and Reuter 1996). The patterns can be correlated with the results of respirometric measurements, which permits an estimation of the individual activity-related energy consumption. Possible causes of the phenotypic variability can then be assessed at the population level and the consequences of such variability for the development of populations more precisely determined (Hölker 2000).

In the case of roach the model predicts two types of mortality. One relates to young fish immediately after winter and is due to bad condition; the other is coupled with the post-reproductive phase, when the females have to cope with a substantial loss in weight due to spawning. Each type leads to a decrease in phenotypic variability, since above all the extreme phenotypes are bound to die. As regards the dependence of growth rates on temperature, the model indicates two contrasting tendencies. In spring with its rich zooplankton supply the growth rate increases with temperature, but in the second half of the year still higher temperatures have the inverse effect, when the intensified turnover cannot be compensated for by a correspondingly proportionally enhanced consumption.

10.5.8 Species Diversity in the Light of Site Conditions and Organismic Motility

Comparative studies of zooplankton assemblages led to the conclusion that their spatial heterogeneity is controlled by a superposition of biotic and abiotic influences whose relative importance varies with scale (the 'multiple forces' hypothesis by Pinel-Alloul 1995). Thus, a high-resolution analysis attributes a distinctly greater importance to biotic processes, while with increasing scale abiotic processes appear to be more essential. The present inquiries have revealed an even greater dominance of abiotic processes with regard to phytoplankton patchiness, which is in agreement with findings of Camarero and Catalan (1991), Verhagen (1994) and Jones et al. (1995). As a consequence, the following generalization of the 'multiple forces' hypothesis can be formulated: With increasing spatial scale abiotic process generally gain preponderance over biotic ones for the development of distinct dispersion patterns of organisms (patchiness) in ecosystems. With increasing size of organisms or enhanced motility, respectively, the relative importance of abiotic processes decreases for a given scale (Schernewski 1999).

The inquiries into the spatio-temporal dispersion patterns summarized in the preceding sections provided comprehensive data sets for more precisely defining the validity of this hypothesis. In fact, the 'equilibrium' between biotic and abiotic control of patch formation appears shifted to both increased organismic size and spatial scale. This implies that a considerable part of the small-scale patterns, traditionally interpreted as biologically controlled, may also be rather or predominantly the result of hydrodynamic processes (cf. Section 10.3.2, Figs. 10.7, 10.8), whose intensity yet remains to be (experimentally) matched with the inherent motility properties of the different components of planktonic and nektonic assemblages. The relevant interrelationships are summarized in Fig. 10.19.

The model, which is open to lake-specific adaptations in dependence on species abundance, biotope size and structure, indicates in the above form that the point of maximum dispersive heterogeneity is scaled down with decreasing organism size.

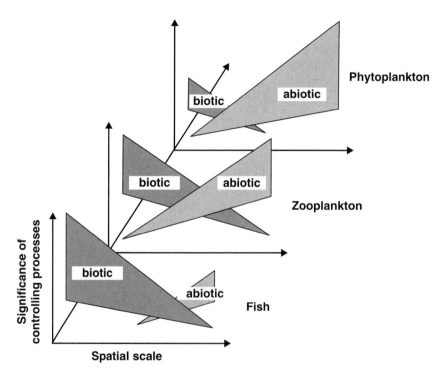

Fig. 10.19 Relative importance of biotic and abiotic processes controlling the scale of phytoplankton, zooplankton and fish dispersion in lakes as dependent on organism size and motility (after Schernewski 1999)

Whence it can be deduced that each population attains a local maximum of heterogeneity under comparable conditions of biotic/abiotic process control. In the light of this model some apparently contradictory findings of several authors may be given a unified interpretation. According to Jones et al. (1995), for instance, the horizontal variability of algae and metazooplankton exceeds that of smaller bacteria and heterotrophic protozoans, while Pinel-Alloul (1995) holds that bigger zooplankton exhibit a more homogeneous dispersion than smaller ones. In fact the basic problem underlying such contrasting views seems to be related to scale, i.e. smaller organisms are likely to exhibit maximum heterogeneity at proportionately smaller scales which are frequently, however, not made subject to high-resolution analyses. Among other things this assumption is corroborated by the findings of Jones and Francis (1982) and Duarte and Vaqué (1992), who observed a marked bacterioplankton patchiness at the centimetre scale, while a lesser spatial resolution provided the picture of a largely homogeneous distribution.

10.6 Conclusions

Lake Belau, as a typical representative of the north German holomictic–dimictic hardwater lakes which originated in the late-Weichselian landscape, has been analysed in its manifold functional relationships with the complex structure of the seasonally varying catchment with an emphasis on the groundwater/lake or lentic ecotones which control in a very specific manner the water and chemical exchange processes (Chapters 8, 9). Detailed macroscale analyses of the riparian and lacustrine biota of Lake Belau point to manifold feedback loops between soil and sediment distribution, nutrient flux patterns between catchment and lake and both the hydrological and floristic character of the ecotones. In dependence on wind speed and plankton motility wind-induced cyclonic circulation systems in the lake bring about a characteristic patchiness of nano- and microplankton assemblages, which is important for sedimentation processes and the development of transient food web structures especially in the central and northern part of the basin. Such short-term phenomena are superimposed on seasonal or interannual fluctuations of the energy budget and material balance of the lake which is typically dimictic but becomes warm-monomictic in years with exceptionally warm winters whose frequency has increased during recent decades.

In contrast to the majority of limnological enquiries which focus either on trace element cycling in the water column or in the bottom sediments, the present study provides for the first time a comprehensive analysis of the intercompartmental spatio-temporal concentration and flux dynamics of dissolved and particulate nutrients (C, N, P) and trace elements (Mn, Fe, Cu, Zn, Mo, Cd, Sb, Ba, rare earth elements, Pb, U) on the basis of high resolution sampling and monitoring. It shows that a detailed knowledge of the redox-controlled seasonal manganese and iron cycles is fundamental to the understanding of the aquatic trace element geochemistry and indispensable for comprehensive restoration purposes. Comparative analyses of all of the lakes of the study area on behalf of such abatement measures show that the eutrophication potential of nutrient sources depends on both the location and spatio-temporal variability of the inputs, and the proportion of phytoplankton-available nutrient fractions. Altogether, the annual averages of nutrient and chlorophyll-a concentrations decrease in the lakes along the Schwentine River, but according to inherent differences in the individual retention and elimination potentials short-term departures from the rule are quite common.

Chapter 11
Ecological Gradients as Causes and Effects of Ecosystem Organization

Felix Müller, Otto Fränzle, and Claus-Georg Schimming

11.1 Introduction

In the course of ecosystem evolution gradients emerge as fundamental features of self-organization processes under the influence of neighbouring geo- and ecosystems (Müller 1998). Thus, analysing ecological heterogeneities (e.g. Kolasa and Picket 1991; Breede 2000; Reiche et al. 2001), ecotones (e.g. Gosz 1992; Fränzle and Kluge 1997; Kluge et al. 2003) or patch dynamics (e.g. Shugart and Urban 1988), implicitly means investigating ecological gradients. In this context two theoretical concepts are fundamental:

1. The thermodynamic *non-equilibrium principle* of Schneider and Kay (1994a, b) is related to the flow and storage of exergy which is a measure of the maximum capacity of the energy content of a system to perform constructive work (see also Jørgensen 2000; Kay 2000).
2. While Schneider and Kay stress the system's degradation capacity, Jørgensen's *exergy optimization principle* puts emphasis on the development of gradients and structures: "If a system receives a throughflow of exergy, the system will utilize this exergy to move away from thermodynamic equilibrium. If the system is offered more than one pathway to move away from thermodynamic equilibrium, the one yielding most *stored exergy*, i.e. with the most ordered structure or the longest distance from thermodynamic equilibrium by the prevailing conditions, will have a propensity to be selected" (Jørgensen 2000, p. 166).

Integrating these principles, the focal hypothesis reads: living systems are degrading and utilizing external gradients by the self-organized formation of a hierarchy of nested internal gradients in correspondence to the energetic environment of the system. This implies:

- Gradients are causes for and results of ecological self-organization. Thus, they can be designated as emergent properties of ecosystems (Nielsen and Müller 2000) and used as indices for the autocatalytic potential of ecological entities.
- Gradients are the result of waxing processes on the one hand (gradient creation) and fundamentals for waning processes on the other (gradient degradation).

O. Fränzle et al. (eds.), *Ecosystem Organization of a Complex Landscape.*   277
Ecological Studies 202.
© Springer-Verlag Berlin Heidelberg 2008

- Gradients operate at different ecological scales. Between these scales, certain mechanisms of self-regulation determine ecosystem development.
- Gradients are ecological orientors. Their size, extent and diversity, and the eco-physiological linkages between them are optimized throughout undisturbed ecosystem dynamics. Therefore, gradients can be used as fundamentals for the assessment of ecosystem evolution.

11.2 Gradients in Ecosystems

In the following analysis two sets of structural and functional gradients are distinguished. While the structural gradients are concentration profiles along a spatial axis, the functional gradients refer to flow schemes between sources and sinks. The general features of these gradient types are:

- *Structural gradients* result from the spatial distribution of ecosystem elements and subsystems which are dynamic components of ecological diversity. Furthermore, ecosystem structure determines the potentials for ecological interactions that operate within the structural patterns.
- *Functional gradients* develop as a consequence of the functional differentiation of ecosystems. Definitions of "function" reach from mathematical arguments to clearly anthropogenic points-of-view (e.g. de Groot 1992). In a conventional perspective, functions are selected relations between ecosystem components, forming patterns of ecosystem processes (Müller and Windhorst 2000). Therefore, in the following paragraphs, functional gradients will be used to elucidate the flows, storage and regulatory processes between ecosystem compartments.

11.2.1 Structural Gradients

Ecosystem structure, i.e. the spatial pattern of the systems' elements, forms an "environmental envelope" (O'Neill 1988; Müller 1992) for ecological processes: all transfers of water, energy or matter follow specific gradients in and between the ecosystem and their environment. Thus, the structural features form an ensemble of environmental constraints, which are related to individual spatial and temporal scales (Breede 2000). They define the potential degrees of freedom for the processes linking the structural elements (O'Neill et al. 1989).

The abiotic parameters, such as the lithological, topographical or geomorphological features of the landscape or the soil pattern (cf. Table 11.1) have developed throughout long periods, and their gradient patterns can be detected on relatively large scales (see Hári and Müller 2000).

These gradients are primarily based on the large-scale abiotic features of a landscape, but at a smaller scale (e.g. the site) they may, vice versa, also exert a high

Table 11.1 Some abiotic structural gradients described in preceding chapters of this volume

Chapter	Authors	Gradient described
2	Blume et al.	Relief formation and erosion processes as causes of externally induced gradient formation processes (Fig. 2.4)
		Geological, lithological and geomorphological gradients in the watershed (Fig. 2.1)
		Horizontal soil distribution (Fig. 2.2)
		Vertical physical, chemical and gradients in the soil profiles (Table 2.1)
10	Fränzle and Schernewski	Bathymetry of Lake Belau (Fig. 10.2)
11	Kluge and Fränzle	Spatial structures and lateral transports (see Fig. 9.1)

Table 11.2 Some biotic structural gradients described in preceding chapters of this volume

Chapter	Authors	Gradient described
2	Blume et al.	Patterns in vegetation structure (Table 2.2)
6	Irmler et al.	Vertical gradients of soil fauna (Fig. 6.1)
		Vertical stratification gradients of Diptera and Coleoptera in the beech forest and in the alder break (Figs. 6.2, 6.3)
		Application of gradient analysis referring to faunistic and floristic features versus various environmental factors (Table 6.2)
		Reproductive success of robins as an effect of structural gradients in the beech forest (Fig. 6.7)
7	Irmler et al.	Species compositions as functions of environmental gradients in and between ecosystems and between them (Fig. 7.2, Table 7.3)
		Distribution of ground beetles between the beech forest and the hedgerows on the agrarian land ecosystems as a function of the land use and land structure gradients (Table 7.4)
		Dynamics of gradients of fish densities in different zones of Lake Belau (Fig. 7.9)
10	Fränzle and Schernewski	Community distribution in reed belts of the lake (Section 10.4.3)
		Horizontal and vertical distribution of algal populations in Lake Belau (Section 10.4.3)

influence on the abiotic constraints. Many examples of such gradient types have been elaborated in the Bornhöved Lakes District (see Table 11.2). An additional case study is exemplarily documented in Fig. 11.1. It shows how the distribution of four plant species depends on the distance from the banks of Lake Belau as a consequence of their autecological setting in terms of soil moisture conditions.

Both the abiotic and the biotic gradients are strongly influenced by human activities, in particular land use measures (cf. Fig. 2.4). As an example, Fig. 11.2 illustrates the distribution of carbon and nitrogen compounds in the soils of the agrarian ecosystem A3 as studied by Mette (1994) in order to more precisely define

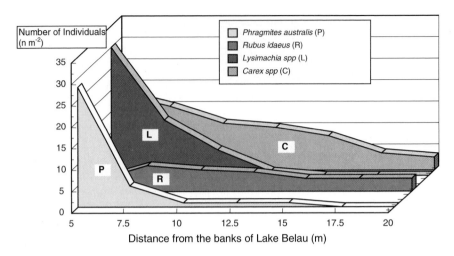

Fig. 11.1 Abundances of four plant species in relation to the distance from the banks of Lake Belau in the alder carr (after Dittert 1994)

the ecological significance of *hedgerows*. The field was evenly managed with regard to ploughing and the application of fertilizer and pesticide.

The hedgerow at the edge of the field affects the nutrient balance in several manners: on the one hand, solar radiation is reduced, especially on the northern flank of the hedgerow. This leads to a reduction of the maize yield in dependence on the distance from the hedge. Furthermore, litter input is higher at the fringes of the field. Because of shading and litter fall, the amounts of carbon and nitrogen stored in the soil exhibit similar gradients. Taking these three factors into account it is easily understandable that, especially in a northern exposure, there is a high surplus of nitrogen in the peripheral soils, which is not taken up by the crops nor by the hedge plants. Hence, a more efficient land use management has to include a reduced agricultural activity at the edges of the fields, which would not only reduce nitrate leaching into the groundwater but would also be a more economic strategy for those sites with high chemical gradients.

Within the framework of a special sub-programme, focusing on concentration profiles in the beech stand (W2), soils were sampled in 1-m, 10-m and 50-m grids, taking nine samples per grid, which were homogenized to represent the respective plot. Some results of this study are shown in Fig. 11.3. Here, the *x*-axis of the gradient triangles represents the spatial extent of the gradients. As an indicator for this variable, the ranges of the respective semi-variograms have been used. The *y*-axis represents the strength of the gradients, which is shown by the percentage of the range between the first and third quartile of the data sets.

In Fig. 11.3, the smallest-scale gradients are mainly due to water and nutrient uptake of the tree roots and their exudates and the input of water and nutrients by stemflow and throughfall. The existence of such gradients is directly dependent on

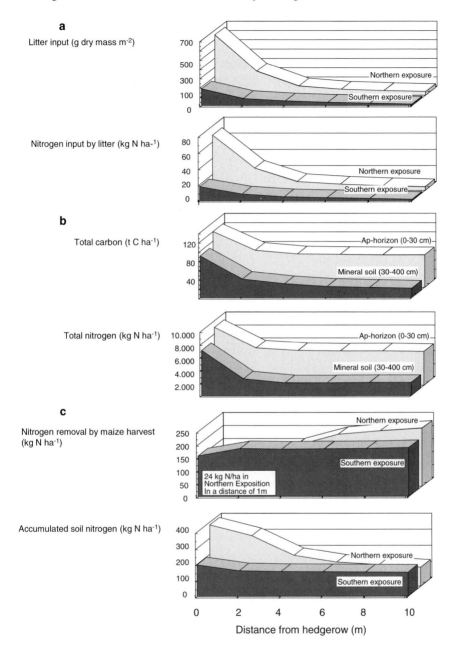

Fig. 11.2 a-c Components of the nitrogen and carbon budgets of the agroecosystem A3 as concentration gradients between a hedgerow and the centre of the field and in relation to the exposure of the hedgerows flanks (after Mette 1994)

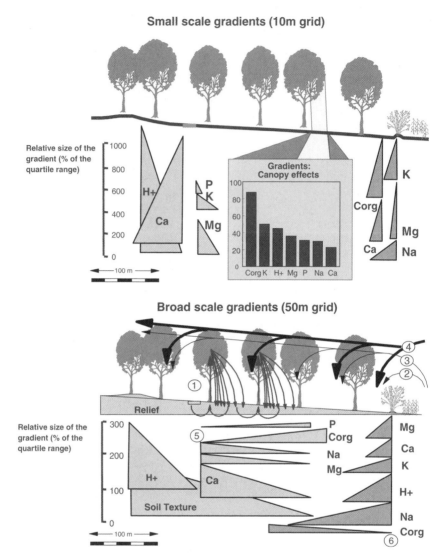

Fig. 11.3 Spatial gradients of different soil parameters in the upper soil layer of the beech forest in the core area of research. The *upper* diagram shows the gradients of the 10×10 m grid (W1.2), including some information about the 1×1 m grid investigations; the *lower* diagram is based upon the 50×50 m sampling pattern. The gradients are presented in the form of triangles, calculated on the base of the ranges of semi-variograms (horizontal axis) and the percentage of the ranges between the first and third quartile values of the whole data set (vertical axis). The numbers in the *lower* diagram refer to (*1*) a marled forest track, (*2*) fertilizer dust inputs from a neighbouring field, (*3*) short-distance transport due to ammonia release from slurry, (*4*) long-distance transport of macro- and micronutrients, (*5*) gradients around a forest track and (*6*) gradients at the forest fringe

the physiological status of the trees and their dynamics. When the trees fall, the gradients will slowly be degraded.

Furthermore, in the upper part of Fig. 11.3, marked gradients appear in the 10×10 m grid. They are dominated by man-made edge effects: On the one hand they originate from the neighbouring fields (right side), and on the other from a linear marling measure in the middle of the forest which took place about 100 years ago and which has exerted a great influence on the pH value and the Ca content (left side).

The lower part of the figure shows large-scale gradients which reflect different relief forms and soil patterns. Besides these extremely slowly changing constraints, the pattern is dominated by the effects of the above marling measures and the influence of fertilization on the neighbouring fields. These fertilizers enhance the chemical buffer capacities of the soils, preserving relatively high saprophagous activities, which leave only small amounts of organic carbon (SOC) at the forest fringe. Between this area and the central forest track, the inputs of ammonia and acid precipitation are effective, reducing the saprophagous activity, which leads to an accumulation of phosphorus and SOC in the soils. On the left side, again the effects of the marling measure are obvious: 100 years ago, this zone was only 2 m wide. This steep gradient has smoothed down very much; today it is 125 m wide. As there are no effective abiotic lateral transport processes in the forest soils (cf. Section 2.2.3), only root uptake of Ca by the beech trees and the subsequent distribution of the cations by leaching and litter fall can account for this gradient pattern (cf. Section 8.2.2). The long-term process has modified the living conditions of the undercover vegetation, too, which is clearly reflected in its specific phytosociological character.

The basic processes influencing the gradient pattern can be arranged in a spatio-temporal order which starts with relief and soil formation and includes marling and fertilization effects. All other processes are developing within this framework. Thus, the abiotic heterogeneity exerts a very high influence on the biodiversity pattern of the forest ecosystems: the higher the heterogeneity of the constraining (mainly abiotic or anthropogenic) gradients, the higher the diversity of potential ecological niches which can be occupied by a correspondingly higher number of species.

In addition to the forest investigations, a similar analysis was carried out by Müller (1998) and Reiche et al. (2001) in the neighbouring agrarian ecosystems A1, A2, A3. While in the forest above all the indirect effects of human activities are constraining the internal development, the land use practices on the neighbouring fields gave directly rise to an extreme degradation of structures and potentials. As a consequence, the gradients of Ca like those of SOC or phosphorus disappeared completely. A set of other gradients can be found in preceding chapters of this volume as summarized in Table 11.3.

11.2.2 Functional Gradients

Ecological functions are analysed as interrelated biotic and abiotic processes that operate in and between ecosystems, thereby connecting the structural elements and

Table 11.3 Some gradients characterizing ecological heterogeneity, described in preceding chapters of this volume

Chapter	Authors	Gradient described
2	Blume et al.	General characterization of spatial patterns on different scales (Section 2.2.3)
6	Irmler et al.	Distribution of soil fauna and ground beetles (Fig. 6.6)
7	Irmler et al.	Distribution of flies and Arachnea (Fig. 7.4)
9	Kluge and Fränzle	Spatial variability of nitrogen and phophorus contents (Fig. 9.11)
10	Fränzle and Schernewski	Small-scale variation of short-wave net radiation (Section 10.3.1)
		Heterogeneity of the physical stratification in Lake Belau (Figs. 10.6–10.8)

changing their qualities as well as their quantities. From this viewpoint, ecological structures characterize the multiple storage compartments of energy, matter, water and information which define the quantities of the corresponding pools, their eco-physiological potentials, and their spatial patterns. As these structural elements are interrelated by exchange processes, the latter can be described by means of functional gradients, making due allowance for the pool-specific conductivities of transfer resistances involved (cf. Prigogine 1976; Fränzle 1994).

11.2.2.1 Energetic Gradients

Energy is of fundamental importance in the framework of ecological functions because all reactions and activities are connected with energy uptake or consumption, and energy export. Therefore, energetic gradients can be found in a wide spectrum of scales, reaching from the large-scale climatic boundary conditions of a site over the site-specific vertical distribution of energy in the atmospheric sub-system, the soils and the organisms down to the level of surface processes, e.g. at the interface between leaves and the atmosphere. Furthermore, the energy balance determines the stratification and mixing of lakes (Chapter 10). Various other energetic considerations concerning the Bornhöved Lake District have been elaborated by Venebrügge (1996), Kutsch et al. (1998), Kutsch et al. (2001b, c), Steinborn (2001) and Svirezhev and Steinborn (2000; cf. Table 11.4).

At the ecosystem scale, the biological energy balance is strictly connected with the transfer, storage and budgets of carbon. Therefore, the following example will focus on a comparison of two ecosystems in terms of carbon budgets. They are usually depicted in the well known form of the carbon cycle, where energy equivalents are transferred between the distinct storage units in an orderly sequence. In the interest of the gradient concept another approach is deemed preferable here since it is based on unfolding the carbon cycle (Patten 1992), as in Fig. 11.4.

The figure differentiates the existing carbon gradients in two groups, the living organic matter of plants, soil animals and micro-organisms and dead organic matter

Table 11.4 Some energetic gradients described in preceding chapters of this volume

Chapter	Authors	Gradient described
3	Dilly et al.	Assimilation and respiration processes as basic components of carbon dynamics (Tables 3.1, 3.2)
		Carbon allocation processes in the plant–environment–subsystem (Fig. 3.7)
		Decomposition processes (Fig. 3.9)
4	Kutsch et al.	Energy balances on diffent scales (Figs. 4.6, 4.9)
		Carbon and energy fluxes (Figs. 4.7, 4.8)
5	Hörmann et al.	Evapotranspiration and its physiological/meteorological conditions as energetic components of the water balance (Figs. 5.8, 5.9)
7	Irmler et al.	Food consumption in the aquatic food web (Table 7.4)
10	Fränzle and Schernewski	Energetic setting of Lake Belau and its drainage basin (Section 10.3.1)
		Limnetic energy cascades (Section 10.5)

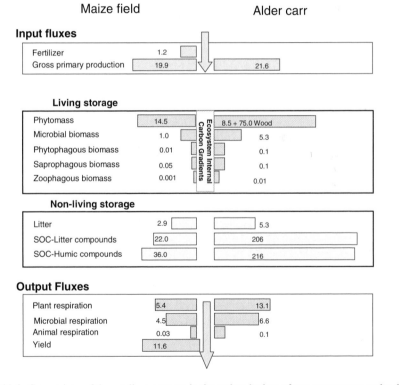

Fig. 11.4 Comparison of the gradient patterns in the carbon budget of two ecosystems under different land use. Carbon flows and storage are presented as: t C ha^{-1} year^{-1}; the figures have been scaled down to the cubic root of the original data. Data compilation by Kappen, Kutsch and Müller (unpublished data)

in the soil and the wood of plants. The incoming energy equivalents are used to build up new structures (raise the existing gradients, e.g. by net primary production) and to provide the necessary energy for the maintenance of existing gradients. The result of these physiological reactions is the system's energy loss by respiration.

The energetic gradients cover a wide range, reaching from $1\,kg\,C\,ha^{-1}$ (invertebrate biomass of the maize field) to a total of $422\,t\,C\,ha^{-1}$ in the soil organic matter (SOC) of the wet alder stand. These different gradients operate on distinct spatio-temporal scales. For example, short-term changes occur especially in the agroeco-system (which is managed on an annual basis) and within the biotic compartments, while long-term storage is found in the wood of alder trees or in the soil organic matter which may have turnover rates of millennia. Generally, these phenomena are dimensionally interrelated. For example, the saprophagous biomass turns out to be an extremely small quantity, while the final product of the pertinent metabolic proc-esses, the soil organic carbon compounds provide for a huge energy storage. Thus, the internal energy flows passing the food web are effectively used to enhance the long-term storage capacity of the whole ecosystem. In contrast to forest ecosys-tems, the efficiency of this process is distinctly reduced in agroecosystems where intensive land use practice reduces storage.

11.2.2.2 Hydrological Gradients

In vegetated areas where the transformation of the available solar energy mostly takes place in the vegetative cover, the major water transfer from soil to air is via the transpiration stream. It involves a number of segments through which the rate of water transport depends on the potential difference causing flow and the resist-ance encountered by the moving water. Water moves from the soil to the root, nor-mally in the liquid phase, through the roots to the conducting vessels of the plant and in them to the leaves where it eventually evaporates (cf. Chapter 5).

Since the water along the whole pathway forms a thermodynamic continuum, the forces involved in its movement and the resistance to its flow in the various segments of the pathway can be analysed in a quantitative way. Water movement from soil to root is initiated by water deficits developed in the plant leaves as a result of evaporation. Through the continuity of water in the transpiration stream, this deficit results in the development of a water-potential gradient which extends progressively through the plant water system and into the soil. The steepness of this gradient depends not only on the magnitude of the evaporation rate at the leaf surface which is dictated by the energy supply there, but also on the resistances to water flow along the transpiration pathway.

As transpiration proceeds and the soil gets drier from day to day, its hydraulic potential (Ψ_{soil}) decreases progressively, and this decrease becomes more and more rapid as the soil-water content is depleted. This has two important effects on the hydraulic potential of the plants (Ψ_{plant}) and hence on internal water deficits. In the first place there is an inevitable decrease in Ψ_{plant} since the degree to which water deficits within plants can be eliminated by absorption at night when transpiration has ceased, is limited to the equilibrium situation $\Psi_{soil} = \Psi_{plant}$. Secondly, as Ψ_{soil} decreases with increasing soil dryness, the hydraulic conductivity of the soil decreases rapidly. The net result is that progressively steeper gradients in water potential between the root and the surrounding soil are required in order to maintain the flow of water from the soil to the root surface. Figure 11.5 illustrates the scale-dependent fluctuations of the volumetric soil-water content of an Eutri-brunic Arenosol (A3) during the 1989–1998 period. A synopsis of the variable climatic situation underlying the fluctuations is provided in Table 2.1.

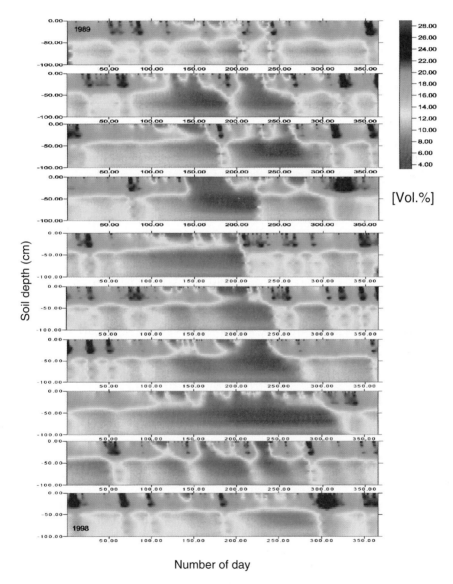

Number of day

Fig. 11.5 Annual and interannual fluctuations of the soil water content (vol%) of crop rotation field A3 during the 1989–1998 period. Data compilation and illustration by G. Hörmann (unpublished data)

Not only in the upper soil layers is the soil water content high, but also below 70 cm the water content increases with depth because the capillary connection to the surface is no longer effective. Another steep gradient can be found at the ploughing depth of 30–40 cm. Here the natural connections between the soil aggregates are mechanically disrupted annually, which brings about a considerable disturbance of the soil moisture regime (Table 11.5).

11.2.2.3 Chemical Gradients

The dynamics of nutrients and other chemical compounds in ecosystems is under the influence of chemical gradients in the ecosystem. Their size and the efficiency of the degrading processes depend on the above hydrological and energetic processes. Chemical compounds are imported by deposition processes and man-made activities. Within the system they are accumulated in certain compartments (e.g. living biomass, litter, different soil horizons), building up chemical gradients. Between the pools, fluxes take place that transfer the compounds to other pools or to adjacent systems. The general features of such processes are described in detail in Chapters 8, 9 and 10 (Table 11.6).

Table 11.5 Some climatic and hydrological gradients described in preceding chapters of this volume

Chapter	Authors	Gradient described
2	Blume et al.	Climatic and hydrological gradients between the investigated ecosystems (Section 2.2.4)
4	Kutsch et al.	Evapotranspiration as a basic element of the ecosystem energy balance (Figs. 4.1, 4.2)
5	Hörmann et al.	Water balance, stomatal conductance, transpiration (Section 5.3)
6	Irmler et al.	Hydrological gradients as basic living conditions for the abundances of animals (Section 6.4)
9	Kluge and Fränzle	Hydrological pathways in a catchment system (Fig. 9.2)
		Vertical groundwater flow patterns (Fig. 9.4)
		Microscale features of water exchange (Fig. 9.8)

Table 11.6 Some chemical gradients described in preceding chapters of this volume

Chapter	Authors	Gradient described
8	Fränzle and Schimming	Element contents in soils (Table 8.4)
		Bioconcentration factors of plants (Figs. 8.12, 8.13)
		Seasonal fluctuations of nutrients in soils (Figs. 8.3, 8.7)
		Average annual element fluxes (Figs. 8.9, 8.10)
9	Kluge and Fränzle	Nitrogen fluxes through ecohydrological shore types (Fig. 9.12)
10	Fränzle and Schernewski	Nutrient distribution, balances, and dynamics in Lake Belau (Figs. 10.9–10.14)

Figure 11.6 shows some chemical gradients in the soils of the beech forest (W1) and agroecosystem A3. In addition to the respective profile parameters (depths and soil horizons), the concentration profiles of Ca, K and N can be found in the figure, whereby the boxes are related to the total nutrient contents in the soil substrates which are due to primary sedimentological inhomogeneities and differences in manuring or fertilization (cf. Sections 2.2.1 and 2.2.3). The two Ca profiles (A) show comparable contents in the BC horizons, but the distribution above this depth is very different. While the forest soil displays a slight gradient in the upper 20 cm, the field soil shows a decreasing concentration beneath the A- and B-horizons. This pattern can be attributed to liming measures which are generally responsible for the higher Ca content of the field soil. The low Ca content between the A and the BC horizons can be attributed to plant uptake, decalcification and – predominantly – to the effects of forest soil acidification. Looking at the Ca fluxes, a slight parallelism with the Ca content becomes obvious, including opposite gradients in the A horizon. In spite of these differences, the seepage outputs (150 cm) from both ecosystems are nearly the same.

Potassium shows different distribution patterns. In the Arenic Umbrisol of the beech forest a gradient is largely lacking, except for the uppermost subhorizon of the mineral soil. As the forest plants seem to be in a critical situation with regard to K supply (cf. Chapter 8); therefore this element is recycled in the biotic compartments of the ecosystem very intensely and efficiently. The root uptake seems to be so high that the K concentrations in soil solution decrease rapidly, already in the A horizon. Thus, there is practically no gradient below 2 cm depth, and the ecosystem output by leaching is very small. The Eutri-brunic Arenosols of field A3 exhibits both higher contents and flows of potassium because this element is supplied by regular fertilization. Due to this high degree of K provision and saturation, the outputs are relatively high, and there are no significant gradients of the K concentration in the soil solution.

Analysing the vertical distribution of the nitrogen content indicates a vertical gradient at both sites. Correlating with the soil organic matter content, there are very high amounts of N in the upper soil and rather small fractions below 1 m. Also the fluxes reflect this tendency in the forest. This applies much more to NH_4^+ due to its marked dependence on microbial activity, while the nitrate gradient is less accentuated because of the high mobility of this anion.

In Fig. 11.6 the concentrations of exchangeable Ca and K in the soil column define vertical gradients which are strictly biota-controlled in the case of K, while the Ca gradient is a reflection of both sedimentary and biotic differentiation. In the sense of a summary the lower right inset illustrates the Ca and K contents of the soil column and the corresponding amounts of the exchangeable proportions of these elements.

Some additional structural indications of the nutrient status are given in Fig. 11.7 on the ecosystem level. Interpreting these illustrations, two points have to be taken into account: (a) the differences between the beech forest and the agroecosystem have developed throughout the past century, because in earlier times both sites had been under tillage; (b) The alder stand (W6) is characterized by large peat reserves

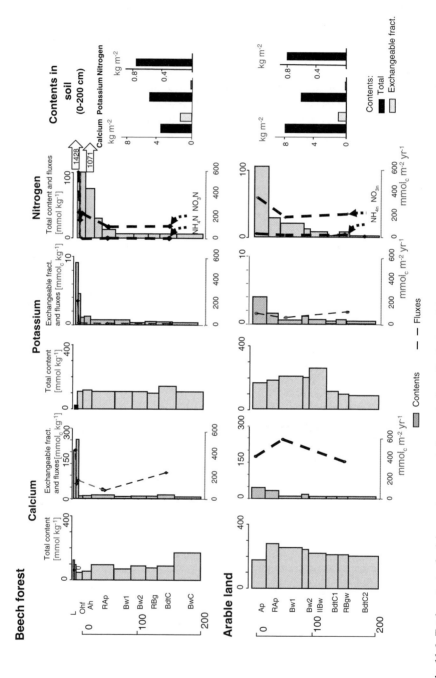

Fig. 11.6 Total contents of calcium, potassium and nitrogen in the soil columns of the beech forest and the agrarian ecosystems and related vertical flux rates of these elements. Data by C.G. Schimming (personal communication) and Wetzel (1998)

on the one hand, and by high soil humidity and high groundwater levels on the other, and by the capacity of the *Alnus–Frankia* symbiosis for nitrogen fixation.

Figure 11.7 illustrates the existing stocks of carbon and nitrogen compounds of three ecosystems which can be taken as indicators for the size of existing gradients and which illustrate the storage capacities of these ecosystems. They differ between beech forest and field by a factor of three concerning the total carbon contents; with regard to the carbon fraction which is bound in living compartments, the difference increases to a factor of nearly six. Thus, the carbon gradients have been developing in very different directions due to the land use change about 100 years ago. The total nitrogen storage does not show comparable differences, but this is due to high fertilizer amounts and high nitrogen exports from the field (cf. Chapter 8). Consequently, the biotic compartments of the forest play a more significant role as nitrogen stores and processors than is the case on arable land. The alder carr accumulates carbon and nitrogen mostly in the soil compartment (see Chapter 8 for a discussion of the dynamics of this function); therefore, the relative importance of the biotic stores is smaller than those of the beech forest.

These ecochemical conditions are coupled with quite distinct structural qualities as can be seen in the lower part of Fig. 11.7. Here, the numbers of plant and soil fauna species of the three ecosystems are compared. The flora attains its highest diversity in the beech forest, and about twice as many soil fauna species as in the field soils are found, too. In contrast, in the alder carr, only 30 plant species are

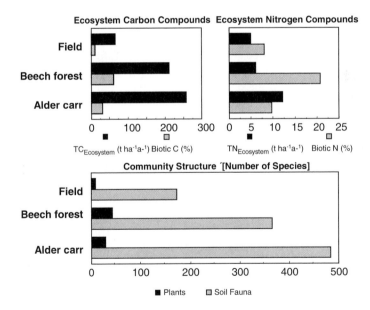

Fig. 11.7 Compilation of ecochemical and community-ecological characteristics of three terrestrial ecosystems in the Bornhöved Lake District (maize field A3, beech forest W2, alder carr W6). Data after Kutsch et al. (1998, 2001)

found (beech forest: 44), while the soil fauna exhibits the highest diversity of all systems studied. This may be caused by the high diversity of niches, which can be found in this nutrient-rich zone with many different humidity conditions and ecotonal situations.

In the light of the above findings the following (preliminary) conclusions can be drawn:

- It is possible to describe empirical results from ecosystem research in various contexts as ecological gradients if the data are understood as concentration profiles in an observer-defined environment.
- Structural gradients are referring to spatial reference systems. They describe the distribution of ecological variables in terms of metric vectors.
- Functional gradients are referring to ecophysiological reference systems. They are related to ecological storage mechanisms and the interactions between them, i.e. ecological flux phenomena. The reference scale can be defined by the dominating flow directions between imports from the environment and exports to neighbouring systems.
- Both structural and functional gradients operate at different spatio-temporal scales, providing constraints for the gradient assemblages on lower hierarchical levels.

11.3 Gradients as Elements of an Integrative Ecosystem Theory

In general, it can be postulated that throughout an undisturbed development ecosystem complexity (or the number of interrelations between ecological gradients) will increase asymptotically to the state of maturity (Odum 1969). Within this development, exergy storage will keep rising (Fig. 11.8) and simultaneously more and more gradients are built up. With this increasing structural diversity also the diversity of flows and the system's ascendency (Ulanowicz 2000) will grow as well as certain network features (Fath and Patten 2000), and therefore also the energy necessary for the maintenance of the developing system must grow. These basic thermodynamic requirements have many consequences for other ecosystem features. For instance, the food web will become more and more complex, heterogeneity, species richness and connectivity will be enhanced (see Fig. 11.4), and many other attributes shown in Fig. 11.8 will follow a similar long-term trajectory.

This deduction is a theoretical principle which can hardly be found in reality due to the continuous effect of disturbances. Especially in the case of high external inputs, the orientor values might decrease rapidly, i.e. in a retrogressive manner. In the following sequence, an adaptive or resilient system will find the optimization trajectory again; while a heavily disturbed ecosystem might no longer be able to improve the orientor values (cf. also Chapter 12). Therefore, the orientors can indicate the robustness of ecosystems as well. Consequently, their values are also suitable to represent

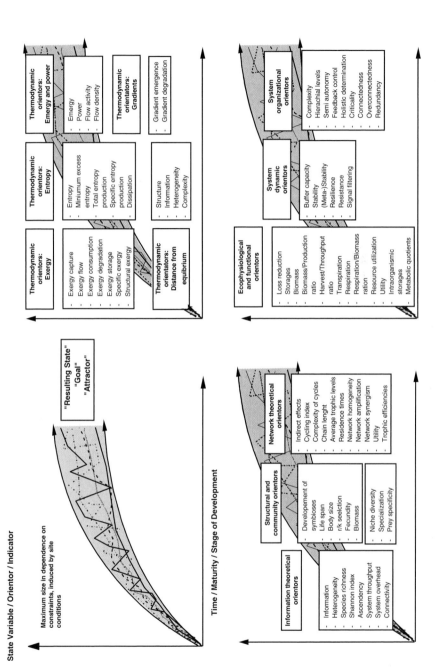

Fig. 11.8 Ecological orientors in the sense of Bossel (2000). The ecosystem properties listed exhibit an optimization during the long-term development in undisturbed situations in the sense of Müller and Fath (1998) and Müller and Jørgensen (2000)

the ecological risk, which is correlated to external inputs or changes of the prevailing boundary conditions. High orientor values, however, do not guarantee high stability or high buffer capacity. Following Holling's ideas on ecosystem resilience and development, in the mature stage complex ecosystems become "brittle", their adaptability decreases because of the high internal connectivity and the respective interrelationships. Thus, the dynamics of external variables can force the mature system to break down and then start with another developmental sequence.

11.4   Conclusions

The present Chapter outlines a concept to integrate basic approaches and results of ecosystem research on a unified basis. Different types of ecological gradients are described, illustrated and finally inserted into the theoretical framework of self-organizing processes. In parallel, there are cross-references to the preceding chapters of this volume where further examples can be found. The basic hypothesis is that all living systems transform gradients, building up an interior hierarchy of nested gradients. They provide potentials for degradation processes, and thus gradients are prerequisites for all flows and transfer processes in ecological systems. In this perspective, the gradient principle provides models which combine structural, functional and organizational approaches in ecosystem analysis. Gradients are important parameters of self-organizing processes, as they are the potentials ("driving forces") as well as the final results of self-organization.

Part III
From Research to Application

Chapter 12
An Indicator-Based Characterization of the Bornhöved Key Ecosystems

Joachim Schrautzer, Felix Müller, Hans-Peter Blume, Uwe Heinrich,
Ernst-Walter Reiche†, Uwe Schleuß, and Klaus Dierssen

12.1 Introduction

The subject-matter of the present Chapter is a hierarchical ecosystem classification of the study area which, in combination with a large-scale map, amalgamates the results of pedological, ecophysiological, phytosociological and geozoological inquiries presented in Chapters 2–7 with the budgeting approaches and ecological stoichiometric analyses described in Chapters 8–10. To this end the orientor approach (Bossel 1998; Müller and Jørgensen 2000; Section 11.3) is applied which considers retrogressive successions of different ecosystem types in order to characterize ecosystem integrity on various spatial levels, which is as important for theoretical considerations in the framework of ecosystem analysis and modelling as it is for practical landscape and land planning use or management purposes (cf. Chapter 13). This target requires a methodology to scale-up findings from the site to the landscape level on the one hand and from short to long-term effects on the other. Thus, the following questions will be discussed:

- Can the ecosystem indicators presented in Chapter 11 be used on the landscape level?
- How do the indicators reflect different degrees of human impact?
- Are there significant interrelations between the landscape-level results and the gradient approach?

In the preceding chapters two significant impacts on the landscape dynamics of the Bornhöved Lake District during the past 30 years have been analysed, namely eutrophication and wetland drainage. Their intensity can be classified by means of the hemeroby approach (Blume and Sukopp 1976) which has been widely applied in Central Europe (Kowarik 1999). Accordingly four levels of increasing human impact can be distinguished, namely oligo-, meso-, eu- and polyhemerobic systems which are associated with different degrees of productivity, structural and functional differentiation. Focusing on the landscape level, the pertinent questions then are:

- Which are the interrelated consequences of such impacts for ecosystem and landscape structures and functions?

O. Fränzle et al. (eds.), *Ecosystem Organization of a Complex Landscape.*
Ecological Studies 202.
© Springer-Verlag Berlin Heidelberg 2008

- How do these impacts influence the nutrient balance of the landscape, and how can the respective changes be assessed by selected functional indicators?
- Which measures can be recommended for sustainable environmental management?

To discuss these issues, the multidimensionally defined ecosystems of the study area are analysed along a gradient of land use intensity (or hemerobic stages), which reflects successional sequences. In the next step, different ecosystem types are combined with the results of a water and substance simulation model operating on the landscape level. The modelling results are then connected with selected indicative output parameters in order to characterize different ecosystem states. On this basis, the ecological integrity of ecosystems can be evaluated, using a multi-parameter amoeba visualization. Finally, nutrient balances for the whole study area are presented and discussed within the scope of sustainable landscape management.

12.2 Methodology

The focal ecosystem types are classified and characterized from a structural point-of-view. The resulting regional data pool is used to indicate the ecosystem states on an integrative level, using a GIS model-coupling approach.

12.2.1 Ecosystem Classification

The study area as depicted in Fig. 12.1. comprises the lowlands around Lake Belau with Histosols and Gleysols and the surrounding uplands with different mineral soils without groundwater influence. In both areas, the land use intensity varies considerably, which involves a set of hemerobic states and succession series. On Histosols and Humic Gleysols the retrogressive successional sequence starts with undrained alder carrs which developed under natural conditions from *Phragmites* reeds and ends with intensely drained grasslands, whereas the development on mineral soils without groundwater contact goes from beech forests to fields. The ecosystem types of these sequences cover 88% of the study area and also larger acreages of Schleswig–Holstein (Dierssen 2004a; Schrautzer 2004).

In greater detail the characteristics of the above ecosystem types are displayed in Table 12.1.

12.2.2 Ecosystem Structure and Diversity

Without human interference, the developmental potential of most lowland biotope complexes in Central Europe would exhibit a propensity to forest systems, reflecting the 'potential natural vegetation' and 'natural' potential of such sites (Bohn et al. 2003).

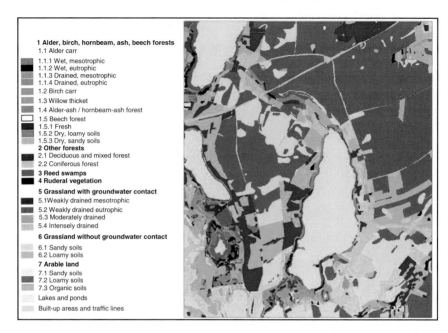

1 Alder, birch, hornbeam, ash, beech forests
1.1 Alder carr
  1.1.1 Wet, mesotrophic
  1.1.2 Wet, eutrophic
  1.1.3 Drained, mesotrophic
  1.1.4 Drained, eutrophic
1.2 Birch carr
1.3 Willow thicket
1.4 Alder-ash / hornbeam-ash forest
1.5 Beech forest
  1.5.1 Fresh
  1.5.2 Dry, loamy soils
  1.5.3 Dry, sandy soils
2 Other forests
2.1 Deciduous and mixed forest
2.2 Coniferous forest
3 Reed swamps
4 Ruderal vegetation
5 Grassland with groundwater contact
5.1 Weakly drained mesotrophic
5.2 Weakly drained eutrophic
5.3 Moderately drained
5.4 Intensely drained
6 Grassland without groundwater contact
6.1 Sandy soils
6.2 Loamy soils
7 Arable land
7.1 Sandy soils
7.2 Loamy soils
7.3 Organic soils
Lakes and ponds
Built-up areas and traffic lines

Fig. 12.1 Ecosystem types of the Bornhöved Lake District

The increasing human impact alters the composition of the actual species pool and the abiotic functions of a site, involving structural and functional changes of primary producers, food webs and their qualitative and quantitative composition and diversity.

In order to analyse the interrelations between species richness, hemerobic levels and succession stages, the mean species number based upon a data set of about 1000 relevés was calculated for each site type. The size of these relevés varies between $25\,m^2$ (herbaceous vegetation) and $100\,m^2$ (forests). Species richness is not closely related to hemerobic or succession stages; therefore, qualitative aspects like life history and functional traits or influences of hemerophilous species are utilized (Kleyer 1999; Dupré and Diekmann 2001; Dierssen 2004b; Kahmen and Poschlod 2004). Thus, the distribution of life strategy types according to Grime (2001) can be calculated for the ecosystem types defined.

12.2.3 Water and Nutrient Budgets

Within the framework of the present budgeting approaches the functional variables were calculated by means of the hierarchical model system WASMOD. This describes processes of the water, nitrogen and carbon budgets at different temporal (from days upwards) and spatial (sites, catchments, landscape units) scales. The model was validated for mineral soils (Reiche 1994; Weiss 2000) and Histosols (Trepel 2000).

Table 12.1 Ecosystem types of the Bornhöved Lake District (cf. Fig. 12.1)

Ecosystem type	Criteria		Soil patterns		Mean ground water level (cm)	Soil potentials		
	Vegetation[a]	Soil variation[b] / Soil unit	Principal type	Inclusions		Available water supply (l m^{-2})	C/N[d] (mass based)	pH[e]
Forests								
Alder carr								
Wet mesotrophic	ALS	Fibric-ombric Histosol	fi-om HS dy (eu)	dy GL s	20 → 0	Unlimited	17–25	3.5–4.5
Wet eutrophic	ALT	Hemi-ombric Histosol	hm-om HS eu	eu GL s (l)	20 → 0	Unlimited	14–17	4.5–5.5
Drained mesotrophic	AGL	Hemi-ombric Histosol	sa-om HS dy	dy GL, hu GL s (l), gl-cu AT	80–35	Unlimited	14–17	<3.5
Drained eutrophic	AGP	Sapric-ombric Histosol, humic Gleysol	sa-om HS dy (eu), hu GL eu	eu GL 1 (s)	50–20	Unlimited	12–14	3.5–4.5
Birch carr	BPT	Hemi-ombric Histosol	hm-om HS	dy GL s	20 → 0	Unlimited	21–27	3.4–3.7
Willow thicket	SC	Hemi-ombric Histosol	hm-om HS dy	hi GL s, hu GL s	n.t.	Unlimited	12–25	3.5–6.5
Alder-ash forest	AFRAX	Sapric-ombric Histosol, Humic Gleysol	sa-om HS hu GL s	eu GL s	40–5	Unlimited	12–14	5.0–6.0
Hornbeam-ash forest	CFRAX	Humic Gleysol	hu GL	hu Gl	105–55	150–250	11–13	4.5–6.0
Beech forest								
	AFC	Dystric Luvisol	dy Lu	dy ST, dy GL	>400	90–150	10–17	3.5–4.5
	AFT	Arenic Umbrisol, Arenic Gleysol (relic)	ar UM, dy-ar GL	cu AT s, hu GL s	>400	50–90, 90–140 (Gleysol)	10–17	3.5–4.5
	AFTD	Dystric Arenosol, Cumulic Anthrosol	dy AR, cu AT s	dy LV, ar UM	>400	50–90, 90–140 (Anthrosol)	14–20	<3.5
Forests (others)								
Deciduous and mixed forest		Cambic Arenosol, Dystric Luvisol, Ombric Histosol	cb AR, dy LV, om HS, cb AR, dy LV, om HS	cu AT, hi GL, hu GL, cu AT	>400	90–250, unlimited (Histosol)	12–30	<3.5–7.5
Coniferous forest								
Others								
Reed swamps and tall sedge reeds	P, MC	Limnic Histosol	lm HS	gt-lm FL, tip-lm FL	5 → 0	Unlimited	14–20	3.5–4.5

Ecosystem	Vegetation[a]	Principal soil type[b]	Soil variation[b]				[d]	pH[e]
Ruderal edges	ARB, TAR, UAE	Loamy and sandy Gleysol	dy Gl, eu Gl s (l)	dy GL, eu GL l (s)		90–140	n.t	3.5–0.5
Grassland with groundwater contact								
Weakly drained Mesotrophic	Cmes	Hemi-ombric Histosol	hm-om HS dy (eu)		20–5	Unlimited	17–20	3.5–5.5
Weakly drained eutrophic	Ceu	Hemi-ombric Histosol	hm-om HS eu	eu GL, ar GL, hu GL	20–5	Unlimited	14–17	4.5–6.5
Moderately drained	RAT	Sapri-ombric Histosol	sa-om HS	ar GL, hu GL	35–20	Unlimited	12–14	4.5–6.5
Intensely drained	RAL, LCA	Histosol, Humic Gleysol	sa-om HS, hu GL	gl-cu AT, eu GL	80–35	Unlimited 90–140 (Gleysol)	10–12	4.5–6.5
Grassland without groundwater contact								
	LCT	Cambic Arensol, Cumulic Anthrosol	cb AR s, cu AT s, eu LV, CM	la AR eu	>400	90–150	10–17	4.5–5.5
	LCT	Luvisol, Cambisol	eu LV, CM	cu AT	>400	90–250	10–17	4.5–5.5
Arable land	LCT	Cambic Arenosol, Cumulic Anthrosol	cb AR s, cu AT s	la AR	>400	90–250	10–17	5.5–6.5
		Luvisol, Cambisol	eu LV, CM	cu AT, eu GL l	>400	90–250	10–17	5.5–6.5
		Sapri-ombric Histosol	sa-om HS	hi GL, eu	n.t.	200–270	n.t.	n.t.

[a] Vegetation types within ecosystems: *ALS Carici elongatae–Alnetum sphagnetosum, ALT Carici elongatae–Alnetum typicum, AGL Alnus glutinosa*-community (subunit of *Lonicera periclymenum*), *AGP Alnus glutinosa* community (subunit of *Poa trivialis*), *BPT Betula pubescens* community (typical subunit), *SC Salicetum cinereae, AFRAX Fraxino–Alnetum, AFC Asperulo–Fagetum* (subassociation of *Circaea lutetiana*), *AFT Asperulo–Fagetum typicum, P Scirpo–Phragmitetum MC Magnocaricion* communities, *ARB Artemisietea* basic community, *TAR Tanaceto–Artemisietum, UAE Urtico–Aegopodietum, Cmes* mesotrophic Calthion communities, *Ceu* eutrophic Calthion communities, *RAT Ranunculo–Alopecuretum typicum, RAL Ranunculo–Alopecuretum* (subunit of *Lolium perenne*), *LCA Lolio–Cynosuretum* (subunit of *Agrostis stolonifera*), *LCT Lolio–Cynosuretum typicum*

[b] Soil variation: (i) principal soil type: *AR* Arenosol, *AT* Anthrosol, *CM* Cambisol, *FL* Fluvisol, *GL* Gleysol, *HS* Histosol, *LV* Luvisol, *PZ* Podzol, *RG* Regosol, *ST* Stagnosol, *UM* Umbrisol, *AFT, CFRAX*; (ii) subunits: *ar* arenic, *ca* calcaric, cambic, *cu* cumulic, *dy* dystric, *eu* eutric, *fi* fibric, *hi* histic, *hm* hemic, *hu* humic, *lm* limnic, *lu* luvic, *om* ombric, *sa* sapric, *sk* skeletic, *sd* spodic, *st* stagnic, *tip* protothionic, *l* loamy, *s* sandy.

[c] For methods see Schlichting et al. (1995).

[d] For Ah and Ap horizons.

[e] Top soil values.

The outputs of the model are arranged in accordance with the indicators employed to evaluate different functions of the systems (Table 12.2). WASMOD is GIS-supported and combines information of a digital landscape analysis with dynamic modelling of water, nitrogen and carbon fluxes (Fig. 12.2). The data sources available for landscape analysis are soil maps based upon a state-wide soil evaluation (Cordsen 1993), land use maps, and a digital elevation model. The data are calibrated according to location and pedological characteristics with data from 1600 soil profiles of the whole study area (Schleuß 1992). Subsequently, in each mapping unit of WASMOD a 30-year simulation run of water and nutrient fluxes was carried out. To transfer these results to the classified ecosystem types, the mapping units of the WASMOD approach and the map of ecosystem types (Fig. 12.1) were combined. Nutrient and water balances of the entire study area were then calculated by multiplying the WASMOD results with the acreages of all the ecosystem types defined.

Table 12.2 Indicanda and indicators used in the landscape study

Indicandum	Indicator(s)
Structural gradients – biota	Number of plant species, hemerobic stages, life strategy types
Energetic gradient – exergy capture	Net primary production (NPP)
Energetic gradients – entropy production	Microbial soil respiration
Energetic gradients – metabolic efficiency	NPP/soil respiration
Hydrological gradients – biotic water flows	Evapotranspiration/transpiration
Chemical gradients – nutrient losses	Nitrogen net mineralization, nitrate leaching, denitrification
Chemical gradients – storage capacity	nitrogen balance, carbon balance

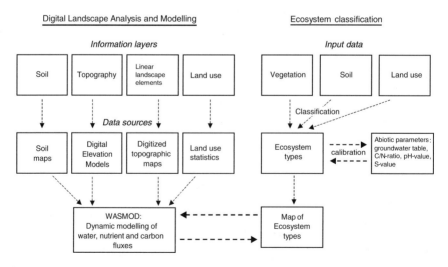

Fig. 12.2 Methodology of digital landscape analysis and ecosystem classification

12.2.4 Indication of Ecosystem Integrity

Ecosystem intregrity can be considered the result of evolutionary optimization processes relating to exergy capture, energy and nutrient storage, reduction of nutrient losses, cycling processes and flow densities, functional efficiency, respiration, transpiration, and site structure. In light of the above, water and nutrient budgets ecosystem integrity can then be defined multidimensionally by a set of indicator variables (Müller et al. 2000; Barkmann et al. 2001; Baumann 2001), which are summarized with the corresponding indicanda in Table 12.2.

12.3 Characterization of Ecosystem Types

The genesis and structure of the study area are described in Chapter 2 of this volume. The comparative analysis of the vegetation cover has provided information about some 500 vascular plants. Although 66% of the species are considered to be idiochorophytes, most of them are hemerophilous and thus are largely missing in oligo- and mesohemerobic sites. Many taxa of the regional species pool prevail in mesic and wet grassland (*Molinio–Arrhenatheretea*, 16.3%), crop fields (*Stellarietea mediae*, 12.8%) and mesic deciduous forests (*Querco–Fagetea*, 12.1%) (Dierssen 2004 a). In the following Section, the basic vegetation features and soil characteristics of the ecosystem types are described.

12.3.1 Successional Series on Histosols

In the wet alder carrs oligo- and mesohemerobic species dominate, whereas indicators for eu- and polyhemerobic sites are missing (Fig. 12.3). The sites are water-saturated for most of the year (mean groundwater table between 20 cm below surface and 0 cm; cf. Table 12.1). According to plant species composition and site conditions the wet alder carrs can be divided into mesotrophic dystric and eutrophic eutric types.

In the drained alder carr, oligohemerobic species decrease while euhemerobic ones increase compared with the previous retrogression steps (Fig. 12.4). The number of vascular plants decreases as well. The mean groundwater table is significantly lower than that of the wet alder carr (between 80 cm and 20 cm below surface; cf. Table 12.1). The drained alder carr can further be subdivided into mesotrophic, dystric and eutrophic subtypes which also differ hydrologically since the sites of the eutrophic alder carr are usually wetter than those of the mesotrophic subtypes (cf. Table 12.1).

Reed swamps characterize the wettest sites investigated (mean groundwater table between 5 cm below surface and soil surface). They originated partly from alder carrs due to deforestation which was coupled with high sedimentation and mineralization rates.

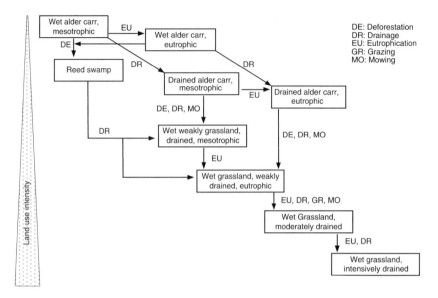

Fig. 12.3 Retrogressive succession sequences of Histosol ecosystems

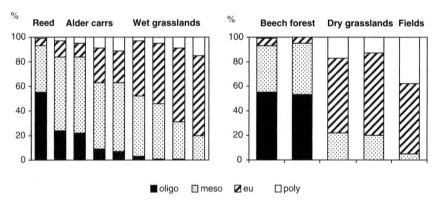

Fig. 12.4 Proportion (percent coverage) of oligo-, meso-, eu- and polyhemerobic species in the ecosystems of the study area

Deforestation, drainage and eutrophication with concomitant mowing or grazing of the systems result in the development of wet grasslands. Oligo- and mesohemerobic vascular plants decrease whereas eu- and polyhemerobic species increase. The mean groundwater table of weakly drained grasslands fluctuates between 20 cm and 5 cm below ground. With regard to the trophic status of the sites mesotrophic grasslands can be distinguished from eutrophic wet ones which differ according to the C/N ratios of their soils (Table 12.1).

Moderately and intensly drained grasslands cover most of the Histosols of the study area (Table 12.1). They originated from primarily weakly drained systems after introduction of intensive drainage measures, which resulted in peat minerali- zation and eutrophication. The hydrological characteristics of these systems differ with regard to situation and temporal fluctuations of the groundwater level (Scholle and Schrautzer 1993).

12.3.2 Successional Series on Mineral Soils

Most of the beech forests of the Bornhöved Lake District originated from crop rota- tion fields (cf. Section 2.2.6) and are characterized by a predominance of oligo- and mesohemerobic plant species (Fig. 12.4). According to species composition and soil units two types are distinguished. On loamy parent material Stagnic and Haplic Luvisols predominate, whereas on sandy material Brunic Arenosols and Arenic Umbrisols occur which have a lower water holding capacity, higher C/N ratios and lower pH values than Luvisols (Table 12.1). Under tillage loamy and sandy units are associated with Colluvic Anthrosols on footslopes and above hedgerows.

Most grasslands on mineral soils are used as pastures with high fertilization rates and high livestock densities (Table 12.1). By analogy with the differentiation of the beech forest types, grasslands without groundwater contact are subdivided into types on loamy and sandy soils. The upper soil horizons of intensively managed grasslands have higher pH values than the beech forest soils (Table 12.1).

High-yield agroecosystems predominate in the study area. The species richness and density of weeds is extremely low (commonly <10 species per $400\,m^2$); ubiq- uitous generalists characterizing eu- and polyhemerobic sites prevail. According to site conditions a distinction between fields on loamy soils and fields on sandy soils is indicated (Table 12.1). On arable land pH values are higher than in forests or grasslands on mineral soils due to fertilization (Table 12.1).

For illustration purposes Fig. 12.5 summarizes pertinent succession sequences on mineral soils.

12.4 Discussion

12.4.1 Patterns of Plant-Species Richness

Among the systems on Histosols reed swamps exhibit the lowest species richness. Most plant species are CS-strategists (Figs. 12.6, 12.7), i.e. perennials with the capacity for lateral vegetative spread by means of rhizomes (e.g. *Phragmites australis*, *Carex acutiformis*). The highest species richness occurs in mesotrophic wet alder carrs and mesotrophic wet grasslands (Fig. 12.6). During the course of

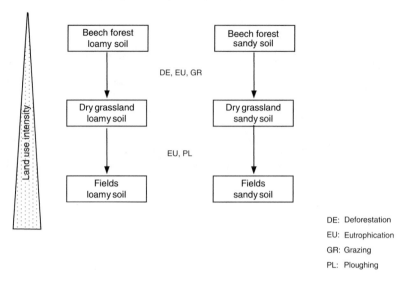

DE: Deforestation
EU: Eutrophication
GR: Grazing
PL: Ploughing

Fig. 12.5 Retrogressive succession sequences of mineral soil ecosystems

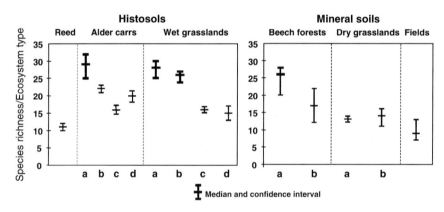

Fig. 12.6 Species richness of the Bornhöved ecosystems. Subunit codes: *Alder carrs*: *a* wet, mesotrophic, *b* wet, eutrophic, *c* drained, mesotrophic, *d* drained, eutrophic; *Wet grasslands*: *a* weakly drained, mesotrophic, *b* weakly drained, eutrophic, *c* moderately drained, *d* intensively drained; *Beech forests*: *a* loamy soils, *b* sandy soils; *Dry grasslands*: *a* loamy soils, *b* sandy soils; *Fields*, arable land (not differentiated in this figure): *a* loamy soils, *b* sandy soils. Size of reference plot is 25 m² in reef stands, grassland and fields, 100 m² in forests and alder carrs

secondary successions the species richness of these systems decreases continuously. The increased human impact affects the distribution of different functional types of plants within the systems, i.e. competitors (C-strategists) decrease and CSR-strategists increase (Fig. 12.7). CSR-strategists are typical for habitats in which competition, stress and disturbance vary temporarily or are constant features of the

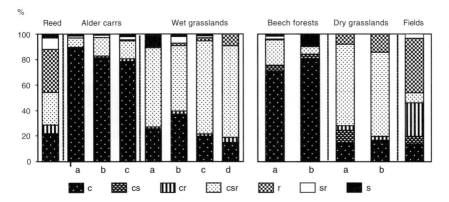

Fig. 12.7 Distribution of functional types of the Bornhöved ecosystems (adapted from Grime 2001). CSR codes: *C* competitors, *R* ruderals, *S* stress-tolerators. *a*, *b*, *c*, *d* as in Fig. 12.6

habitat (Grime 2001). The highest rates of CSR-strategists can be found in moderately and intensively mowed and grazed wet grasslands. In these ecosystems the vegetation is affected by high external (fertilization) and internal (mineralization) nutrient availability promoting competition, by temporary stress situations (oxygen deficiency in the root zone caused by flooding after heavy rainfall; Schrautzer and Trepel 1997), and by major disturbance due to intensive grazing. The dominant CSR-strategists of moderately and intensively drained wet grasslands (*Agrostis stolonifera*, *Alopecurus geniculatus*, *Glyceria fluitans*, *Ranunculus repens*) are well adapted to such conditions (Crawford 1996). During the succession from weakly drained mesotrophic to intensively drained wet grasslands the proportion of S-strategist decreases and that of CR- and R-strategists increases. Representative species of S-strategists are, e.g. *Carex echinata* and *Carex panicea*, which get on well with the low nutrient availability of mesotrophic wet grasslands because of the high longevity of their leaves and their elevated growth rates. The increase in CR- and R-strategists in intensively drained wet grassland is caused by enhanced disturbance of the habitats due to grazing.

A decrease in species richness can also be observed during secondary successions on mineral soils as shown above. Species richness, however, is not only influenced by human impact but also by site conditions. Thus, in beech forests on loamy soils with a fairly high base saturation more species exist than in beech forests on acid sandy soils (Härdtle 1995). One explanation of this phenomenon is the low nutrient availability in the upper horizons of forests soils producing stress for the vegetation of the herb layer (cf. Chapter 8). Only a few species are adapted to these conditions, e.g. the S-strategists *Deschampsia flexuosa*, *Carex pilulifera* and *Luzula campestris*. The transition from beech forest to dry grasslands is characterized by a decrease in C-strategists and an increase in CSR-strategists which are mostly typical grassland species with wide ecological amplitudes. In the field systems most species are R- and CR-strategists.

12.4.2   Comparative Carbon Budgets of Ecosystems and their Successional Phases

On Histosols the simulated net primary production (NPP) is high in near-natural reed swamps and alder carrs (cf. Fig. 12.8). Simulation results for alder stands show that a drainage of these systems leads to a weak increase in NPP. Kutsch et al. (2001c) presented pertinent NPP data of the eutrophic drained alder carr. Based upon two-year measurements, the authors calculated a mean NPP of 8430 kg C ha^{-1}

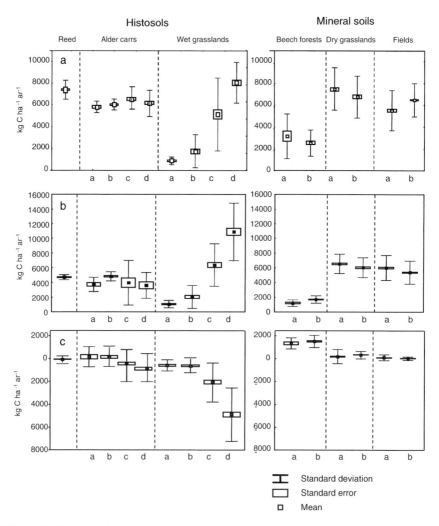

Fig. 12.8 Carbon budgets of the Bornhöved ecosystem types. **a** Net primary production, **b** microbial soil respiration, **c** C balance. *a, b, c, d* as in Fig. 12.6

year^{-1} for this stand (cf. Chapter 4), which is about 10–20% beyond the WASMOD-modelled NPP range. One possible reason for this difference is the specific location of the narrow alder strip on the banks of Lake Belau, which may cause light conditions different from those of more extensive closed stands.

NPP is low in weakly drained wet grasslands on Histosols and increases on intensely drained sites, which is in accordance with the results of phytomass investigations from other areas (Schrautzer 2004) and can be explained by external (fertilization, nutrient input via groundwater or surface water) and/or internal (accelerated mineralization due to drainage) nutrient accumulation (Grootjans et al. 1985; Koerselman and Verhoeven 1995). Kutsch et al. (2001c) measured a mean NPP of 6400 kg C ha^{-1} year^{-1} on a moderately drained wet grassland which is within the range of the simulated data (cf. Chapter 4).

On mineral soils an increase in NPP during secondary successions occurs. The extensively used mesic grassland systems exhibit a low productivity, and the beech forests have an even lower NPP rate (ca. 2000–4000 kg C ha^{-1} year^{-1}; Kutsch et al. 2001c). The simulated NPP of the intensely used mesic grasslands varies in the range 6000–9000 kg C ha^{-1} year^{-1}. Sach (1999) collected data on this ecosystem type which vary between 3500 kg and 7000 kg C ha^{-1}year^{-1}. On arable land, besides climatic variables, NPP primarily depends on fertilizer input and crop selection. The simulated NPP varies between 5000 kg and 7000 kg C ha^{-1} year^{-1} and represents the typical range of high-yield cereal production. The highest NPP of this agrosystem type can be expected for wheat (10 000 kg C ha^{-1} year^{-1}) or maize cultures (12 700 kg C ha^{-1} year^{-1} after Kutsch et al. 2001c).

In both successional series the simulated rates of microbial soil respiration (MSR) are in the same order of magnitude as the results of the NPP (Fig. 12.8; cf. Section 3.4). Furthermore, the simulated and the measured data presented by Kutsch et al. (2001c) are in good agreement. The simulated MSR of different alder carr stands varies between 4000 kg and 5000 kg C ha^{-1} year^{-1}, whereas Kutsch et al. (2001c) determined an average MSR value of 4780 kg C ha^{-1} year^{-1} for the alder carr as a result of two-year investigations. Finally, the measured MSR values of moderately drained wet grasslands (8350 kg C ha^{-1} year^{-1}), beech forests (3070 kg C ha^{-1} year^{-1}) and the maize field (5090 kg C ha^{-1} year^{-1}) are also in the range of the simulated data.

The storage capacity of ecosystems can be assessed by their C balances. Growing fens on sites with a high groundwater table throughout the year act as carbon sinks due to peat formation (Dierssen and Dierssen 2001). According to the simulation results of the present study, the C balances of reed swamps and wet alder carrs vary between weakly negative and weakly positive values (Fig. 12.8). Kutsch et al. (2001c) determined a positive C balance of 3560 kg C ha^{-1} year^{-1} of an alder carr (cf. Section 4.2.4.). However, this amount was caused by timber increment and not by peat accumulation. The simulation results reveal that drained alder carrs as well as wet grasslands have negative C balances. Furthermore, in wet grasslands a continuous increase of the system's function as a carbon source can be observed with increasing drainage intensity, caused by accelerated peat decomposition (Fig. 12.8). The simulated C balances of the ecosystems on mineral soils are positive or well balanced. According to both modelled and measured data the C balance is most positive in mature beech stands (cf. Chapter 4).

12.4.3   Comparative Nitrogen Budgets of Ecosystems and their Successional Phases

The net nitrogen mineralization (NNM) can be used as an indicator for nitrogen surplus of the systems, because a high N supply often cannot be matched by a correspondingly increased plant uptake. Regarding the ecosystems on Histosols, low NNMs were simulated for reed swamps, wet alder carrs and weakly drained wet grasslands (Fig. 12.9). The modelled values for these systems correspond with pertinent measurements of various authors, e.g. Janiesch (1981) and Döring-Mederake (1991). Moreover, Grootjans (1985) and Schwartze (1992) determined NNM rates of weakly drained wet grassland which are comparable with the simulations of the present enquiries. As shown in Fig. 12.9, drainage leads to a continuous increase of NNM in alder carrs (up to more than 200 kg N ha^{-1} year^{-1}) as well as in wet grasslands (up to 350 kg N ha^{-1} year^{-1}). Measured NNM rates of drained ecosystems on Histosols can be considerably higher (Wiebe 1998; Sach 1999). These differences can be presumably attributed to biased measurements since systematic comparisons revealed that NNM rates determined by the polyethylene bag method (applied by Janiesch 1981 or Sach 1999) may yielded twice the rate found by means of PVC tubes (Münchmeyer et al. 1998; Schrautzer 2004).

Among the ecosystems on mineral soils, the beech forests have the lowest modelled NNM (Fig. 12.9). Considerably higher NNM rates between 100 kg and 250 kg N ha^{-1} year^{-1} result for intensely used mesic grasslands (fertilization: 130 kg N ha^{-1} year^{-1}). Sach (1999) measured NNM rates of these systems (fertilization: 100 kg N ha^{-1} year^{-1}) which vary between 102 kg and 205 kg N ha^{-1} year^{-1}. Relatively low NNM rates were modelled for fields, which can be explained by the fact that the major N input to arable land results from fertilizer application (cf. Chapter 8).

Simulated N leaching increases in both successional series with elevated land use intensity. However, N leaching from Histosols is relatively low and even in intensely drained wet grassland this process seldom exceeds 50 kg N ha^{-1} year^{-1} despite high NNM rates (Fig. 12.9). The reason is that on drained Histosols most of the nitrogen leaves the system by denitrification (Tschirsich 1994; Davidsson and Leonardson 1997). In contrast, on mineral soils higher N leaching and proportionally lower denitrification rates result in simulation runs. N leaching is also high in unfertilized beech forests, which can be related to relatively high air-borne input of nitrogen compounds (cf. Chapter 8). The simulated low denitrification rates of the beech forests were empirically confirmed by Mogge (1995), who measured denitrification rates of 6.6 kg to 19.0 kg N ha^{-1} year^{-1} (Fig. 12.9). In agreement with modelling results only reed swamps prove to be weak nitrogen sinks or sources, whereas weakly or moderately drained wet grasslands are strong N sources. The simulated N balances prove all ecosystems on mineral soils to be weak nitrogen sinks or sources.

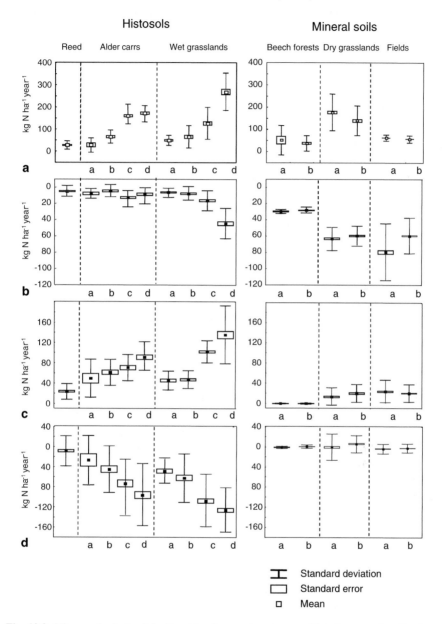

Fig. 12.9 Nitrogen budgets of the Bornhöved ecosystem types. **a** Net nitrogen mineralization, **b** leaching, **c** denitrification, **d** N balance. *a, b, c, d* as in Fig. 12.6

12.4.4 Efficiency Measures

In both successions modelled metabolic efficiency figures decrease with increasing
land use intensity (Fig. 12.10). Regarding the ecosystems on Histosols, this development can be mainly attributed to high C mineralization rates due to drainage. Among
the ecosystems on mineral soils the beech forests assimilate more efficiently carbon
than dry grasslands and field crops with the exception of maize. The very high metabolic efficiency of this species is likely to be due to its character as a C4 plant
(Kutsch et al 2001c). In both successional series the simulation results of the water
budgets exhibit a tendency towards decreasing biotic water use (lower proportion of
transpiration in evapotranspiration) with increasing land use intensity.

12.4.5 Integrative Characterization of Ecosystem Evolution

The transition of wet to drained alder carrs leads to a decrease in species richness
and an increase in N mineralization, N leaching, denitrification and microbial soil
respiration, whereas NPP remains on the same level. The N and C balances show

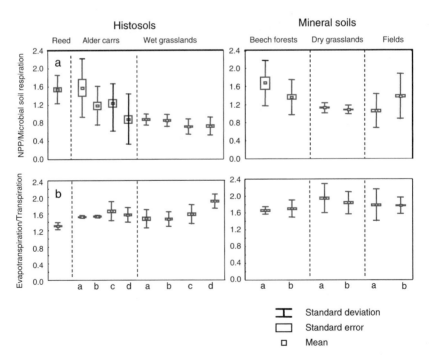

Fig. 12.10 Efficiency measures of the Bornhöved ecosystems. **a** Metabolic efficiency, **b** biotic
water use. *a, b, c, d* as in Fig. 12.6

that drainage induces a shift of the systems functions from C sinks to C sources and a corresponding one from weaker to stronger N sources. The efficiency indicator (NPP/soil respiration) of the C budget decreases throughout this development. A comparison of the mesotrophic and eutrophic stages of the alder carrs indicates that the effects of eutrophication are smaller than those of drainage measures (Fig. 12.11).

The development of drained alder carrs to weakly drained wet grasslands results in a rise of the water table, which is due to lower transpiration rates of wet grasslands in comparison to humid forests. Higher water tables reduce the intensity of C and N processes and diminish the function of the systems as nutrient sources despite considerably lower NPP or nutrient uptake, respectively. The number of plant species increases during this development, which can be explained by the moderate disturbance intensity of wet grasslands (Wheeler and Shaw 1991). A characteristic change in the indicators can be observed, however, if wet grasslands are drained. In this case, there is a rapid and marked decrease in plant species richness associated with major changes of the primary productivity. In parallel, the efficiency measures are decreasing. Moreover, intensely drained wet grasslands become strong N and C sources.

A further comparison of the ecosystems on mineral soils reveals that the relatively high structural diversity of beech forests is correlated with low rates of primary productivity (cf. Sections 8.4 and 8.5). Due to lacking fertilization net N mineralization, N leaching and denitrification of these ecosystems have the lowest values of all the systems analysed. As a consequence, soil respiration is distinctly reduced and the system has a positive carbon balance. Furthermore, the efficiency measures of the beech forest are higher than those of the more intensively managed ecosystems (Fig. 12.12).

The higher productivity of agroecosystems is coupled with a loss in species diversity and a reduction in efficiency on the one hand and, despite fertilization measures, an increase in nutrient losses on the other, by leaching and denitrification and carbon loss by soil respiration. With regard to the carbon balance, the sink function of beech forests is much higher than that of the agroecosystems (cf. Chapters 4 and 8).

12.4.6   Relationships between Species Richness and Ecosystem Functioning

There is agreement that the abundances of many species are predictable from the extent of their habitats. Whereas physical site conditions have played a prominent role in controlling the abundance of plant species in the past, current changes in land use patterns and management intensity have now become the dominant control mechanisms of species richness, rarity and abundance (Hodgson 1986a, b; Hodgson et al. 1998). As a rule, species-rich plant communities lack

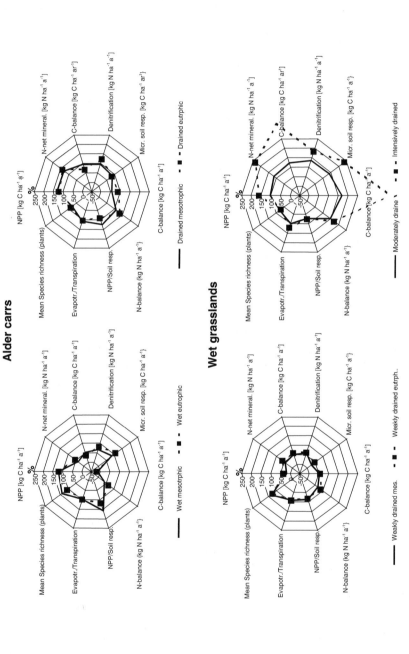

Fig. 12.11 Comparison of indicator values of four types of alder carrs and wet grasslands on different sites. The reference value (100%) represents the average of all ecosystem types defined

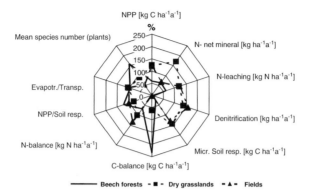

Fig. 12.12 Comparison of ecosystems on mineral soils. The reference values (100%) represent the average of all beech forests, dry grasslands and fields of the Lake Belau catchment

aggressive dominants (Huston and Smith 1987). Yet species richness may be only a weak indicator for ecosystem integrity and functioning, where species with a subordinate frequency and cover occur in higher numbers. Chiefly the dominant species may strongly determine ecosystem functioning, whereas subordinates or other scarce species may gain in importance in the event of ecosystem change (Díaz and Cabido 2001). Moreover, the potential consequences of reduced biodiversity for ecosystem functioning and services are only insufficiently characterized by using models which are limited to a single trophic level (Loreau et al. 2002) or change by shifting from the local to the regional level (Hughes and Petchey 2001; Loreau et al. 2001; Cardinale et al. 2004). Thus, additional functional attributes will be more informative for characterizing the potential role of a species in an ecosystem:

- Plant strategy types differ in nutrient and water uptake, growth, resistance and resilience to stress and perturbations (cf. the CSR paradigm of Grime 2001) and are suitable indicators for the interactions between species trade-offs and environmental constraints which vary in the course of succession processes and disturbances ('intermediate disturbance hypothesis' of Connell 1978).
- A different sensitivity of species to human impact in their habitats can be used to classify the hemerobic degree as a main driving force of ecosystem change.
- The regional and coenological state of idiochorophytes of primary and secondary sites ('apophytes') and invading agriophytes and ephemerophytes characterize floristic changes according to land use.
- A shift in physical site characteristics like hydrology, climate as well as soil acidity and nutrition can be indicated by species with a stenoecious behaviour in regard of these environmental factors (Ellenberg et al. 2001; Fränzle 2003).

The fairly high species richness in meadow and pasture communities compared with some forest sites and reed swamps is the result of a high frequency of CSR-strategists that are well adapted to sites with frequent disturbance by mowing and grazing.

Table 12.3 Nutrient budgets of the terrestrial ecosystems of the study area

Process	Histosols (157 ha)	Mineral soils (939 ha)
	Transfers (kg ha^{-1} year^{-1})	
NPP(C)	5432	5994
N net mineralization	194	70
N leaching	2	58
Denitrification	104	22
Microbial soil respiration	6629	5503

The shift in plant species composition from forests to grasslands and field systems is well documented by an increasing frequency of haemorophilous apophytes and agriophytes in the intensely managed plant communities (Dierssen 2004b).

12.4.7   Nutrient Balances of the Study Area

Most of the fixed carbon is assimilated by the ecosystems on mineral soils (Table 12.3). When the relative values of NPP (kg ha^{-1}) are taken into consideration, however, stands on Histosols reveal almost the same productivity as those on mineral soils. More than 30% of the total mineralized nitrogen is produced by ecosystems on Histosols which cover only 14% of the study area. The total N losses by leaching are much higher in ecosystems with mineral soils than in those on Histosols. In contrast, denitrification is almost five times higher in Histosol systems. Despite much smaller acreages, the total denitrification of Histosols matches almost that of mineral soils, because most of the nitrate produced is immediately denitrified owing to a direct coupling of nitrification/denitrification processes. As a consequence, N leaching from peatlands is relatively low. In contrast, the denitrification potential of mineral soils is low (Mogge 1995) and the N surplus is mainly transported away by seepage water.

Also the higher microbial soil respiration of Histosols in comparison with that of mineral soils is the result of an accelerated mineralization due to drainage. According to C balances, the drained peatlands of the study area act as strong carbon sources, whereas the mineral soils with their different vegetation cover are carbon sinks. Thus, the study area as a whole reacts as a nutrient source and therefore sustainable landscape management has to focus on measures to reduce nutrient losses by restoration of fens or a careful management of field systems on mineral soils in order to reduce N leaching substantially (cf. Chapters 8, 9, 13).

12.5   Conclusions

The orientor concept underlying the above ecosystem map states that certain system variables are optimized throughout the undisturbed development of ecosystems. With regard to the retrogressive successions analysed, this means that the respective indicator values will predominantly decrease with increasing human impact. The data presented provide a significant test for this hypothesis: fertilization as well as drainage are practised in order to enhance the productivity of ecosystems. All system comparisons show that these agricultural measures are successful, but all other indicators illustrate just as well that this type of optimization has far-reaching consequences. With growing productivity, i.e. an increase in exergy capture, species richness decreases and the composition of life-strategy types changes significantly. Furthermore, microbial soil respiration (which is an important path for entropy production) increases as well. In our case studies this process is due to high mineralization rates in the disturbed wetland systems. Thus, the stress component of ecosystem respiration is much higher than the energy demand of the heterotrophic system components (Chapter 8). So far, entropy production seems to be a consequence of stress rather than a function of the energetic demands of the complex flux patterns of ecosystems.

With regard to the gradient concept the results presented indicate that retrogression is coupled with a long-term degradation of both structural and functional gradients and a concomitant loss of energy and nutrients. What is the issue for environmental management? The material and energetic balances presented clearly indicate that the actual land use pattern of the study area is only in parts a reflection of the sustainable development paradigm, since there has been a loss of nutrients, a loss of energy and a loss of structural and biotic diversity. These losses not only represent a reduction of systems' capital but are also associated with a substantial impairment of other environmental qualities. Thus, it could be shown that nitrate loss leads to eutrophication in the upper groundwater domain and consequently in the corresponding aquatic ecosystems as well (cf. Chapter 9). Analogously, the carbon compounds which are lost may contribute to global warming effects. Therefore, landscape management should rapidly concentrate on measures to reduce the CO_2 production of wetland ecosystems, e.g. by restoration measures. Analogously, the agricultural practices should be modified, aiming at a reduction of N losses as a focal management target.

Chapter 13
Ecosystem Research and Sustainable Land Use Management

Jan Barkmann, Hans-Peter Blume, Ullrich Irmler, Winfried Kluge,
Werner L. Kutsch, Heinrich Reck, Ernst-Walter Reiche[†],
Michael Trepel, Wilhelm Windhorst, and Klaus Dierssen

13.1   Introduction

Today's society is concerned with a multitude of events affecting the ecosystems of our planet, ranging from natural processes such as weather anomalies and climate changes to man-made phenomena such as the impact of environmental chemicals or occupational and outdoor noise. To better understand and assess the importance of these events and to resolve the related problems, the basic approach of systems ecology is indicated, appropriately coupled with societal considerations in a comprehensive, transdisciplinary way.

Thus, in its capacity as an international pilot project of UNESCO's Man and the Biosphere Programme, the Bornhöved Project also focused on applied aspects of ecosystem research, in particular on sustainable landscape/land use management. Two kinds of problems have to be distinguished in this respect, which will be discussed in Section 13.2 of the present Chapter as a theoretical background to a collection of illustrative examples of applied ecosystem research in Section 13.3. The first issue is how to solve particular, well defined 'technical' problems, e.g. in conservation biology or resource protection, the second is how to provide instrumental knowledge for societal decision-making at large.

13.2   Ecosystem Research and Land Use Strategies

13.2.1   Concepts of Sustainable Landscape Management

Most simply, the interactions between humans and the environment are represented by the flow of ecosystem goods and services whose utilization usually has an environmental impact on ecological systems (Fig. 13.1).

In terms of the DPSIR (Driving Forces, Pressures, State, Impact, Response) approach of the European Environment Agency (EEA 1999), human needs 'drive' the use of ecosystem structures and processes. The utilization causes 'pressures' on the 'state' of the ecological systems. These pressures result: (a) in further changes

O. Fränzle et al. (eds.), *Ecosystem Organization of a Complex Landscape.*
Ecological Studies 202.
© Springer-Verlag Berlin Heidelberg 2008

Fig. 13.1 Simple model of human and biosphere interactions (after de Groot 1992, modified)

of the system and (b) in changes of the ecosystems' capacity to supply ecosystem services. This last step feeds back into the human sphere. Humans rely on ecosystem goods and services and thus evaluate their sustainable provision.

The societal 'response' to changes in the state of the environment differs to a large extent with the degree that the human actors consider essential ecosystem services at risk. These considerations, in turn, depend on risk awareness and risk attitude of the actors. From a legal point of view, the precautionary principle in German and European law demands a careful, risk-averse environmental planning and management. Because both should rest on sound knowledge, applied ecosystem research must investigate to which extent the capacity of ecological systems to deliver goods and services is endangered. In particular, it has to be analysed:

- On which ecosystem states, processes or structures the ecosystem services depend in detail;
- Which pressures threaten the relevant ecosystem states, processes or structures;
- How the long-term capacity of ecological systems to provide essential services can be protected and developed.

The importance of the various preconditions of sustainable development justify to put up formal policy goals that secure the sustained functioning of the ecological, the social and the economic systems. In order to differentiate these instrumental goals from the *ultimate* social goals of sustainable development, they can be dubbed *proximate* goals. In order to highlight the interrelatedness of the three spheres, the German Environmental Advisory Council coined the term *retinity* (SRU 1994). Retinity demands that all relevant decisions have to account for the interrelatedness of economy, ecology, and the social sphere (also, see section "Merging environment and economics in decision making" in WCED 1987, p. 62ff).

Combining the demand for retinity with the proposed hierarchy of goals, multi-dimensional bundles of ultimate and proximate objectives are to be taken into account in decision-making. The complexity of the problems excludes, however, that a "single blueprint for sustainability can be found" (WCED 1987, p. 40). The normative multi-dimensionality in decision-making is most apparent when the trade-offs and potential conflicts between ecology, economy and the social sphere are considered. Frequently, even different *ecological* objectives are in conflict (Fränzle 1991). Thus, ecological objectives investigated in the Bornhöved Project that can be in conflict include, e.g.:

- Conservation goals including the protection of autochthonous biological diversity;
- The protection of ecological processes or structures that generate environmental goods or services;
- The implementation of the precautionary principle.

There is no ecological reason to believe that such diverging objectives can be maximized *simultaneously* with a single set of management guidelines maximizing 'the right' ecosystem structures or processes. If in conflict, however, there is no ecological guideline from which the conflicting objectives, for example between biodiversity conservation and protection of the economically productive resource base, could be prioritized (Barkmann 2001a). Because of this principal inability of science to give clear advice in development decisions within the sustainable development paradigm, no generally valid 'way to sustainability' can be defined. In effect, the analysis leads to *discursive* sustainable development strategies in which the goals of sustainable landscape development are a matter of informed societal choice (Herzog 2002).

Understanding sustainable development as the regulative idea of the discourse on environmental and developmental impacts applied sciences have to re-formulate their tasks. Applied ecosystem research, for example, is called upon to prepare the knowledge and capabilities for optimal use by diverse groups of stakeholders in decision-making processes. It must make its expertise 'discourse-able'. In this model, it is the role of systems ecology to introduce ecological knowledge into the discourse in order to facilitate rational social decision-making in accordance with the essentials of sustainable development (Barkmann 2001a, b).

13.2.2 The Demand for an Ecosystem Approach in National and International Regulations

13.2.2.1 International Conventions, Directives and Action Plans

The regulatory framework for an ecosystem approach to sustainable land use in Germany is set by international, national and regional law including acts of the federal states and municipal ordinances. The first pertinent international convention was the European Soil Charter of the Council of Europe in 1972 which pointed out that "Soil is one of humanity's most precious assets. ... Soil is a limited resource, which is easely destroyed ... A regional planning policy must be conceived in terms of the properties of the soil and the needs of today's and tomorrow's society. ... Governments and those in authority must purposefully plan and administer soil resources. Soil is an essential but limited resource. Therefore, its use must be planned rationally, which means that the competent planning authorities must not only consider immediate needs but also ensure long-term conservation of the soil while increasing or at least maintaining its productive capacity."

In the following decades, in particular since the 'Rio Summit' in 1992, all international conventions dealing with environmental or development issues refer to the sustainable development concept. The detailed regulations of these conventions must be regarded as the result of international political bargaining and compromise. Nevertheless, the issues treated reflect an increasingly integrated, global long-term perspective. An important example is the changing perception of environmental problems in the international and national debate, for example from point sources to non-point source pollution problems, and from media-orientation to cross-cutting issues of environmental management (e.g., Yaffee 1999; WBGU 2000; Hartje et al. 2002). Pollution control, originally mainly orientated towards (human) health protection, becomes integrated in more comprehensive and cross-media concepts concerning natural resources use including potential climate effects. A more integrated 'retinity' perspective can be recognized in the Convention on Biological Diversity (CBD) and the United Framework Convention on Climate Change (UNFCCC), both adopted in 1992.

In May 2000 the Fifth Conference of the Parties (COP 5) in Nairobi agreed on a primary framework for action under the CBD, the so-called Ecosystem Approach (ESA). The ESA consists of 12 principles and five operational guidelines that widely reflect sustainable development concerns, largely including social and economic considerations. The UNFCCC and the resulting Kyoto Protocol focus on the greenhouse gas concentration of the atmosphere. However, they are motivated by mitigating the impact of climate change on essential ecosystem services. Since the Marrakesh Accord, land use, land use change, and forestry (LULUFC) are supposed to be monitored and integrated into the national carbon balance – with potentially significant influences on landscape planning and agricultural practice. Yet, these more comprehensive approaches do not solve the complex optimizations tasks, for example, between carbon sequestration by monoculture plantations of fast-growing tree species and the maintenance of more diverse, but less carbon-fixing plant communities (cf. Chapter 4).

Other international and European agreements or directives that must be considered in landscape planning procedures include:

- Environmental Impact Assessment Directive (EIA Directive 85/337 EWG);
- EU Habitat Directive 92/43, EU Birds Directive, RAMSAR Convention;
- EU Directive 2000/60/(EC) Establishing a Framework for Community Action in the Field of Water Policy; Kyoto and Montreal Protocols, Water Framework Directive.

While the EIA Directive concerns the instruments of landscape planning, the second group of regulations has a clear sectoral focus on conservation, and the third is rather clearly media-oriented.

13.2.2.2 Relevant National Acts and Regulations

The Federal Nature Conservation Act (BNatSchG) was revised in 2002. Now it includes, inter alia, more precise provisions for farming practices. Additionally, the creation of a network of linked biotopes covering at least 10% of the German territory is demanded to preserve biological diversity. Some of the main principles and goals of the Federal Nature Conservation Act [§2(1)] are:

- To safeguard, develop, or restitute ecosystem complexes including their specific biotic functions, matter and energy flows as well as their landscape structures;
- To protect soils in a way that they can fulfill their ecological functions;
- To preserve and develop biological diversity of habitats, communities, species and the genetic diversity within populations;
- To secure and develop the variety, identity and beauty of the landscape as a source of recreational facilities.

Important supplements, such as the Federal Soil Protection Act (1985), amended by the 1999 Soil Rehabilitation Act (*BBodSchG*) and the fertilization regulation (DüV 2006) contain further specifications on the proper utilization and protection of nature and natural resources. The acts provide protection against soil destruction, soil compaction, erosion, and require the preservation of soil structure, a high biological activity and a site-specific humus content of soils under tillage. Werwer (2003) described ways to appropriately consider such pedoecological requirements in landscape and rural planning.

Two peculiarities of the German system of environmental regulation are to be underlined (see Fig. 13.2). The first task is the coherent, hierarchical approach for *landscape planning* (German: Landschaftsplanung) that parallels the more comprehensive *land use planning* (German: Raum- und Flächennutzungsplanung). The Federal Nature Conservation Act stresses the autonomous position held by landscape planning as the pertinent instrument of nature conservation and landscape management in comparison with other planning instruments and administrative procedures. In landscape planning on the different administrative levels, the respective authorities are obliged to follow a number of specific procedures which include:

- Mapping and description of the current ecological conditions as a comprehensive information base on landscapes and ecosystems;
- Evaluation of present and likely future ecological features (structural as well as functional);
- Assessment of environmental changes resulting from current land use practices, enacted planning regulations and likely land use scenarios;
- Development of regional/local environmental guidelines and quality targets;
- Proposals for future land use, for environmentally compatible agriculture, for special conservation measure-ments, etc.

To accomplish these tasks, scientifically valid indicator systems have to be developed in order to meet the demands of various valuation procedures. These indicators shall reflect the financial limits that constrain the collection of environmental data.

The impact regulation and compensation procedures according to BNatSchG §§ 19 and 20 apply whenever the shape of the ground or the land use type are significantly altered and if serious or lasting disruptions of the attributes of an ecological system are expected. These procedures take place as an integrated part

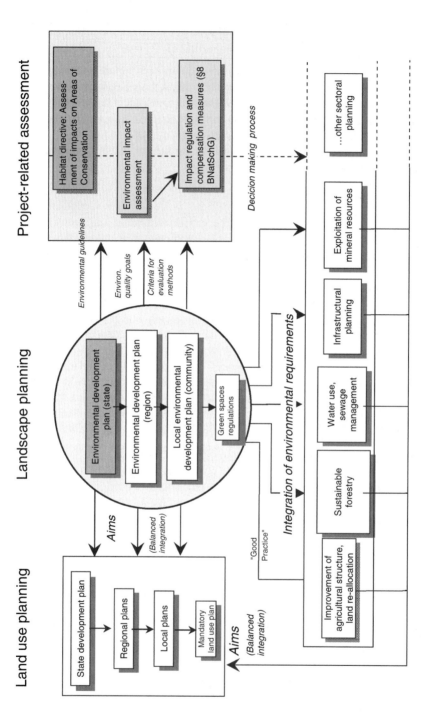

Fig. 13.2 Environmental planning in the Federal Republic of Germany. Instruments related to landscape planning according to the Federal Nature Protection Act (after Jessel and Reck 1999)

of the general planning, permit, and approval processes for any development project. It is the most important objective of the environmental impact regulation and compensation process to minimize any impact or to calculate the compensation for the impact, which the developer has to implement on his/her cost. Compensation must match the 'ecological value' and the function of the disturbed ecological features. Furthermore, the compensation must fit the general goals of the Federal Nature Conservation Act and the specific goals of pertinent regional landscape planning provisions. If full ecological compensation cannot be achieved with reasonable efforts, the project permit can be denied or *replacement measures* or *payments* can be required by the competent authorities. Thus, the impact regulation and compensation procedures demand ecological investigations similar to the Environmental Impact Assessment (EIA) mandated by European law. There are two important differences, however:

- A higher spatial resolution is required than by EIA;
- Instead of a pure risk assessment, the assessments are to be based on *valid predictions* of the impact at hand. The scope of the predictions has to cover effects of the project on the functional capacity of the affected ecosystems and on the appearance of the landscape. For decision-making and for the calculation of the size of the compensation or replacement measures, data are required concerning the predicted effect and the value of the affected features and functions. Methodologically, elements of ecological landscape description and prediction have to be amalgamated with valuation procedures (cf. Fig. 13.3).

Because sampling and analysis of ecological data are expensive, the role of scientific methods for *predictive* impact assessment is controversial. Is a consensus view on the value of a certain ecosystem type and a general assessment of its sensitivity to the impacts at hand sufficient (so-called biotope-valuation procedures; e.g., Dierssen and Reck 1998) or are detailed, specific predictions necessary? The importance of this question is illustrated by the vastly differing results of both approaches. Even the different calculation methods of biotope-values and impact-severity values result in calculated compensation needs for identical impacts that differ by more than 100% (Oles 2001).

13.2.3 Ecosystem Integrity – Protection in the Face of Unspecific Risks to the Human–Environment Interaction

Several key problems of sustainable development in general and of landscape planning and management in particular are not yet resolved. Among these problems are the increasing reduction of ecosystem integrity by the depletion or degradation of non-renewable ecological resources such as fertile, uncontaminated soil, or biological diversity on the genetic, the species, and the ecosystem levels, and of "unspoiled" landscapes suitable for physical and spiritual recreation. Growing fragmentation, in

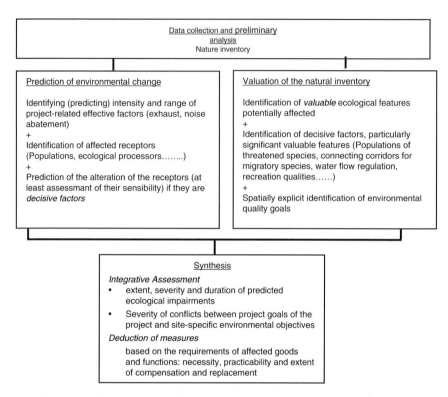

Fig. 13.3 Impact assessment and definition of compensation/replacement measures by ecologically and socio-politically based valuation procedures

particular the suburbanization of rural areas, and an increasing uniformity of landscapes continue to threaten biological diversity because they severely reduce the habitats of specialist species. Pollution, e.g., eutrophication and acidification, facilitates the spread of robust generalist species (EEA 1999; SRU 2000, 2002) while it destroys the habitats of less competitive species adapted to low nutrient substrates (Dierssen 2002; Dierssen and Hoffmann-Müller 2004). The ongoing reduction in the habitats of many species proves the insufficiency of current environmental legislation and management and compromises a key factor of ecosystem resilience (Holling 1986; Perrings 1995).

In the face of these problems, long-term plans for a 'sustainable' future have to be developed. The precautionary principle demands preemptive actions to safeguard the capacity of the ecological systems to provide ecosystem services. One way to deal with this challenge is the application of a functional interpretation of ecosystem integrity. In the following, it is understood as a guideline for the protection against unspecific risks to the human–environment interaction within the *sustainable development* paradigm. Ecological integrity aims at the long-term maintenance of the abstract ability of ecosystems to provide goods and services. This is achieved by protecting those biotic and abiotic structures and processes that maintain the self-organizing capacity of ecosystems. The above *Kiel interpretation of ecosystem integrity* (Barkmann et al. 2001) differs from other accounts of the integrity concept as it explicitly combines a *social values approach* within the framework of sustainable development with insights from ecosystem science (e.g., Miller 1995).

Ecological integrity aims at a pro-active provision against unspecific ecological risks that endanger the capacity of ecological systems to provide ecosystem services. In order to cope with these unspecific risks, an orientation at the most general, thermodynamically justified properties of ecological systems constitutional for ecological self-organization is suggested. Unspecific risks result from our insufficient knowledge about the long-term development of the man–environment system concerning social, stochastic and epistemic uncertainties:

- *Social uncertainties*, because we do not know in which way the demands of further generations on global environmental systems will change;
- *Stochastical uncertainties*, because we cannot predict the direction, speed and dimension of changes of global environmental systems because of their inherent complexity and non-linearity;
- *Epistemic uncertainties*, because we do not know in detail which potential risks and influences may force changes in environmental systems and whether we have the models and theories to estimate potential risks in space and time with sufficient precision.

Based upon the scientific approaches applied in the Bornhöved-Project, the Kiel approach therefore connects an attempt to identify the "integrity" of ecosystems with a guideline for sustainable landscape management based on the precautionary principle.

13.2.4   The Role of Biological Diversity – Losing Insurance Benefits

Biological diversity is difficult to investigate because of the following reasons:

- At the species level, it is impossible to determine all taxa of various animal and plant groups at a given site in an operational way.
- The number of species strictly depends on the site scale, but for various species groups in a different way, e.g., in relation to the motility of the population members.
- Diversity indices have been developed to convey the extent to which individuals are distributed evenly among species.
- Temporal turnover is of importance.
- Below the species level, the knowledge of phylogenetic and genetic variability is less developed. The number of genetic populations in the world has been esti-

mated between 1.1 and 6.6 milliards (Hughes et al. 1997). Moving above the species level, the estimation of higher taxa richness is a less data-demanding surrogate for species richness. This can only be a very preliminary solution, though.

- The evaluation of species functional groups may be a more sophisticated approach to characterize the relationship between biological diversity and ecosystem functioning than species density (Gaston 1996; Purvis and Hector 2000).
- Species/area curves are based on the presence of one or more individuals of species and do not necessary assure viable populations.

Many hypotheses try to explain the spatial and temporal changes of species richness (Palmer 1994; Gaston 1996; McCann 2000). Overall species richness, for instance, increases to some extent exponentially with area. It also depends on parameters such as annual amount of precipitation, summer temperature, sea surface temperature or potential evapo-transpiration. It decreases with elevation above sea level, and from the tropics to arctic and antarctic areas. Local species richness is correlated with regional species richness. Niche differentiation as well as phytomass normally increase with increasing productivity (Tilman et al. 1997b; Grime 2001). Nutrient enrichment and abandonment of traditional land use practices are key factors threatening biodiversity in cultural landscapes (e.g., Bollens et al. 1998; Sala et al. 2000; Dierssen 2004). Derived from regression analysis and based on computer simulations with cellular automates, the increasing grain size of landscape pattern or the increasing aggregation in intensively cultivated areas appear strongly negatively correlated with various diversity measures (α-, β-, γ-, δ-diversity, sensu Whittaker 1977; Steiner 2002). Sala et al. (2000) point out that land use change is the most severe driver of change in biodiversity. For many systems, such as forests in northern nemoral areas, nitrogen deposition is the key factor.

Before the 1970s, ecologists believed that more diverse communities enhanced community stability. This perspective changed, as the search for a general diversity-stability relationship widened its scope. Some investigators found that ecosystem function can be more directly related to functional diversity, e.g., the presence or absence of Poaceae or nitrogen-fixing Fabaceae (Schulze and Mooney 1993; Hooper and Vitousek 1997; Tilman et al. 1997a; Schläpfer and Schmid 1999). A similar relationship was found for mutualistic arbuscular mycorrhizal fungi (van der Heijden et al. 1998).

Huston (1997) re-evaluated experiments on the relationship between diversity and productivity. These experiments provide no evidence that increasing biodiversity improves ecosystem functions – if not all physical conditions that might effect the treatment are carefully monitored. Recent investigations and experiments focus on the diversity and interaction strength of food web structures in ecosystems and on their importance for system resilience (Mason et al. 2003). Taken together, the results indicate that diversity can limit energy flows in consumer–resource interactions. It inhibits excessive consumption which in turn destabilizes food web dynamics (McCann 2000).

Nevertheless, disappointing gaps remain between the alarming rates of human-induced decline in biological diversity and their effect on system functioning. Regardless the outcome of these mechanistic debates, the conservation of biodiversity is essential because we rarely know a priori which species are critical to current ecosystem functioning or which species provide resilience and restistance in the face of environmental changes (Hooper and Vitousek 1997; Chapin et al. 2000). Thus Mooney et al. (1996, p. 476) conclude inter alia:

- The loss of genetic variability within a population can reduce its flexibility to adjust to environmental change.
- The addition or deletion of a species can affect the capacity of an ecosystem to provide services.
- Certain 'keystone' species, when deleted, have profound effects of ecosystem functioning. Due to a lack of a generally applicable theory, these effects can only be assessed by direct experiments.
- The capacity of ecosystems to resist changing environmental conditions is positively related to species richness.
- The simplification of ecosystems in order to produce greater yield of individual products comes at the cost of the loss of ecosystem 'stability'.
- Fragmentation and disturbance of ecosystems, especially on the landscape level, have profound effects of the services provided.
- Currently, we are more successful in simplifying ecosystems than in reconstructing complex ones.

Ecosystem process rates may show no simple correlation with species richness, and diversity may not be the driver of such processes. Nevertheless, species diversity indicates an ecosystem's ability to buffer perturbations. For example, if there is functional redundancy in the 'key' species of an ecosystem, the system is considerably less susceptible to species deletion and species invasion. The greater the variance of potential species responses in a system, the better a possible insurance effect (Lawton and Brown 1993; Yachi and Loreau 1999). At the global scale, it is the *functional* integrity of the earth that is at issue, not biodiversity sensu strictu (Woodwell 2002): The progressive impoverishment of natural communities and ecosystems through human-induced chronic disruption and cumulative environmental dysfunction only in the later stages also lead to species extinction. Because ecosystem resilience is the precondition for the capacity of ecosystems to provide human society with essential services, the ecological insurance that is provided by functionally diverse systems entails important economic insurance benefits.

13.3 Beyond Sectoral Planning – Outline of the Precautionary Ecosystem Approach to Sustainable Landscape Planning

13.3.1 From Sectoral Planning to an Ecosystem-Oriented Approach

The management of rural landscapes from an applied ecosystem research perspective requires an optimization of a multi-dimensional set of ecological goals within a framework of social and economic demands (cf. Section 13.2). In the present Section several examples of an ecosystem-oriented approach to landscape management, as developed in the framework of the Bornhöved Project, are presented.

In a first case the indicator set of ecosystem integrity (Chapter 12) is applied to representative ecosystems of the Bornhöved Lake District. As these indicators were designed to quantify the capacity for ecosystem self-organization and ecosystem resilience, this issue is related to a precautionary approach to landscape planning. Then selected tools and strategies for the optimization of landscape management procedures are described which comprise integrative modelling concepts that can be used for prognoses and/or scenario assessment. The models were developed in the study area and thereafter successfully applied in other landscapes of the North German Plain for comparison purposes.

In this connection, three major environmental issues were chosen in order to draw general conclusions for the management of rural landscapes:

- The need of carbon sequestration as a tool for mitigating climate change;
- The protection of water bodies against nitrogen load;
- The preservation of biodiversity.

13.3.1.1 Ensuring Ecosystem Services from Forest Ecosystems

In the core area of research growing forests act as carbon sinks. Among the forest types analysed beech forests and (currently unmanaged) alder carrs are of particular interest. The net ecosystem production (NEP) varies between 300 g and 400 g C m^{-2} year^{-1} (Kutsch et al. 2001a), which is in the normal range of broadleaved forests in the nemoral zone (Granier et al. 2000; Pilegaard et al. 2002). Unmanaged or old forests can develop a higher carbon sink capacity up to 600 g C m^{-2} year^{-1} (Knohl et al. 2003). The reason may be an optimized canopy structure that enhances light use efficiency and gross primary productivity.

The investigations of the nitrogen balance showed that at least the beech forest of the core area is stressed by high atmospheric nitrogen deposition (Schimming et al. 2001; Chapter 8) which has resulted in nitrogen over-saturation. Yet it still acts as a buffer between atmospheric deposition and groundwater, because about 60% of the atmospheric deposition are stored in the biomass and the organic matter of the forest soil. This is due to the capacity of beech trees to adapt pedospheric uptake to the actual demand especially by cycling amino compounds between shoot and roots, whichin involves fairly high emission rates of NO and N$_2$O and nitrate leaching (Rennenberg et al. 1998; Geßler 2001). Due to drainage and atmospheric N deposition, the soils of the alder carrs have frequently shifted from ammonium to nitrate dominance as the main N source of the plants as is indicated by an increasing amount of *Fraxinus excelsior* in the tree layer (Janiesch 2003). Indicator-species analyses have provided ample evidence that also in other parts of northern Germany such changes in the floristic structure of deciduous forests due to eutrophication and acidification are the rule (Diekmann and Dupré 1997).

Forests generally provide habitats for autochthonous and partly endangered species (Ellenberg 1998). Overall, 560 vascular plant species in Germany are more or less restricted to forests (v. Oheimb 2003), and a fraction of 21% (1371 species) out of 6492 beetle species are restricted to dead wood (Köhler 2000). The species richness of temperate forests in Central European forests is the result of both long-term historical processes and recent anthropogenic impacts. The geographical situation – the Ice Age barrier function of the Alps impeding large-scale migratory processes of flora and fauna – in connection with long and repeated glacial periods affected a number of tree species. Additionally, a strong change in species composition and decrease in richness is caused by forest management which prefers even-aged stands in the optimal growing state. Moreover, the above-mentioned input of nutrients and acidification influence the composition of plant and animal populations.

The forests in the Bornhöved area, like everywhere in Schleswig–Holstein, are seldom in an oligohemerobic state because of:

- Their even-age tree structure according to forest management;
- Their low age and former agricultural use;
- An artificially increased game species density.

For edaphic reasons and the even-aged tree canopy structure, the Bornhöved beech stands are comparatively poor in plant species (20–23 plant species per 0.4 ha). The alder carrs have 17–44 plant species per 0.1 ha (Wiebe 1998) with species-poorer phytocoenoses occurring mostly on soils with an artificially lowered water table. The amount of epiphytic cryptogams in the forests is low compared with unpolluted forest systems as in some mountainous areas in southern Germany and southern and western Europe.

Species richness of animals is comparatively high like in other forests of Middle Europe (Weidemann and Schauermann 1986), i.e., in the beech forest and the alder carr 907 and 1229 species, respectively, were recorded (Chapter 6), which is not least a reflection of the relative importance of the soil moisture factor.

13.3.1.2 Ensuring Ecosystem Services from Agrarian Ecosystems

Agroecosystems are an example of how people have accepted the decline of services by oligohemerobic ecosystems, such as erosion prevention, nutrient and water regulation, in favour of the production of fodder for domestic animals and crops for humankind. This is closely associated with considerable energy costs and a deterioration of the functions of neighbouring ecosystems, and frequently with a loss of biological diversity at the landscape level (Basedow et al. 1988; Mooney et al. 1996).

The Kyoto Protocol opens new commercial possibilities for using the biosphere as a carbon sink. Using agroecosystems as carbon sinks may be the most appropriate practice from both environmental and socioeconomic points of view (Smith et al. 2001). Degraded agroecosystems in Africa may benefit significantly from the improved land management that would be part of a carbon sequestration programme. One may agree with UNEP that there are potentially important synergies to be made between the Convention on Climate Change, the UN Convention to Combat Desertification and the UN Convention on Biodiversity (Olsson and Ardö 2002).

There is a widespread belief that low-yielding organic agricultural systems are more sustainable than high-yielding farming systems. Nevertheless, also adapted conventional mixed farming in smaller plots providing more field margins or farming based on the traditional lay system maintain conventional yields at low costs. Mechanical weeding by 'organic farmers' moreover includes a high use of fossil fuels including pollution by nitrogen oxides (Trewavas 2001).

Land use intensity and field border structures determine the vegetation diversity (Arx et al. 2002). Ubiquitous nitrophytic species prevail on arable land. As fallow sites show, the seed banks contain a fairly high amount of seeds, but again only of ubiquitous weeds with a low site specificity (Jödicke and Trautz 1994). If not cultivated, the seed bank of specific species may be depleted within a time-span of about 20 years (Waldhardt et al. 2001). Many cultivated plants and common allochthonous weeds, especially self-pollinators, have developed local races more rapidly than stenoecic autochthonous ones (Allard 1999).

13.3.1.3 Aquatic Systems and Wetlands

At various spatial scales human activity has degraded freshwater bodies and their specific vegetation and fauna more severely than other biota by impairing the water quality of the catchment. Continuous eutrophication results in changes in the food web structure and alterations of the trophic cascades in an often unpredictable way. The study of functional interrelations between hydrology and biota at the catchment scale is a new approach to achieving sustainable water management (Zalewski 2000). Thus, nutrient load, limnological processes, and their spatial and temporal variability have been intensively investigated in eutrophic Lake Belau (Pöpperl et al. 2001; Schernewski 2003).

The assessment of the current status of Lake Belau involved the trophic conditions (Chapter 10) which are mainly characterized by concentrations of TP (unfiltered) during spring circulation as well as mean summer concentrations of TP (unfiltered) from May to September, mean Chlorophyll-a concentrations in the epilimnion and a mean Secchi depth in summer (LAWA 1999). With mean N/P ratios of 15 in Lake Belau, N-limitation in the lake water body with N/P < 7.2 was observed only during exceptional cases from July to September (Naujokat 1997; Schernewski 1999). For a more sophisticated evaluation of the current trophic status of the predominantly P-limited Lake Belau and the development of integrated management programmes it is necessary to combine the external nutrient input with internal conditions of the lake itself. Thereby special attention must be paid to lake morphometry, interactions at the water/sediment interface and the nutrient budget of the littoral zone including macrophytes (cf. Chapters 9, 10).

Quantifying the origin of water entering Lake Belau allows a first assessment of the external nutrient inputs. The upstream catchment area totals 28.4 km^2 and the surrounding subterranean catchment 4.5 km^2. Thus, the inflowing water originates to 84% from groundwater recharge at the land areas (with 95% terrestrial areas and 5% wetlands), to 13% from precipitation on the lakes water surface, to >1.5% from overland flow from adjacent slopes, drainage of settlements, and to 2% from direct inflow of purified waste water.

Thus, integrated water management begins with groundwater protection, where the agricultural area of the catchment (80% of the total catchment with arable and pasture land) is of particular importance (Dibbern 2000; Schimming et al. 2001). Despite a low to medium vulnerability of the upper aquifer of the Bornhöved Lake District, the agricultural land use has resulted in an increase in nitrogen concentrations of the upper groundwater domain with mean values of 10–25 mg DIN l^{-1} (cf. Table 9.3) which is nearly one order of magnitude higher than the nitrogen concentration currently measured in the water body of Lake Belau. This emphasizes the importance of a precautionary area-wide groundwater protection (Schenk and Kaupe 1998; Dibbern 2000).

Currently measured P concentrations in the upper groundwater with values of 0.05–0.1 mg TP (unfiltered) l^{-1} nearly match the concentrations in the lake (cf. Table 9.4). For phosphorus dissolved in the groundwater, a partitioning into the natural background concentration (or geogenic fraction) and an anthropogenic fraction due to increased P fertilization is difficult. Increased P concentrations on the arable land, and the persistence of P compounds in particulate form in terrestrial soils, together with a high proportion of groundwater with hydraulic travel times of 2 years to 50 years, nearly rule out a short-term effect of reduced fertilizer application on the external phosphorus input. In contrast to diffuse P outputs from the terrestrial catchment via groundwater which amount to nearly 85% of the input to the riparian and shore wetlands, the P input via surface runoff and interflow is generally overestimated (cf. Section 9.4). Only a short-term effect can be brought about if the adjacent slopes, as already common practice around Lake Belau, are used as grassland and manuring of slopes and wet grasslands with sewage is ended.

In future, a quantification of external nutrient inputs has to consider explicitly the potentially high retention of near-natural riparian zones and lake banks. Case studies, where the effect of stepwise measures for a restitution of riparian zones at Lake Belau was estimated, showed that closing of drainage ditches and a complete restitution of mesohemerobic alder carrs in the wetland and of reed belts in the littoral zone would reduce the non-point input of nitrogen by 20% and that of phosphorus by 30% compared with the 1989–1998 situation (Kluge 2004). These figures depend mainly on the local geo- and ecohydrological bank types and thus cannot be directly transferred to the banks of other lakes.

At present, Lake Belau acts as an efficient nitrogen sink with a retention coefficient of $R = 40\%$ and a lesser sink for phosphorus ($R = 10\%$). With regard to the actual trophic state of the lake defined in terms of the Vollenweider approach (OECD 1982), this requires a reduction of external P inputs from the present-day 1300 kg TP year^{-1} to approximately 250 kg TP kg year^{-1} (or 2 kg TP ha^{-1} year^{-1} lake area) to achieve at least in the long run mesotrophic conditions (Pöpperl et al. 2001). This estimate deliberately neglects P resuspension from the sediments of Lake Belau ('internal fertilization') and the upstream lakes whereby the shallower Schmalensee is particularly at risk.

Several sediment cores from the deepest part of Lake Belau with a maximum length of 30 m allow to estimate sedimentation rates and to unravel the development of the trophic state of the water body during the last 9000 years. The concentrations of selected elements in the warved sediments increased considerably since Mesolithic times (7200–3800 B.C.), e.g. TOC from 1.1 wt% to 10 wt%, or Pb from 0.25 ppm to 48 ppm (Garbe-Schönberg et al. 1998), which reflects a waxing human impact, as can also be deduced from detailed palaeobotanical data of the sediment cores (Wiethold 1998).

A successive restoration of eutrophic lakes, as proposed in the Water Framework Directive (EU 2000; Keitz and Schmalholz 2002), is only possible if the lakes, together with their catchment areas, become part of an integrative protection programme which aims rigorously at the reduction of nutrient input (Hupfer and Scharf 2002; Becker and Lahmer 2004). This requires controlled interventions during all stages of nitrogen transport including best land use practice to reduce general nutrient losses in the surrounding catchment. The success of wetland restoration clearly depends on site selection to achieve the specific restoration goals. Process-based ecohydrological lake models, however, which are able to describe the long-term development of trophic conditions integratively by considering the external and internal nutrient inputs as well as nutrient retention in riparian ecotones and internal nutrient transformations, are still a target ahead.

13.3.1.4 Biotope Fragmentation and Matrix Biotope Interactions in Rural Landscapes

In intensely used landscapes, only few oligo- and mesohemerobic sites have been left among the cultivated areas. In the Bornhöved Lake District typical landscape features with predominant linear or reticular structures are hedgerows ('Knicks'), forest fringes and road or field verges. They differ from the predominant matrix systems by their structure and largely the absence of direct land use. All these systems have a comparatively high species density in relation to the intensively used matrix biotopes, yet ubiquitous nitrophytes dominate the vegetation. At least four questions arise which are important for the evaluation of the quality of these remaining habitat fragments in agricultural landscapes:

- Are they able to maintain their actual species composition and ecosystem functioning?
- How are these systems able to form efficient buffers of oligo- and mesohemerobic sites against the input of xenobiotics or nutrients?
- Do these systems form efficient connections/corridors for the dispersal of propagules and individuals between 'habitat islands' in rural landscapes?
- How can the quality of these sites be improved for the benefit of ecosystem functioning and biodiversity conservation?

In dependence on their structural aspects, especially hedgerows and their margins act as sinks for airborne nutrients, fertilizers from the ajacent fields and xenobiotics like heavy metals or organics (Marxen-Drewes 1987; Krinitz et al. 1996; Asman et al. 1998). Hedgerows parallel to the banks of lakes and rivulets may act as barriers to particulate phosphorous input into the surface water systems (Schernewski et al. 1996).

The recent decline in species richness in meso-hemerobic biotopes has been attributable to biotope destruction, and to subsequent limitation in seed and small animal dispersal. Some authors seek evidence for livestock as vectors for propagules at various spatial scales in the past cultural landscape (Bruun and Fritzbøger 2002). Extensively used grassland, fallow land or tracks are of importance for the survival of rare and sporadically distributed plant (and animal) populations. Zacharias (1990) especially mentions the importance of forest fringes as habitats of less frequent species. Simmering et al. (2001) and Arx et al. (2002) examined, how far 'sharp edges' between meadows and fields contribute to floristic species diversity. GIS-based habitat models were developed and evaluated for those species considered to be actually endangered by extinction and listed in Annex II of the European Union Habitats Directive. These habitat models are in increasing demand for conservation purposes and planning issues (Kuhn 1997; Hunger 2002).

Other authors try to establish a methodological framework in order to optimize networks or areal systems to secure conservation aims (Altmoos 1999). Thus, roadside verges in mesohemerobic landscapes are considered important for plant communities and pollinator guilds (Stottele 1991), while scattered species appear to decrease in abundance and to be replaced by nitrophilous generalists (Sykora et al. 2002). In fact, some decades ago, the edge systems of forests and hedgerows in Schleswig–Holstein still harboured a great amount of species (Tischler 1948, Weber 1967), but the intensive use and fertilizer input into contiguous fields reduced the floristic and faunistic diversity of these communities.

To summarize: the dynamics and management of the euhemerobic matrix systems should be seen as the driving forces for the development and quality of habitat fragments and mesohemerobic sites between these matrix ecosystems (Jules and Shahani 2003); therefore their importance for the dispersal of sensitive and stenoecious species can hardly be overestimated (Irmler et al. 1996; Borkowsky and Schmalhaus 2004).

13.3.2 Digital Landscape Analysis and Modelling as Tools for an Integrative Landscape Management

The integrative approach of ecosystem research and thus the challenge to take a large variety of processes, functions and structures into appropriate account, requires developing tools which allow to study the biological relationships with the abiotic environment and the use of natural resources. For heterogeneous systems, the approach to show the specific pattern and extent of structures is of crucial importance (Allen and Hoekstra 1991; Müller et al. 1997a, b; Breckling et al. 2005). In particular, the conflict between the necessity of abstraction and the menace of neglecting essential details must be dealt with. In such situations the development of simulation models is indicated to investigate the effect of single or complex processes on the whole system as well as a top-down view analysing the impact of the system's dynamics and properties on the subsystems and low-level processes as forcing functions or boundary conditions. On the population level, methods and perspectives of meta-population dynamics and habitat modeling based on statistical and expert-based models have been developed and proved to be efficient (Kleyer 1995; Kuhn and Kleyer 2000). With these models it is possible to evaluate the habitat requirements of selected species by quantifying the quality of biotopes, while likely relationships between biodiversity and landscape structure can be more precisely defined by recourse to guilds or functional groups (Kleyer et al. 2000).

This biological approach is complemented by a modular system of geo-ecological models which have been elaborated to study water and matter flows at different temporal scales covering sites, plots as well as catchments with a size of up to several hundreds of square kilometres. The application of both approaches allows to include a wide multi-scale knowledge. Thus, it is possible to develop scenarios in order to detect possible future changes due either to global climatic changes or to altered land use schemes, which provide a sound basis for discussing phase space trajectories of ecosystem development in the light of the sustainability paradigm.

13.3.3   Modelling Biotic Interactions with Individual-Based Models

Ecological systems are known for their complex interactions covering several spatial and temporal scales. It is challenging to study those effects which may be: (a) caused on the local level by changes of the global climatic systems, or (b) to investigate the dynamics of populations on broader spatial scales than their habitats, based upon selected interactions of species on the individual level (Müller et al. 1997b; Müller and Nielsen 2000). Unravelling the bewildering complexity of food web structures is another essential part of an ecosystem analysis. Furthermore, individual organisms act as co-ordinating entities, which integrate internal processes into a coherent behaviour and physiological reaction to the environment. This coincides with the concept of nature protection focusing on target species (e.g. Reck 2003).

For a detailed presentation of the methodology and case studies that deal with the dispersal of ground-living arthropods and the complex role of the environment for reproduction success of bird species that depend on diverging climatic constraints, the reader is referred to Breckling et al. (2005).

13.3.4   Ecohydrological Modelling of Wetland Systems

While investigating the interactions between Lake Belau and its drainage basin it became clear that special attention must be paid to wetlands, because they act as ecotones between terrestrial and aquatic ecosystems and control the water and nutrient exchange between these major systems (Chapter 9). Wetland ecosystems have in general a high potential for the transformation and accumulation of carbon and nutrients as well as for maintaining the local biodiversity (Naiman and Décamps 1990; Pollock et al. 1998). However, when quantifying the effects of wetlands on the water and nutrient dynamics at the catchment scale, most frequently only rough estimates are avaiable. This uncertainty is caused:

- By the specific abiotic site conditions of wetlands, where analytical procedures developed for terrestrial ecosystems (or mineral soils) do not work well under wet and partly anaerobic conditions;
- Because most wetlands receive additional water and nutrient input via lateral or longitudinal inflow pathways.

While small-scale process-oriented research must address wet conditions and extremely sharp oxygen and redox gradients in the soil profile, at the landscape scale the quantification of inflowing substances is determined by the geohydrological conditions of the wetlands (Fränzle and Kluge 1997) and their position within a drainage basin (Groffman et al. 1988; Bedford 1999; Trepel and Palmeri 2002). In the Lake Belau catchment, therefore, wetland modelling activities started with a cartographic analysis of the geo-hydrological conditions between the mineral uplands and the lake basin (Piotrowski and Kluge 1994). With the 2D-groundwater flow model FLOTRANS (University of Waterloo, Centre of Groundwater Research, Canada), the water flow was simulated (cf. Chapter 9). The modelling results clearly underline that the wetland systems which surround Lake Belau receive lateral water and nutrient inputs both from the mineral uplands and from the lake. However, the exchange direction changes in the course of the year.

To obtain more precise water budgets for these wetland systems the nonlinear box-model FEUWA was developed (Kluge et al. 1994). The basic assumption of this model approach is that the water budget of a single site in a wetland can be calculated on the basis of climatic data, soil profile descriptions, geohydrological structure, and the water levels of the upper and lower surrounding groundwater systems and the lake, respectively. With this approach, the physical parameters needed for a process-based modelling system could be fitted with an automatic parameter acquisition algorithm. An automatic parameterization was chosen because measured soil physical properties are limited by the volume of the soil sample itself and, as a rule, do not reflect the geohydrological spatial heterogeneity of wetland systems. This concept was further developed into the FEUWANET multi-box model which can be used for modelling the surface and subsurface water exchange and water balance of heterogeneous riparian wetland sites in line with the main lateral groundwater flow direction (Dall'O et al. 2001).

Additionally, the WASMOD/STOMOD model system (Reiche 1996; Trepel 2000) was applied to quantify the nitrogen budget of peatlands used as wet meadows (Trepel 1999). Modelling the nitrogen budget with a process-based model at the site-scale identified denitrification as the main pathway for nitrogen loss from drained peatlands (Davidsson et al. 2002; Trepel and Kluge 2002a). Field methods failed in the quantification of these denitrification processes from peatland soils, however, because under wet conditions the acetylene inhibition method does not block denitrification completely due to insufficient diffusion of the acetylene into the soil-water phase (Mogge 1995).

At the same time the competent authorities of the State Agency for Nature and Environment of Schleswig–Holstein (LANU) began to develop a peatland restoration programme which aimed at restoring the nutrient retention capacity of riparian wetland systems. Based on the experience with water and nutrient exchange in lentic ecotones gained during the ecosystem research project, the wetland research team developed a scientifically founded and user-friendly methodology for quantitatively modelling water and nitrogen exchange between minerotrophic peatlands and their surroundings. To this end a decision support system was conceived with a reduced amount of input data, in particular basic geological and hydrological data such as the occurrence of impermeable layers of gyttja or clay, and the size of the upstream basin.

This WETTRANS model is based on a path-oriented budget approach which had primarily been developed for the quantification of nutrient flow processes between Lake Belau and its contiguous terrestrial ecosystems (Kluge 2004, Trepel and Kluge 2004). It considers inflow pathways for the transport of water and nitrogen to the wetland and outflow pathways for the transport of water and nitrogen from the wetland to the neighbouring water body. The different inflow pathways are characterized by different nutrient concentrations whose sum determines the nutrient load of a specific wetland. The outflow pathways differ in their physical and chemical conditions which can be connected with different nutrient transformation coefficients. By changing the drainage conditions of the wetland the resultant effects on nitrogen transformation are quantified. This information is important for environmental authorities when planning wetland restoration projects. The model identifies the sensitive pathways to and from a wetland on the basis of both the geohydrological structure and the management strategies and is useful to guide wetland managers successfully in the restitution of hydrological conditions (Trepel and Kluge 2002b).

13.3.5   Process-Oriented Modelling on the Landscape Level

The following modelling concepts try to develop scenarios for a future sustainable land use based on the development of scenarios of actual and future useful planning measures. Their importance for landscape planning will increase, although their implementation in planning procedures to date is scarce. The development of such scenarios for decision-makers requires a steady and careful study of realized measures in order to learn from possible mistakes. Improvements include the development of GIS-supported scenarios of land use measures and process-oriented simulations to assess the consequences of economic decisions for the ecological integrity of the sites involved (Reiche et al. 1999; Meyer 2000).

13.3.5.1 Improvement of Planning Proposals by Means of Macro-Scale Landscape Scenarios

The intensity and manner of agricultural land use determines structures and processes of ecosystems and their interactions in rural landscapes. Landscape planning according to the German federal conservation act tries to harmonize various demands in order to enable options for recent and future sustainable land use concepts. Actual shortcomings in landscape planning processes are, for instance:

- An insufficient consideration of the environmental media: soil, water, and atmosphere;
- a missing integrative approach connecting ecological and economical aspects;
- for the above reason, a low acceptance of the relevant proposals by landowners.

Based on the actual land use pattern, rule-supported scenarios of possible landscape changes were developed, which involved novel GIS techniques. One scenario simulates the impact of an intensified agriculture, the other a more 'ecological' way of agricultural production, thus reflecting two developmental extremes. The landscape analysis is based on the software package DILAMO (Reiche et al. 1999). The simulation model component of this package, WASMOD/STOMOD, reacts sensitively to changes of land use and permits to quantify different ecological indicators and provides economically relevant information. Without the application of these efficient computer-aided methods neither small-scale nor large-scale land use changes could be evaluated in high qualitative and quantitative resolution. The application of such scenario techniques is especially useful in simulating multiple uncertain factors concerning the potential outcomes of prospective land use schemes. In a case study for the Bornhöved Lake District, Meyer (2000) assumed three different strategies for the agricultural sector, i.e., expecting no changes (recent situation with many examples of excessive fertilizer application), or giving priority to the principles of organic farming (scenario B) and allowing agricultural management schemes as they could be expected due to further globalization processes (scenario A, agro-industry). A change to an 'ecological' land use (scenario B) may decrease the nutrient load of contiguous ecosystems and should be considered in sensitive areas. But also the results of scenario A reveal that part of the recent nutrient losses can be avoided, if existing regulations and proposals are applied rigorously (cf. Fig. 13.4).

A further example of the advantages and prospects of the scenario approach is provided for the Hohenschulen experimental farm of the Kiel University by Weiss (2000). Here again the comprehensive WASMOD/STOMOD model system proved superior to other, and generally distinctly less differentiated, modelling approaches in its capacity to precisely predict nitrogen losses in dependence on climate and soil properties.

13.3.5.2 Identification of Areas Liable to Soil Erosion

The long-lasting fertilization of agro-ecosystems for centuries has induced a considerable nutrient accumulation of the top soil. The erosion and correlative

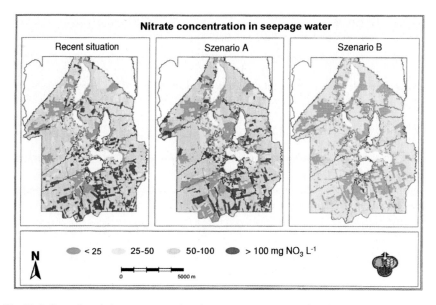

Fig. 13.4 Scenarios of nitrate concentrations in seepage water, comparing the actual situation in the Bornhöved Lake District with variants: *Scenario A* agro-industry, *Scenario B* biological farming

sedimentation of this surface material is an important component in the evaluation of the sensivity of water bodies with regard to eutrophication. For more than a century soil scientists, geomorphologists, and hydrologists have tried to describe these processes in an adequate form, and a considerable number of erosion models has been developed, but their practical use is often limited by time-consuming procedures to determine the model parameters. From the authors' point of view, at present only the USLE Model (Bork 1991) lends itself for convenient larger-scale use. The software package DILAMO by Reiche et al. (1999) comprises modules to estimate long-term means of soil erosion based upon the 'Universal Soil Losses Equation (USLE)', taking complex shapes of slopes into account as well as linear landscape elements (e.g. hedgerows) mitigating surface runoff and soil erosion. Thereby, DILAMO is solely based on data provided by German agencies. The calculations model different land use scenarios, and results are available at the level of landscapes and for selected water bodies. This approach was applied to the Farver Au catchment (32 km²), 50 km southeast of Kiel on behalf of the State Agency for Nature and Environment of Schleswig-Holstein (LANU). As in the Bornhöved Lakes District relief and soils in the Farver Au catchment are formed from moraines of the Weichselian which display a considerable relief energy and a maximum difference in altitude of 140 m. The target of the study was to identify the areas liable to soil erosion and complementarily those where changes in land use could achieve a maximum reduction of soil losses.

Three different land use regimes for the whole catchment were taken into account:

1. Typical crop rotation: winter-rape or green fodder, winter-wheat, winter-barley;
2. modified crop rotation: root crop, wheat, barley;
3. black fallow for the totality of agricultural land.

While (1) and (2) represent realistic land use schemes, (3) was designed to highlight those areas where highest soil loss rates would result from unsuitable land management practices. As expected, the calculated values for annual soil losses differ significantly for the land use regimes: (1) 1773 t year^{-1}, (2) 3043 t year^{-1}, and (3) 21.757 t year^{-1}.

The major advantage of the procedure is the combined use of grid-based and polygon-based spatial data as well as information on linear landscape elements. This allows not only to account for barrier effects of the latter, but also to identify those areas which contribute by direct soil loss to sedimentation processes in the receiving waters (cf. Fig. 9.6). This is achieved by indicating the grid-squares with direct surface runoff into the neighbouring receiving water. Since all waters can be identified by GIS-procedures, it is possible to indicate the sediment-load for each sector of the water courses. In most cases waters with extraordinarily high sediment loads are small brooks or ponds adjacent to arable land. The overall freshwater systems in the watershed studied receive 77 t year^{-1} in variant (1), 129 t year^{-1} in variant (2), and 950 t year^{-1} in (3), respectively. The importance of small ponds as sediment sinks should not be underestimated, especially when they are situated in arable land on large slopes. The DILAMO results also reveal that 13% of the overall soil erosion in the Farver Au watershed is held back by linear landscape elements like hedgerows, thus indicating their relevance, when landscape management close to nature with low soil losses is intended.

13.3.5.3 Potentials for Precision Farming

The use of the Global Positioning System (GPS) to precisely drive agricultural machines on (larger) fields allows adjusting management schemes according to small-scale site-specific requirements. This option permits to differentially account for environmental targets on small plots within fields which are are otherwise and habitually treated uniformly. To study the applicability of GPS-supported precision agriculture as well as the environmental and economic effects have been the goals of the 'pre-agro' project. Again DILAMO was used to provide for high spatial resolution and to run model-based scenarios in order to reveal potential environmental benefits by site-specific agricultural management strategies. While economic benefits are closely related to yields which can be measured comparatively well, the identification of environmental effects not only requires extensive estimates of multiple ecological indicators but also the consideration of longer time-spans and larger spatial units. Thus, the surrounding landscape has to be studied in order to identify off-site effects. Since in 'pre-agro' only model studies have been conducted, the results are currently limited to scenario-based trend analyses based upon plausible assumptions and the application of landscape models which had been calibrated in previous projects. Thus, as a part of the 'pre-agro' project the potential environmental effects of site-specific farming have been assessed in two regions, namely Thumby (Schleswig–Holstein) and Kassow (Mecklenburg–Vorpommern) which have both an acreage of several thousand hectares (Windhorst et al. 2002).

13.3.5.4 Proposals

The results show that the software package DILAMO developed during the "Bornhöved Project" and no less successfully applied in quite a few other projects allows to set up scenarios in order to analyse the effects of precision farming at different spatial scales. The results of the scenarios indicate that a significant reduction of nutrient losses cannot be expected per se from precision farming. Only if the reduction of nutrient losses becomes an additional target besides the production of harvestable biomass site-specific farming proves superior to conventional management practices. Furthermore the results show that the most beneficial effects of precision farming might be to achieve environmental goals, e.g., to optimize the multi-functionality of agricultural production schemes. To which extent optimized placing of seed, better timing of fertilization, etc. is able to cover the additional costs of site-specific farming or to increase the variable gross margin is still uncertain. However, distinct economic advantages could be achieved, if precision agriculture were subsidized as a means to reduce agricultural emissions. The implementation of this advanced technique based upon global positioning techniques and digital maps would also create a new basis for organizing the justification and payment of environmental services rendered to the society by the agricultural sector. While the official bodies could endow the farmers with needed digital site information and maps with special environmental targets, the farmers could provide evidence for meeting the official requirements by documenting the whole production process on a digital basis in space and time. Provided a close cooperation between primary production, nature protection and environmental bodies could be brought about precision farming could become a modern methodological tool to achieve an efficient multifunctional landscape management.

As the presented scenarios are still limited to wheat production, future developments of the software will have to account for crop rotations including all typical cash crops, thus allowing to analyse the suitability of precision farming for other areas or organic farming on a wide spectrum of spatial scales.

13.3.6 Ecological Economics and Scenario-Guided Adjustment of Control Systems

High nutrient loads and changes in land use may seriously impair the integrity of ecosystems with regard to soil buffer capacity, nutrient pools as well as the structural diversity of landscapes including their specific abiotic and biotic components. Therefore Dibbern (2000) analysed the ecological and economic feedback mechanisms of varying land use measures in the Bornhöved District. The key item was to develop integrative evaluation tools by coupling economic planning approaches with ecological simulation models of land use. Thus, the approach includes:

- Analyses of the Bornhöved farm structures by empirical spot-tests and statistical data evaluation;
- regionalization of the data obtained;
- data bank administration and GIS-coupling;
- formulation of a site-dependent economical planning model as a precondition for a coupling with ecological simulation runs;
- evaluation of the procedure in order to develop sociological and economical scenarios for the Bornhöved area;
- synoptic characterization of the economical and ecological feedbacks on ecosystem integrity induced by a change in site management practices.

The economic model system is based on linear optimization procedures which reflect the influence of a successive (five-step) increase of nitrogen fertilizer prices from 0.45 ~ kg^{-1} to 2.50 ~ kg^{-1} on the spatial and technical organization of representative model farms. In comparison with the actual situation (AS) the extreme,

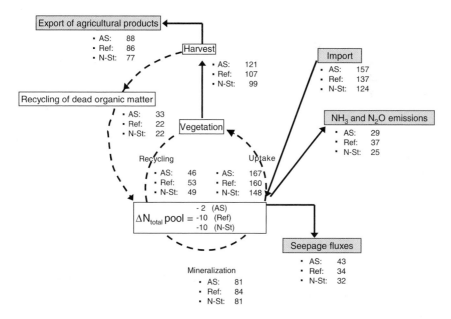

Fig. 13.5 Nitrogen balance of the study area, calculated for different scenarios of the 1992–1996 period. (all data in kg N ha^{-1} year^{-1}). *AS* Actual situation. *Ref* Reference scenario, i.e., an economically optimized production structure. *N-St* nitrogen tax scenario

and most illustrative, variants of the price spectrum are called reference scenario (Ref = 0.45 ~ kg^{-1} N fertilizer) and N-St (nitrogen tax scenario with 2.50 ~ kg^{-1} N fertilizer). The output of these economic model runs serves as an input to the subsequent ecological model system WASMOD/STOMOD which depicts the consequences of fertilizer-dependent adjustment strategies of the model farms in terms of land use patterns and related nitrogen leaching rates. Figure 13.5 summarizes the results of this ecologico-economic scenario technique, in the sense of a nitrogen balance for the entire study area. It shows that profits mostly decrease, and regionally an expansion of monocultures (silage maize) takes place on better soils.

Within the economically optimized production scenarios different variants of crop rotation are possible, e.g., crop rotations with mainly rape and oat or (increasingly) also silage maize monocultures. In grassland systems, a marked difference exists between extensively used pastures on sites with less than 30 land-value points and pastures intensively used for grazing and mowing purposes. Since farm extensions such as capacity enhancements or stable extension are not formulated in the model, both cattle number and the proportion of grassland, cropland, forests, and settlements remain nearly stable in the variants. Therefore, the highest deterioration in marginal utility occurs in farms with exclusive crop production as a consequence of expected rising nitrogen prices, while the lowest changes occur in the finishing sector. From the ecological results of the simulation it follows that a reduction in nitrogen seepage by optimized economic land use can only be achieved at the expense of a latent all-over N storage in the soil.

Chapter 14
Conclusions: Perspectives for Integrative Landscape Planning, Management and Monitoring

Klaus Dierssen and Jan Barkmann

14.1 Introduction

Integrative landscape ecology in cultural landscapes encompasses complex systems, long-term processes, uncertainties, limited but ill-defined carrying capacities, and external effects. One of its prime areas of application is rural landscapes, i.e. areas that are primarily managed for biological productivity. The emphasis on agricultural productivity comes with serious trade-offs in respect to non-agricultural landscape functions and ecosystem services.

A major challenge to implement more sustainable land use practises results from the differing time-scales of ecological, economic and social processes. These time-scales make it more difficult to integrate social and economic research into the environmental sciences. Due to the complexity of ecosystems and the long-term cycles of the processes involved, strict prognoses are rarely possible. Frequently, planners have to rely on the construction and analysis of (computer-)modeled scenarios. The appropriate modeling tools are being developed by the new ecosystem and landscape sciences. These tools should be fairly easy to understand for planners, and – supported by visualizations – contribute to communicate effectively with stakeholders and decision makers (cf. Chapter 13).

A certain land use pattern and intensity can only be judged effectively, if both the targets and time-span are sufficiently quantified. Still, if the societal preference patterns for the targets change during the time-span, if the understanding of the desirability of certain ecosystem states, structures or processes improves, or our perception of the ecological and social impact of the land use patterns changes, we may have to deal with considerably moving targets. However, there is no alternative to a continuously adjusting management process into which new information is iteratively fed and evaluated.

O. Fränzle et al. (eds.), *Ecosystem Organization of a Complex Landscape.*
Ecological Studies 202.
© Springer-Verlag Berlin Heidelberg 2008

14.2   Definition of Goals and Valuation Procedures

For the procedures of target formulation and aggregation based on the detailed planning process, the incorporation of the public is necessary in order to develop participative proposals for goals and adequate measures for target realization.

A conceptual model that integrates ecological as well as normative and applied demands on sustainable land use management was developed in the third phase of the Bornhöved Project. Existing concepts on decision-making processes with strong stakeholder participation (Werner et al. 1997; Wiegleb 1997; Horlitz 1998) were cast into a discourse ethics framework (Habermas 1992; Ott 1994, 1996). The resulting approach aims at meeting three demands:

- The definition of regional objectives for sustainable land use has to be discursive.
- Ecosystem models should be employed to support decision-making processes.
- The complexity of knowledge on potentially relevant states of the socio-ecological systems of the planning region must be reduced in order to be intelligible.

The method is based on a discursive and iterative assessment of management scenarios in reference to the value systems of the concerned stakeholders. A "unified" socio-ecological model provides data on likely system states that correspond to different hypothetical management scenarios. An integrated representation of the projections of the states of the socio-ecological model onto the value systems of the stakeholders can be achieved in an abstract *valuation space* that comprises the basic norms and values relevant to environmental decision making (Heiland 1999; Barkmann 2000; Barkmann and Windhorst 2000). The abstract representation of management scenarios by modelled system states in the valuation space must be translated into measures that can be communicated to non-expert stakeholders. Indicators of sustainable development are an instrument to facilitate that communication. They can be defined as: *a form of constructed environmental knowledge that serves the assessment and/or documentation of ecological, economic, and social system states in reference to the objectives of the sustainable development paradigm as locally applicable employing (quantifiable) phenomena* (Barkmann 2002).

Once sufficiently quantified targets are defined, indicators can be used to prioritize conflicting land use proposals or conflicting regional management strategies. In these cases, the indicators are employed to assess systematically the merits of differing options by comparing their putative effects assessed by a socio-ecological model ("MODENA": *mo*del-aided *d*iscursive *en*vironmental *a*ssessment; Barkmann 2001b).

In the ideal case, a consensually legitimized system of aggregated indicators assigns a unique ranking number to the set of development options. In such an indicator system, the consensus norm and value system of all stakeholders is encoded. If quantified targets are not defined in advance, the definition of the indicator system amounts to the same task as the previous section. Technically, these objectives are quantified and spatially disaggregated *environmental quality standards*.

If a discursive procedure is required for the definition of regional sustainable development objectives, it is thus also often required for the definition of a system of sustainable development indicators. There is a tight mutual relationship between the definition of objectives for complex socio-ecological systems, which is facilitated by complexity-reducing indicators, and the definition of indicators for decision-support, which must reflect the objectives. In order to account for this relationship, model-aided discursive environmental assessment is complemented by a *d*iscursive *i*ndicator *d*evelopment (DID; Barkmann and Windhorst 1999). This procedure is in line with the demand of the German Advisory Council on the Environment which requires an objectives-driven indicator development (SRU 1996). The definition of environmental quality targets, in turn, should be based on properly represented scientific information.

14.3 Appropriate Planning Procedures – Integrative Analysis

Ecosystem research is guided by a hierarchically constructed system of theories and hypotheses which in turn can be represented by hierarchically nested models (cf. Section 1.2). The application of theoretical ecosystem concepts in ecosystem planning and management should be guided in a comparable way by a hierarchical model system. Landscapes as complexes of ecosystems have multiple uses. The preservation of non-renewable landscape functions and services is a precondition for their unimpeded long-term use. Ecosystems can produce, accumulate, store, convert, and deliver energy and substances or act as buffers. Their services include the maintenance of soil fertility, climate regulation and natural pest control. They provide flows of ecosystem goods such as food, timber and fresh water, but also intangible benefits such as aesthetic and cultural values (Daily 1997; Chapin et al. 2000).

There are numerous examples to valuate ecosystem services and assess ecosystems from a natural capital point of view (cf. Costanza et al. 1989, 1997). Based on an earlier publication of van der Maarel and Dauvellier (1978), de Groot (1992, 1994) proposed a functional assessment of the natural environment for planning purposes. He distinguished:

1. *Regulating functions* are related to the capacity of ecosystems to maintain essential ecological processes, which *inter alia* contributes to an environment providing clean air, water and soil, and potentials for biological preservation and restoration.
2. *Carrier functions* provide space and suitable substrates for human activities such as habitation, infrastructure, cultivation, removal of refuse, recreation and sport.
3. *Production functions* provide natural resources ranging from those that are based on biological productivity (agriculture, forestry) to mineral and energy supply.
4. *Information functions* deal with aesthetical and emotional potentials (history, spirituality, symbolization, beauty, authenticity, summarized as 'intrinsic appeal') which contribute to the maintenance of mental health by providing opportunities for cognitive development, reflection, spiritual enrichment and aesthetic experience.

Transferred to the sustainable development concept, it seems wise to preserve these functions and the actual and potential benefits to human society *at the same site* for current and future generations. This approach in parts conforms with former concepts used in German landscape planning (Bierhals 1980).

Most procedures in landscape and conservation planning processes and a subsequent target monitoring are and will increasingly be supported by IT and database technology. Planning decisions will incorporate computer-based decision support systems (DSS), processing comprehensive knowledge and experience from experts in landscape management and planning (Asshoff 1999; Dibbern 2000; Herzog 2002; Sodtke 2003). Nonetheless, land use and restoration planning needs the following fundamental items:

1. An analysis that includes the actual ecological design and a historical review of the key changes;
2. A search for potential landscape functions and services as well as for remnants of less intensively changed ecosystems;
3. A detailed description of point and non-point pollution sources and pathways;
4. A combined ecological and economic risk assessment that includes an analysis of the potentials for source buffering or elimination;
5. A definition of the ecological, economic and social quality goals for the landscape ('region') and their inhabitants, including a hierarchy of management measures;
6. Developing differentiated and integrated management and restoration concepts;
7. Developing an integrated monitoring based on the degree of quality goal realization defined in point 5, considering ecological and economic needs and social acceptance.

14.4   Realization of Planned Measures

The key problems for recent landscape planning and management procedures are: (a) the often insufficient knowledge on the abiotic and biotic state of a landscape or its parts, and (b) the lack of a precise definition of the desired landscape state. In such a situation, an appropriate integrative monitoring of project impacts as well as on conservation successes is impossible.

Landscape planning actually includes two parts. The first part is a legal framework that includes material and procedural provisions for the definition of quality targets and standards. The second is an 'instrumental' part that deals with technical standards of the systematic planning procedure itself. In an evaluation of 40 current landscape plans in Schleswig–Holstein, even a generous analysis concluded that complete and balanced target systems can hardly be found (Scholtissek 2000). The development of such target systems is a challenging enterprise in the face of huge amounts of sectoral, supposedly integrative planning targets. Additionally there is considerable expert argument on these targets, and their application to different sites, landscapes and ecosystem types (e.g. Wulf 2001). Ideally, the rational discourse on targets and standards should be informed by scientific knowledge, reproducible, internally consistent, in line with superior legal or moral norms – and thus comprehensively justified (Bechmann 1988; Schröder 1998; Romahn 2003). For the target formulation and aggregation based on the detailed planning process, the incorporation of the public is necessary in order to develop participative proposals for goals and consistent measures for target realization.

Restoration measures in agriculturally misused sites in northwest Germany through eco-technical restorations can result in significant improvements to ecosystem functioning nearby intensively cultivated landscapes. Meanwhile it seems a strategy to put the cart before the horse in primarily establishing parks and reserves for conservation without preserving the functional integrity of landscapes in general and at various spatial scales. An unalterable precondition is the adjustment of cultivation to the carrying capacity of a catchment area (e.g. Schuller et al. 2000).

14.5 Ecosystem-Oriented Monitoring and Feedback to Planning and Management Processes

A target monitoring can support three key functions:

- Identification and definition of recent and future environmental problems;
- Preparing programme objectives, priorities and selections of environmental politics and management decisions;
- Evaluating impacts on environmental quality resulting from current land use, ecosystem management and political programs.

The aims of inventorying assessment and sustainable use of ecosystem and biological resources require the elaboration of multilevel and multiscale concepts of various intensity. The ecosystem approach is only useful with aggregated indicators, which are calibrated in reference sites. Concerning biological diversity, the calibration should be undertaken for different selected species groups in the same reference areas. Commendable concepts and approaches exist (di Castri et al. 1992; Lugo 1996; Dierssen 2006), but a detailed elaboration and general acceptance are still lacking.

In Germany, the implementation of an ecosystem monitoring in connection with an ecological area sampling (EAS) is proposed. On EAS level I, biotope types and features are mapped on some 800 plots (size 1 km²) to characterize landscape and biotope quality. On level II, phanerogams and selected animal species groups (first step: grasshoppers, carabides) will be recorded, and birds on some 270 plots (Dröschmeister 2001). Functional macro-indicators of ecosystem quality and services should be implemented.

Summing up, the development of an integrated environmental monitoring based on the degree of quality goal realization is proposed, considering ecological and economic needs and social acceptance (GDV 2000; SRU 1991, 1996). An exemplary implementation of this concept in the Biosphere Reserve Rhön has been conducted during 1997–2001 (Schönthaler et al. 2003).

14.6   Transfer of Knowledge to the Community, to Stakeholders and to Decision Makers

Society, stakeholders and politicians today normally react to present-state problems and try to solve them sectorally and commonly with a retarded response. For a precautionary and in that way sustainable policy advice, there is a need of indicators and impact assessment tools going beyond a pure description of the state of the currently highlighted problems. Mechanistic impact management should, step by step, be replaced by an adaptive management that incorporates feedback cycles of varying length. By developing 'intelligent' overall trans-disciplinary models characterizing the environmental, economic and social services of landscape complexes, a structured evaluation will be possible to define, e.g. budgetary adoptions on a set of sustainability indicators. In practice, as a first step, relevant environmental data in a regularly updated form will be increasingly available, e.g. from the State Agency for Nature and Environment of Schleswig–Holstein (LANU) in order to make environmental information become faster and to increase the transparency of public administration measures and procedures (Bornhöft 2003; Barkmann 2004).

References

Aerts R, Chapin FS III (2000) The mineral nutrition of wild plant revisited: a re-evaluation of processes and patterns. Adv Ecol Res 30:1–67

AG Bodenkunde (1982) Bodenkundliche Kartieranleitung. Schweizerbart, Stuttgart

AG Bodenkunde (1994) Bodenkundliche Kartieranleitung. Schweizerbart, Stuttgart

Alcamo J, Bennet E (2003) Ecosystems and human well being: a framework for assessment (Millennium Ecosystem Assessment). Island Press, London

Alexander M (1982) Most probable number method for microbial populations. In: Page AL, Miller RH, Keeney DR (eds) Methods of soil analysis. Part 2. Am Soc Agronomy, Madison, pp 815–820

Allard RW (1999) History of plant population genetics. Annu Rev Genet 33:1–27

Allen TF, Hoekstra TW (1991) Role of heterogeneity in scaling ecological systems under analysis. In: Kolasa J, Pickett STA (eds) Ecological heterogeneity. (Ecological studies 86) Springer, Berlin Heidelberg New York, pp 47–68

Allen TFH, Hoekstra TW (1992) Toward a unified ecology. Columbia University Press, New York

Allen TFH, Starr TB (1982) Hierarchy: perspectives for ecological complexity. University of Chicago Press, Chicago

Altmoos M (1999) Netzwerke von Vorrangflächen. NatSchutz LandschPlan 31:357–367

Ambsdorf J (1996) Phytophage Arthropoda in verschiedenen Erlenbeständen (*Alnus glutinosa*) der Bornhöveder Seenkette unter besonderer Berücksichtigung des Kronenraumes. Faun Oekol Mitt Suppl 20:77–110

Anderson JM (1975) Succession, diversity and trophic relationships of some soil animals in decomposing leaf litter. J Anim Ecol 44:475–495

Arakaki T, Morse MJ (1993) Coprecipitation and adsorption of Mn(II) with mackinawite (FeS) under conditions similar to those found in anoxic sediments. Geochim Cosmochim Acta 57:9–14

Arx G von, Bosshard A, Dietz H (2002) Land-use intensity and border structures as determinants of vegetation diversity in an agricultural area. Bull Geobot Inst ETH 68:3–15

Asman WAH, Sutton MA, Schjørring JK (1998) Ammonia: emission, atmospheric transport and deposition. New Phytol 139:27–48

Asshoff M (1990) Die Mollusken des Belauer Sees und seines Abflusses (Schleswig-Holstein) unter Berücksichtigung produktionsbiologischer Aspekte. MSc thesis, University of Kiel, Kiel

Asshoff M (1999) Die Erschließung und Modellierung ökologischen Wissens für das Management von Feuchtwiesenvegetation – ein Beispiel für die Aufbereitung ökologischen Wissens und den Transfer mit hybriden Expertensystemen. EcoSys Suppl 27

Asshoff, M. Pöpperl R, Böttger K (1991) Ökosystemforschung im Bereich der Bornhöveder Seenkette: Vergleichende Untersuchungen zur Habitatpräferenz und Produktion der Molluken im Belauer See und seinem Abfluß (Schleswig-Holstein). Verh Ges Oekol 20:223–228

Atlas RM, Bartha R (1998) Microbial ecology: fundamentals and applications. Benjamin Cummings Science, Menlo Park

Aue CH (1993) Die Bedeutung der Stoffkonzentration in der Bodenlösung für den Stofffluss in forstlich und agrarisch genutzten Böden einer Norddeutschen Jungmoränenlandschaft im Bereich der Bornhöveder Seenkette. SchR Inst Pflanz Bodenkde 23

Baar HJW de, Saager PM, Nolting RF, Meer J van der (1994) Cadmium versus phosphate in the world ocean. Mar Chem 46:261–281

Baars MA (1979) Patterns of movement of radioactive carabid beetles. Oecologia 44:125–140

Baccini P (1985) Phosphate interactions at the sediment–water interface. Stumm W (ed) Chemical processes in lakes. Wiley, New York, pp 189–205

Baccini P, Suter U (1979) MELIMEX, an experimental heavy metal pollution study: chemical speciation and biological availability of copper in lake water. Schweiz Zeitschr Hydrol 41:291–314

Bach HJ (1996) Bakterielle Populationen und Stoffumsatzpotentiale in Acker-, Grünland- und Waldböden einer Jungmoränenlandschaft in Schleswig–Holstein. EcoSys Suppl 15

Balistrieri LS, Murray JW, Paul B (1992) The biogeochemical cycling of trace elements in the water column of Lake Sammamish, Washington: response to seasonally anoxic conditions. Limnol Oceanogr 37:529–548

Balistrieri LS, Murray JW, Paul B (1994) The geochemical cycling of trace elements in a biogenic meromictic lake. Geochim Cosmochim Acta 58:3993–4008

Bar-Yosef B (1991) Root excretions and their environmental effects: influence on availability of phosphorus. In: Waisel Y, Eshel A, Kafkafi (eds) Plant roots: the hidden half. Dekker, New York, pp 529–557

Barkmann J (2000) Eine Leitlinie für die Vorsorge vor unspezifischen ökologischen Gefährdungen. In: Jax K (ed) Funktionsbegriff und Ungewissheit in der Ökologie. Lang, Bern, pp 139–152

Barkmann J (2001a) Angewandte Ökosystemforschung zwischen Biodiversitäts-, Landschafts- und Ressourcenschutz. Petermanns Geogr Mitt 145:16–23

Barkmann J (2001b) Ökologische Integrität. In: Fränzle O, Müller F, Schröder W (eds) Handbuch der Umweltwissenschaften, Kapitel VI-3.8.2. Ecomed, Landsberg

Barkmann J (2002) Modellierung und Indikation nachhaltiger Landschafts entwicklung. EcoSys 9

Barkmann J (2004) Entwicklung von "angemessenen" Indikatoren für eine Nachhaltige Entwicklung – Beispielfall Schleswig-Holstein. In: Wiggering H, Müller F (eds) Umweltziele und Indikatoren: Wissenschaftliche Anforderungen an ihre Festlegung und Fallbeispiele. Springer, Berlin Heidelberg New York, pp 573–605

Barkmann J, Windhorst W (1999) A discursive approach to defining regionalised sustainability indicators. Paper presented at the 5th auDes-Conference, April 15–17, Zurich

Barkmann J, Windhorst W (2000) Hedging our bets: the utility of ecological integrity. In: Jørgensen SE, Müller F (eds) Handbook of ecosystem theories and management. Lewis, Boca Raton, pp 497–517

Barkmann J, Baumann R, Meyer U, Müller F, Windhorst W (2001) Ökologische Integrität: Risikovorsorge im nachhaltigen Landschaftsmanagement. Gaia 10:97–108

Barkmann S (1998) Ökologische Untersuchungen zur Bedeutung des photoautotrophen Picoplanktons in der Planktonbiozönose des eutrophen Belauer Sees (Schleswig-Holstein). PhD thesis, University of Hamburg, Hamburg

Barkmann S, Fleckner W, Zimmermann H, Kausch H (1994) Nahrungsbeziehungen im Pelagial des Belauer Sees. Interne Mitteilungen zur Ökosystemforschung im Bereich der Bornhöveder Seenkette, April. pp 225–238

Basedow TH, Rzehak H, Liedtke W (1988) Die Bedeutung von mehrjährig wiederholten großflächigen Pestizidanwendungen auf die Stabilität und die Produktivität von Agrarökosystemen. Ergebnisse vergleichend experimenteller Freilanduntersuchungen zur Frage der Belastbarkeit von Agrarökosystemen, 1981–1986. Spez Ber Kernforschungs Jüleich 9:223–386

Bauch G (1966) Die einheimischen Süßwasserfische. Neumann–Neudamm, Melsungen

Baumann R (2001): Konzept zur Indikation der Selbstorganisationsfähigkeit terrestrischer Ökosysteme anhand von Daten des Ökosystemforschungsprojekts Bornhöveder Seenkette. PhD thesis, University of Kiel, Kiel

Baumgarten A, Kinzel H (1990) Mikrozonen im Stammfußbereich von Buchen: Untersuchungen der bodenbiologischen Aktivität. In: Albert R, Burian K, Kinzel H (eds) Zustandserhebung Wienerwald. Verlag der Österreichischen Akademie der Wissenschaften, Vienna, pp 247–288

Bazzaz FA, Carlson RW (1982) Photosynthetic acclimation to variability in the light environment of early and late successional plants. Oecologia 54:313–316

BBodSchG (1999) Gesetz zum Schutz vor schädlichen Bodenveränderungen und zur Sanierung von Altlasten. BGBl. I, G5702, 16:502–510

Bechmann A (1988) Grundlagen der Bewertung von Umweltauswirkungen; die Nutzwertanalyse. Handbuch UVP 1.IX/88

Beck L, Dumpert K, Franke U, Mittmann HW, Römbke J, Schönborn W (1988) Vergleichende ökologische Untersuchungen in einem Buchenwald nach Einwirkung von Umweltchemikalien. In: Scheele B, Verfondern M (eds) Auffindung von Indikatoren zur prospektiven Bewertung der Belastbarkeit von Ökosystemen. KFA, Jülich, pp 548–701

Becker A, Lahmer W (eds) (2004) Wasser- und Nährstoffhaushalt im Elbegebiet und Möglichkeit zur Stoffaustragsminimierung. Forschungsverbund Elbe-Ökologie: Bd. 1. Weissensee, Berlin

Bedford BL (1999) Cumulative effects on wetland landscapes: links to wetland restoration in the United States and southern Canada. Wetlands 19:775–788

Beese F (1986) Parameter des Stickstoffumsatzes in Ökosystemen mit Böden unterschiedlicher Azidität. Göttinger Bodenkde Ber 90

Behre GF (1983) Die Sieb-Flotations-Methode – Bau und Erprobung eines ökologischen Arbeitsgerätes zur mechanischen Auslese von Bodenarthropoden. BSc thesis, University of Bonn, Bonn

Behrendt H, Brüggemann R (1993) Modelling the fate of organic chemicals in the soil plant environment: model study of root uptake of pesticides. Chemosphere 27:2325–2332

Beinhauer R, Hörmann G, Piotrowski JA, Schernewski G, Spranger T (1991) Ökosystemforschung im Bereich der Bornhöveder Seenkette: Auswirkung des extremen Sturms vom 28.8.1989 auf den Wasser- und Stoffhaushalt des Einzugsgebiets "Belauer See" (Schleswig–Holstein). Verh Ges Oekol 20/21:173–180

Berner RA (1981) A new geochemical classification of sedimentary environments. J Sediment Petrol 51:359–365

Bertram C (2002) Verteilung, Wachstum und Nahrungsökologie der Larven und Jungfische im Belauer See, Schleswig–Holstein. PhD thesis, University of Hamburg, Hamburg

Bezzel E, Prinzinger R (1982) Ornithologie. Ulmer, Stuttgart

Beyer L, Wachendorf C, Köbbemann C (1993a) A simple wet chemical extraction procedure to characterize soil organic matter. Comm Soil Sci Plant Ann 24:1645–1663

Beyer L, Blume H-P, Hinß B, Peters M (1993b) Soluble aluminium-and-iron-organic complexes and carbon cycles in hapludalfs and haplorthods under forest and cultivation. Sci Total Environ 138:57–76

Bierhals E (1980) Ökologische Raumgliederungen für die Landschaftsplanung. In: Buchwald K, Engelhardt W (eds) Handbuch für Planung, Gestaltung und Schutz der Umwelt 3.BLV Verlagsgesellschaft, Wien, pp 80–104

Bleeker A, Draajers GPJ, Klap JM, Jaarsveld JA van (2000) Deposition of acidifying components and base cations in Germany in the period 1987–1995. RIVM Report 7221080027. RIVM, Bilthoven

Bloem J, Hopkins D, Benedetti A (2006) Microbiological methods for assessing soil quality. CABI, Wallington

Blume H-P (1984) Definition, Begrenzung und Benennung von Bodenlandschaften. Mitt Dtsch Bodenkde Ges 40:196–176

Blume, H-P, Schimming C-G (1991) Böden Schleswig-Holsteins als Quellen und Senken für Nähr- und Schadstoffe. Forstw Cbl 110:221–227

Blume H-P, Sukopp H (1976) Ökologische Bedeutung anthropogener Bodenveränderungen. SchR Vegetationskde 10:75–89

Böhm W (1979) Methods of studying root systems. In: Billings WD, Golley F, Lange OL, Olson JS (eds) Ecological studies 33. Springer, Berlin Heidelberg New York

Bohn U, Gollub G, Hettwer C, Neuhäuslová Z, Schlüter H, Weber H (eds) (2003) Karte der natür-lichen Vegetation Europas Maßstab 1:2 500 000, 655 S. + Kartenanhang. Bundsamt für Naturschutz, Bonn-Bad Godesberg

Bollens U, Güsewell S, Klötzli F (1998) Zur relativen Bedeutung von Nährstoffeintrag und Wasserstand für die Biodiversität in Streuwiesen. Bull Geobot Inst ETH 64:91–101

Bork HR (1991) Bodenerosionsmodelle. Ber Landwirtsch NF 205:51–67

Bork HR, Schröder A (1996) Quantifizierung des Bodenabtrags anhand von Modellen. In: Blume H-P, Felix-Henningsen P, Fischer WR, Frede H-G, Horn R, Stahr K (eds) Handbuch der Bodenkunde. Ecomed, Landsberg

Borkowsky O, Schmalhaus U (2004) Geobotanische Untersuchungen von Hecken und heckenähnli-chen Biotopen in Mecklenburg-Vorpommern. Nat Natursch Mecklenburg–Vorpommern 37:3–35

Bornhöft D (1993) Untersuchungen zur Beschreibung und Modellierung des Bodenwasserhaushalts entlang einer Agrar- und einer Wald-Catena im Bereich der Bornhöveder Seenkette (Schleswig-Holstein). EcoSys Suppl 6

Bornhöft D (1994) A simulation model for the description of the one-dimensional vertical soil water flow in the saturated zone. Ecol Model 75/76:269–278

Bornhöft E (2003) Das Geodatenmanagement im Natur- und Umweltinformationssystem. Jb Landesamt Nat Umwelt 2002:26–28

Bortmann I (1996) Heterogenitäten der Besiedlung durch Laufkäfer (Col.: Carabidae) in einem Buchenwald. Faun-Oekol Mitt Suppl 22:87–126

Bossel H (1998) Ecological orientors: emergence of basic orientors in evolutionary self-organization. In: Müller F, Leupelt M (eds) Eco targets, goal functions and orientors. Springer, Berlin Heidelberg New York

Bossel H (2000) Sustainability: application of systems theoretical aspects to societal development. In: Jørgensen SE, Müller F (eds) Handbook of ecosystem theories and management. Lewis, Boca Raton, pp 519–536

Brabrand A, Faafeng BA, Nilssen JPM (1990) Relative importance of phosphorus supply to phyto-plankton production: Fish excretion versus external loading. Can J Fish Aquat Sci 47:364–372

Branding A (1996) Die Bedeutung der atmophärischen Deposition für die Forst- und Agrarökosysteme der Bornhöveder Seenkette. EcoSys Suppl 14

Braun-Blanquet J (1964) Pflanzensoziologie. Grundzüge der Vegetationskunde. 3rd ed Springer, Wien New York

Brechtel HM, Lehnhardt F (1982) Einfluss der Grundwasserabsenkung auf Waldstandorten. Fortbildungslehrgang Grundwasser Darmstadt, DVWK Fortbildung 4

Breckling B (1996) An individual-based model for the study of pattern and process in plant ecol-ogy. In: Breckling B, Asshoff M (eds) Modellbildung und Simulation im Projektzentrum Ökosystemforschung. EcoSys 4:241–254

Breckling B (1998) The application of the object-oriented programming paradigm to study eco-logical processes. ASU Newsl 24[Suppl]:15–26

Breckling B (ed) (2005) Emergent properties in individual-based ecological models. Case studies from the Bornhöved Project (Northern Germany). Ecol Model 186:376–507

Breckling B, Reuter H (1996) The use of individual based models to study the interaction of dif-ferent levels of organization in ecological systems. Senckenberg Mar 27:195–205

Breckling B, Reuter H, Middelhoff U (1997) An object oriented modelling strategy to depict acti-vity patterns of organisms in heterogeneous environments. Environ Model Assess 2:95–104

Breckling B, Müller F, Reuter H, Hölker F, Fränzle O (2005) Emergent properties in individual-based ecological models. Introducing case studies in ecosystem research context. Ecol Model 186:376–388

Breede S (2000) Die Anwendung der Hierarchitätstheorie in der Ökosystemforschung. Ecosys Suppl 30

Bruland KW (1980) Oceanographic distributions of cadmium, zinc, nickel, and copper in the North Pacific. Earth Planet Sci Lett 47:176–198

Bruun HH, Fritzbøger B (2002) The past impact of livestock husbandry on dispersal of plant seeds in the landscape of Denmark. Ambio 31:245–231

Bryant DM (1978) The factors influencing the selection of food by the House Martin (*Delichon urbica*). J Anim Ecol 42:539–564

Brüning H (2004) Umweltverträglichkeitsprüfung – Vom Königsweg zum Holzweg der Umweltpolitik? In: Wiggering H, Müller F (eds) Umweltziele und Indikatoren. Springer, Berlin Heidelberg New York, pp 479–516

Burdige DJ (1993) The biogeochemistry of manganese and iron reduction in marine sediments. Earth Sci Rev 35:249–284

Burel F, Baudry J (1994) Reaction of ground beetles to vegetation changes following grassland derelictation. Acta Oecol 15:401–415

Butterweck MD (1998) Metapopulationsstudien an Waldlaufkäfern (Coleoptera: Carabidae) – Einfluß von Korridoren und Trittsteinbiotopen. Wissenschaft und Technik, Berlin

Caldwell MM, Meister H-P, Tenhunen JD, Lange OL (1986) Canopy structure, light microclimate and leaf gas exchange of *Quercus coccifera* in a Portuguese macchia: measurements in different canopy layers and simulations with a canopy model. Trees 1:25–41

Camarero L, Catalán J (1991) Horizontal heterogeneity of phytoplankton in a small high mountain lake. Verh Int Ver Limnol 24:1005–1010

Cardinale BJ, Ives AR, Inchausti P (2004) Effects of species diversity on the primary productivity of ecosystems: exceeding our spatial and temporal scales of inference. Oikos 104:437–450

Carpenter SR, Kraft CE, Wright R, He X, Soranno PA, Hodgson JR (1992) Resilience and resistance of a lake phosphorus cycle before and after a food web manipulation. Am Nat 140:781–798

Castri F di, Vernhes JR, Younès T (1992) Inventory and monitoring biodiversity. Biol Int Sp Iss 27:1–28

Chaabane K, Loreau M, Josens G (1997) Growth and egg production in *Abax ater* (Coleoptera, Carabidae). Pedobiologia 41:385–396

Chapin FS III, Zavaleta ES, Eviner VT, Naylor RL, Vitousek PM, Reynolds HL, Hooper DU, Lavorel S, Sala OE, Hobbie SE, Mack MC, Díaz S (2000) Consequences of changing biodiversity. Nature 405:234–242

Charrier S, Petit S, Burel F (1997) Movements of *Abax parallelepipedus* (Coleoptera, Carabidae) in woody habitats of a hedgerow network landscape: a radio-tracing study. Agric Ecosyst Environ 6:133–144

Cheng W, Zhang Q, Coleman DC, Carroll CR, Hoffmann CA (1996) Is available carbon limiting microbial respiration in the rhizosphere? Soil Biol Biochem 28:1283–1288

Chvàla M (1983) The Empidoidea (Diptera) of Fennoscandia and Denmark II: general part. The family Hybotidae, Atelestidae and Microphoridae. Faun Entomol Scand 12:1–279

Clemen T, Hörmann G (1996) MURKEL – Ein 3-dimensionales Objekt-orientiertes Wasserhaushaltsmodell für Buchenwälder. In: Breckling B, Asshoff M (eds) Modellbildung und Simulation im Projektzentrum Ökosystemforschung. EcoSys 4:165–177

Cole GA (1975) Textbook of limnology. Mosby, Saint Louis

Collinge S (2000) Effects of grassland fragmentation on insect species loss, colonization, and movement patterns. Ecology 81:2211–2226

Connell JH (1978) Diversity in tropical rain forest and coral reefs. Science 199:1302–1310

Cordsen E (1993) Böden des Kieler Raumes. Untersuchungen der Böden natürlicher Lithogenese unter Verwendung EDV-gestützt ausgewerteter Daten der Bodenschätzung. SchrR Inst Pflanz Bodenkde 25

Cornelsen R, Irmler U, Paustian D, Rieger A, Welsch H (1993) Effizienz von Uferrandstreifen als Elemente des Biotopverbunds. Lauf- und Kurzflügelkäfer, Spinnen und Schwebfliegen in Schleswig–Holstein. Natursch Landschaftspfl 25:205–211

Costanza R, Daly HE (1992) Natural capital and sustainable development. Conserv Biol 6:37–46

Costanza R, Faber SC, Maxwell J (1989) Valuation and management of wetland ecosystems. Ecol Econom 1:35–361

Costanza R, d'Arge R, De Groot R, Faber S, Grasso M, Hannon B, Limburg K, Naeem S, O'Neill RV, Paruelo J, Raskin RG, Sutton P, Van Den Belt M (1997) The value of the world's ecosystem services and natural capital. Nature 387:253–260

Côté B, Carlson RW, Dawson JO (1988) Leaf photosynthetic characteristics of seedlings of actin-orhizal *Alnus* spp and *Elaeagnus* spp. Photosynth Res 16:211–218

Council of Europe (1972) European soil charter. Brussels (German reprint in: Rosenkranz D, Bachmann G, König N, Einsele G (1988; chapter 8902) Bodenschutz. Schmidt, Berlin)

Crawford RMM (1996) Whole plant adaptions to fluctuating water tables. Fol Geobot Phytotax 31:7–24

Csermak K, Csermak A, Máté F (1992) Methanbildung im Sediment des Balaton (Plattensee, Ungarn). Limnologia 22:277–282

Csermely P (1998) Stress of life from molecules to man. Ann NY Acad Sci 851

Daily GC (ed) (1997) Nature's services: social dependence on natural ecosystems. Island, Washington, D.C.

Dall'O M, Kluge W, Bartels F (2001) FEUWAnet: a multi-box groundwater level and lateral exchange model for riparian wetlands. J Hydrol 250:40–62

Daunicht W, Salski A, Nöhr P, Neubert C (1996) Ein fuzzy-wissensbasiertes Modell zur Reproduktion von Feldlerchen (*Alauda arvensis*) im Ackerland. In: Breckling B, Asshoff M (eds) Modellbildung und Simulation im Projektzentrum Ökosystemforschung. EcoSys 4:99–106

David JF, Ponge JE, Delecour F (1993) The saprophagous macrofauna of different types of humus in beech forests of the Ardenne (Belgium). Pedobiologia 37:49–56

Davidsson TE, Leonardson L (1997) Production of nitrous oxide in artificially flooded and drained soils. Wetlands Ecol Managem 5:111–119

Davidsson TE, Trepel M, Schrautzer J (2002) Denitrification in drained and rewetted minerotrophic peat soils in Northern Germany (Pohnsdorfer Stauung). J Plant Nutr Soil Sci 165:199–204

Davies JA, McKay DC, Luciani G, Abdel-Wahab M (1988) Validation of models for estimating solar radiation on horizontal surfaces. (Final report, IEA Task IX, IEA-SHCP-9B-1, 1) Downsview, San Francisco

Davies NB (1977) Prey selection and the search strategy of the Spotted Flycatcher (*Muscicapa striata*): a field study on optimal foraging. Anim Behav 25:1016–1033

Davison W (1993) Iron and manganese in lakes. Earth Sci Rev 34:119–163

Delettre Y, Tréhen P, Grootaert P (1992) Space heterogeneity, space use and short-range dispersal in Diptera: a case study. Landscape Ecol 6:175–181

Díaz S, Cabido M (2001): Vive la différence: plant functional diversity matters to ecosystem proc-esses. Trends Ecol Evol 16:646–655

Dibbern I (2000) Ökologisch–ökonomische Modellierung von Landnutzungssystemen – Ein Beitrag zur verbesserten Bewertung dauerhaft umweltgerechter Entwicklungen in der Flächenutzung. EcoSys Suppl 33

Didden W, Born H, Domm H, Graefe U, Heck M, Kühle J, Mellin A, Römbke J (1995) The relative efficiency of wet funnel techniques for the extraction of Enchytraeidae. Pedobiologia 39:52–57

Diekmann M, Dupré C (1997) Acidification and eutrophication of deciduous forests in northwestern Germany demonstrated by indicator species analysis. J Veg Sci 8:855–864

Dienemann P (1997) Autökologische, synökologische und produktionsbiologische Untersuchungen an den Larven der Chironomidae (Diptera, Nematocera) des Belauer Sees (Schleswig–Holstein) EcoSys Suppl 23

Dierschke H (1989) Kleinräumige Vegetationsstruktur und phänologischer Rhythmus eines Kalkbuchenwaldes. Verh Ges Oekol 17:131–143

Dierschke H (1996): Pflanzensoziologie. Ulmer, Stuttgart

Dierssen K (1988) Rote Liste der Pflanzengesellschaften Schleswig-Holsteins. 2. Aufl Schriftenr Landes Nat Landschaftspfl Schleswig–Holstein 6

Dierssen K (1990) Einführung in die Pflanzensoziologie. Akademie, Berlin

Dierssen K (2000) Conservation biology. In: Jørgensen SE, Müller F (eds) Handbook of ecosys-tem theories and managment. Lewis, Boca Raton, pp 475–485

Dierssen K (2002) Was ist Erfolg im Naturschutz? SchrR Dt Rat Landespfleg 73:91–95

Dierssen K (2004a) Vegetation Schleswig–Holsteins. EcoSys Suppl 41:36–60

Dierssen K (2004b) Erhaltung der Artenvielfalt – Qualitative Aspekte. Artenschutzreport 15:44–50

Dierssen K, Dierssen B (2001) Moore. Ulmer, Stuttgart

Dierssen K, Hoffmann-Müller R (2004) Naturschutzziele, Naturschutzplanung und Indikatoren für den Zustand der Natur aus der Ökologischen Flächenstichprobe. In: Wiggering H, Müller F (eds) Umweltziele und Indikatoren, Technische Anforderungen an ihre Festlegung und Fallbeispiele. Springer, Berlin Heidelberg New York, pp 267–308

Dierssen K, Reck H (1998) Konzeptionelle Mängel und Ausführungsdefizite bei der Umsetzung der Eingriffsregelung im kommunalen Bereich. Natsch LandschPlan 30:341–345, 373–381

Dilly O (1994) Mikrobielle Prozesse in Acker-, Grünland- und Waldböden einer norddeutschen Moränenlandschaft. EcoSys Suppl 8:1–127

Dilly O (2001) Metabolic and anabolic responses of four arable and forest topsoils to nutrient addition. J Plant Nutr Soil Sci 164:29–34

Dilly O, Irmler U (1998) Succession in the food web during the decomposition of leaf litter in a black alder (*Alnus glutinosa* (Gaertn.)L.) forest. Pedobiologia 42:109–123

Dilly O, Munch J-C (1996) Microbial biomass content, basal respiration and enzyme activities during the course of decomposition of leaf litter in a black alder (*Alnus glutinosa* (L.) Gaertn.) forest. Soil Biol Biochem 28:1073–1081

Dilly O, Munch J-C (1998) Ratios between estimates of microbial biomass content and microbial activity in soils. Biol Fertil Soils 27:374–379

Dilly O, Nannipieri P (2001) Variation of enzyme activities in the A horizon of an arable and a forest soil induced by nutrient addition. Biol Fertil Soils 34:64–72

Dilly O, Mogge B, Kutsch WL, Kappen L, Munch J-C (1997a) Aspects of carbon and nitrogen cycling in soils of the Bornhöved Lake District. I. Microbial biomass content, microbial activities and in situ emissions of carbon dioxide as well as nitrous oxide of arable and grassland soils. Biogeochemistry 39:189–205

Dilly O, Eckhardt FEW, Blume H-P (1997b) Mikrobielle Prozesse in der Humusauflage und dem Oberboden eines Buchen- und eines Erlenwaldes einer norddeutschen Moränenlandschaft. EcoSys 6:15–30

Dilly O, Blume H-P, Kappen L, Kutsch WL, Middelhoff U, Wötzel J, Buscot F, Dittert K, Bach H-J, Mogge B, Pritsch K, Munch J-C (1999) Microbial processes and features of the microbiota in Histosols from a black alder (*Alnus glutinosa* (L.) Gaertn.) forest. Geomicrobiol J 16:65–78

Dilly O, Bach H-J, Buscot F, Eschenbach C, Kutsch WL, Middelhoff U, Pritsch K, Munch J-C (2000) Characteristics and energetic strategies of the rhizosphere in ecosystems of the Bornhöved Lake District. Appl Soil Ecol 15:201–210

Dilly O, Bartsch S, Rosenbrock P, Buscot F, Munch JC (2001a) Shifts in physiological capabilities of the microbiota during the decomposition of leaf litter in a black alder (*Alnus glutinosa* (Gaertn.) L.) forest. Soil Biol Biochem 33:921–930

Dilly O, Winter K, Lang A, Munch J-C (2001b) Energetic eco-physiology of the soil microbiota in two landscapes of southern and northern Germany. J Plant Nutr Soil Sci 164:407–413

Dittert K (1992) Die stickstoffoxidierende Schwarzerle-Frankia-Symbiose in einem Erlenbruchwald der Bornhöveder Seenkette. Ecosys Suppl 5

Dobler G (1985) VIII. Abundanzdynamik und Entwicklungszyklen von Zikaden (Homoptera, Auchenorrhyncha) im zentralalpinen Hochgebirge. In: Janetschek H (ed) Ökologische Untersuchungen an Wirbellosen des zentralalpinen Hochgebirges (Obergurgl, Tirol). Alpin Biol Stud 18

Dokulil M, Hamm A, Kohl J-G (2001) Ökologie und Schutz von Seen. UTB 2110. Gacultas University, Vienna

Döring-Mederake U (1991) Feuchtwälder im nordwestdeutschen Tiefland. Gliederung-Oekologie-Schutz Scripta Geobot 19

Driescher E, Gelbrecht J (1990) Phosphat im unterirdischen Wasser. Acta Hydrophys 34:79–95

Dröschmeister R (2001) Bundesweites Naturschutzmonitoring in der "Normallandschaft" mit der Ökologischen Flächenstichprobe. Nat Landschaft 76:58–69

Duarte CM, Vaqué D (1992) Scale dependence of bacterioplankton patchiness. Mar Ecol Progr Ser 84:95–100

Duchin F, Lange G-M (1994) Strategies for environmentally sound economic development. In: Jansson M, Hammer C, Folke C, Costanza R (eds) Investing in natural capital: the ecological economic approach to sustainability. Island, Washington, D.C., pp 250–265

Dunkan A (1997) Quantifying the fish–zooplankton interaction as an ecosystem response – a historical account of the 16th PEG meeting 1984. Arch Hydrobiol Spec Iss Adv Limnol 49:139–152

Dupré C, Diekmann M (2001) Differences in species richness and life history traits between grazed and abandoned grassland in southern Sweden. Ecography 24:275–286

DVWK (1996) Ermittlung der Verdunstung von Land- und Wasserflächen. Dtsch Ver Wasser Kulturbau Merkbl Wasserwirtsch 238

Eber W (1981) Struktur und Dynamik der Bodenvegetation im *Luzulo–Fagetum*. In: Dierschke (ed) Struktur und Dynamik von Wäldern. Ber Int Sympos IVV Rinteln 1982:495–511

EEA (1999) Environment in the European Union at the turn of the century. (Environmental assessment report 2) European Environment Agency, Luxembourg

Ellenberg H (1977) Stickstoff als Standortfaktor, insbesondere für mitteleuropäische Pflanzengesellschaften. Oecol Plant 12:56–67

Ellenberg H (1998) Biologische Vielfalt – Ein Indikator für nachhaltige Entwicklung der Wälder? Forschungsrep 1:25–28

Ellenberg H, Fränzle O, Müller P (1978) Ökosystemforschung im Hinblick auf Umweltpolitik und Entwicklungsplanung. Forschung im Bereich Umweltgrundsatzangelegenheiten, Abschnitt Ökologie, MAB. Bundesministerium des Innern, Bonn

Ellenberg H, Mayer R, Schauermann J (1986) Ökosystemforschung. Ergebnisse des Sollingprojektes 1966–1986. Ulmer, Stuttgart

Ellenberg H, Weber HE, Düll R, Wirth V, Werner W, Paulissen D (2001) Zeigerwerte von Pflanzen in Mitteleuropa. Scripta Geobot 18

Elschenbroich C, Salzer A (1988) Organometallchemie. Teubner, Stuttgart

Enquete-Kommission (1998): Konzept Nachhaltigkeit. Vom Leitbild zur Umsetzung. Abschlussbericht der Enquete-Kommission "Schutz des Menschen und der Umwelt – Ziele und Rahmenbedingungen einer nachhaltig zukunftsfähigen Entwicklung" des 13. Deutschen Bundestages. Zur Sache 4/98. Deutscher Bundestag, Referat Öffentlichkeitsarbeit, Bonn

Erisman JW (1992) Atmospheric deposition of acidifying compounds in the Netherlands. PhD thesis, University of Utrecht, Utrecht

Erisman JW, Draaijers GPJ (1995) Atmospheric deposition in relation to acidification and eutrophication. Stud Environ Sci 63

Erlenkeuser H (1998) Die absolute Zeitstellung der Warvensequenz in der Sedimentfolge Q300 aus dem Belauer See/Schleswig-Holstein. In: Wiethold J. Studien zur jüngeren postglazialen Vegetations- und Siedlungsgeschichte im östlichen Schleswig-Holstein. Universitaetsforsch Praehistorisch Archaeolog 45:355–365

Eschenbach C (1995) Zur Physiologie und Ökologie der Schwarzerle (*Alnus glutinosa*). PhD thesis, University of Kiel, Kiel

Eschenbach C (1996a) Modellierung der Primärproduktion der Schwarzerle (*Alnus glutinosa*). EcoSys 4:195–206

Eschenbach C (1996b) Zur Ökophysiologie der Primärproduktion der Schwarzerle (*Alnus glutinosa* (L.) Gaertn.). Verh Ges Oekol 26:89–95

Eschenbach C (1998) Modelling growth, development and architecture of black alder trees with an object-oriented approach. ASU Newsl 24 Suppl:75–86

Eschenbach C, Kappen L (1996) Leaf area index determination in an alder forest – a comparison of three methods. J Exp Bot 47:1457–1462

Eschenbach C, Kappen L (1999) Leaf water relations of black alder (*Alnus glutinosa* (L.) Gaertn.) growing at neighbouring sites with different water regimes. Trees 14:28–38

Eschenbach C, Herbst M, Vanselow R, Kappen L (1996) Transpiration eines Erlenwaldes (*Alnus glutinosa*) und eines Buchenwaldes (*Fagus sylvatica*) an benachbarten Standorten. Verh Ges Oekol 26:97–103

Eschenbach C, Middelhoff U, Steinborn W, Wötzel J, Kutsch W, Kappen L (1997) Von Einzelprozessen zur Kohlenstoffbilanz eines Erlenbruchs im Bereich der Bornhöveder Seenkette. EcoSys Suppl 20:121–132

Esteves FDA (1979) Die Bedeutung der aquatischen Makrophyten für den Stoffhaushalt des Schöhsees. 1. Die Produktion an Biomasse. Arch Hydrobiol Suppl 57:117–143

Fahrig L, Jonson I (1998) Effect of habitat patch characteristics on abundance and diversity of insects in an agricultural landscape. Ecosystems 1:197–205

Falge E, Tenhunen J, Baldocchi DD, Aubinet M, et al (2002) Phase and amplitude of ecosystem carbon release and uptake potentials as derived from FLUXNET measurements. Agric For Meteorol 113:75–95

FAO (1998) World reference base for soil resources. (World soil research report 84) FAO, Rome

FAO (2006) Guidelines for soil description. FAO, Rome

Farquahr GD, Firth PM, Wetselaar R, Wier B (1980) On the gaseous exchange of ammonia between leaves and the environment: determination of the ammonia compensation point. Plant Physiol 66:710–714

Fassbender HW (1977) Modellversuch mit jungen Fichten zur Erfassung des internen Nährstoffumsatzes. Oecol Plant 12:263–272

Fath B, Patten BC (2000) Ecosystem theory: network environ analysis. In: Jørgensen SE, Müller F (eds) Handbook of ecosystem theories and management. Lewis, Boca Raton, pp 345–360

Feger K-H (1993) Bedeutung von ökosysteminternen Umsätzen und Nutzungseingriffen für den Stoffhaushalt von Waldlandschaften. Freiburger Bodenkde Abh 31

Fitts CR (1995) TWODAN 2-D analytic groundwater flow model v. 4.0. manual. TWODAN, Scarborough

Fleck W (1986) Bodenwasserbilanz, Streuverdunstung und Wasserverbrauch von Buche und Fichte auf Standorten und in Einzugsgebieten des Schönbuchs. In: Einsele G (ed) Das landschaftsökologische Forschungsprojekt Naturpark Schönbuch. VCH, Weinheim, pp 133–160

Flohn H (1957) Zur Frage der Einteilung der Klimazonen. Erdkunde 11:161–175

Fowler D (1984) Transfer to terrestrial surfaces. Philos Trans R Soc Lond 305:281–297

Fränzle O (1978) The structure of soil associations and cenozoic morphogeny in Southeast Africa. In: Nagl H (ed) Beiträge zur Quartär- und Landschaftsforschung. Hirt, Vienna, pp 159–176

Fränzle O (1981) Erläuterungen zur Geomorphologischen Karte 1: 25 000 der Bundesrepublik Deutschland GMK 25 Blatt 8, 1826 Bordesholm. Geo Center, Stuttgart

Fränzle O (1988a) Glaziäre, periglaziäre und marine Reliefentwicklung im nördlichen Schleswig–Holstein. Schr Naturwiss Ver Schleswig–Holstein 58:1–30

Fränzle O (1988b) Umweltbelastung und Umweltschutz in der Bundesrepublik Deutschland. Geogr Runds 40:4–11

Fränzle O (1993) Contaminants in terrestrial environments. Springer, Berlin Heidelberg New York

Fränzle O (1994) Thermodynamic aspects of species diversity in tropical and ectropical plant communities. Ecol Model 75/76:63–70

Fränzle O (1998a) Ökosystemforschung im Bereich der Bornhöveder Seenkette. In: Fränzle O, Müller F, Schröder W (eds) Handbuch der Umweltwissenschaften, Kap V-4.3. Ecomed, Landsberg

Fränzle O (1998b) Sensitivity of ecosystems and ecotones. In: Schüürmann G, Markert B (eds) Ecotoxicology – ecological fundamentals, chemical exposure, and biological effects. Wiley, New York, pp 75–115

Fränzle O (2001) Alexander von Humboldt's holistic world view and modern inter- and transdisciplinary ecological research. Northeast Nat 8[Spec Iss 1]:57–90

Fränzle O (2003) Bioindicators and environmental stress assessment. In: Markert BA, Breure AM, Zechmeister HG (eds) Bioindicators and biomonitors. Elsevier, Amsterdam, pp 41–84

Fränzle O, Killisch W (1994) Die Biplot-Technik als Analyseinstrument komplexer Datenmatrizen. In: Schröder W, Vetter L, Fränzle O (eds) Neuere statistische Verfahren und Modellbildung in der Geoökologie. Vieweg, Braunschweig, pp 129–143

Fränzle O, Kluge W (1997) Typology of water transport and chemical reactions in groundwater/ lake ecotones. In: Gibert J, Mathieu J, Fournier F (eds) Groundwater/surface water ecotones: hydrological interactions and management options. (International hydrological series) Cambridge University Press, Cambridge, pp 127–134

Fränzle O, Kluge W (2003) Analyse der Energie- und Stoffflüsse von See–Umlandsystemen. Bochumer Geogr Arb Sonderheft 14:35–46

Fränzle O, Killisch W, Mich N (1986) Die regionale Differenzierung und zeitliche Veränderung der Emissionssituation in der Bundesrepublik Deutschland. Kieler Geogr Schr 64:31–77

Fränzle O, Bruhm I, Grünberg K-U, Jensen-Huß K, Kuhnt D, Kuhnt G, Mich K, Müller F, Reiche E-W (1987a) Darstellung der Vorhersagemöglichkeiten der Bodenbelastung durch Umweltchemikalien. Forschungsbericht 106 05 026 im Umweltforschungsplan des BM Umwelt, Naturschutz und Reaktorsicherheit. Geogr Inst Univ Kiel

Fränzle O, Kuhnt D, Kuhnt G, Zölitz R (1987b) Auswahl der Hauptforschungsräume für das Öko-systemforschungsprogramm der Bundesrepublik Deutschland. Forschungsbericht 101 04 043/02 im Umweltforschungsplan des Bundesministers für Umwelt, Naturschutz und Reaktorsicherheit. Geogr Inst Univ Kiel

Fränzle O, Kluge F, Jelinek S (1996) Die Bedeutung von Uferökotonen für den Wasser- und Nährstoffhaushalt von Ökosystemen. Heidelberger Geogr Arb 104:450–459

Fränzle O, Reiche EW, Windhorst W (2001) Conceptual framework of the Bornhöved Lake District research. In: Tenhunen JD, Lenz R, Hantschel R (eds) Ecosystem approaches to landscape management in Central Europe. (Ecological Studies 147) Springer, Berlin Heidelberg New York, pp 40–48

Fränzle S, Markert B (2000) Das Biologische System der Elemente (BSE): Eine modelltheoretische Betrachtung zur Essentialität von chemischen Elementen. Umweltwissensch Schadstoff-Forsch 12:97–103

Fränzle S, Markert B (2006) Metals in biomass. Environ Sci Pollut Res 10

Fuchs M, Tanner CB (1967) Evaporation from drying soil. J Appl Meteorol 6:852–857

Galvez JA, Niell FX, Lucena J (1988) Description and mechanism of formation of a deep chlorophyll due to *Ceratium hirundinella*. Arch Hydrobiol 112:143–155

Gansert D (1994) Die Wurzel- und Sproßrespiration junger Buchen (*Fagus sylvatica* L.) in einem montanen Moder-Buchenwald. Cuvillier, Göttingen

Garbe-Schönberg C-D, Wiethold J, Butenhoff D, Utech C, Stoffers P (1998) Geochemical and palynological record in annually laminated sediments from Lake Belau (Schleswig–Holstein) reflecting palaeoecology and human impact over 9000 a. Meyniana 50:47–70

Garniel A (1991) Weichselzeitliche Morphogenese im nördlichen Mittelholstein unter besonderer Berücksichtigung der Eisabbauvorgänge. Schr Naturwiss Ver Schleswig–Holstein 61:25–54

Gash JHC, Morton AJ (1978) An application of the Rutter model to the estimation of the interception loss from Thetford forest. J Hydrol 38:49–58

Gaston KJ (ed) (1996) Biodiversity: a biology of numbers and differences. Blackwell, Oxford

GDV (ed) (2000) Katastrophe Natur? Strategien zur Bewältigung von Naturkatastrophen. (GDV-Essays und Fakten Bd 4) Gesamtverband der Deutschen Versicherungswirtschaft GmbH, Karlsruhe

Gehrmann J, Andreae H, Fischer U, Lux W, Spranger T (2001) Luftqualität und atmosphärische Stoffeinträge an Level II-Dauerbeobachtungsflächen. Bundesministerium für Verbraucherschutz, Ernährung und Landwirtschaft. Arbeitskreis B der Bund-Länder-Arbeitsgruppe Level II, Bonn

Geiger R (1961) Das Klima der bodennahen Luftschicht. Vieweg, Braunschweig

Geßler A (2001) Der Stickstoffhaushalt von Buchen in einem stickstoffgesättigten Waldökosystem. Forstarch 72:118–122

Gessner M, Schieferstein B, Müller U, Barkmann S, Lenfers UA (1996) A partial budget of primary organic carbon flows in the littoral zone of a hardwater lake. Aquat Bot 55:93–105

Geyer B, Jarvis PG (1991) Preface to "a review of models of soil-vegetation-atmosphere-transfer schemes (SVATS): a report to the TIGER III Committee". TIGER III Committee, Edinburgh

Giller PS (1996) The diversity of soil communities, the 'poor man's tropical forest'. Biodivers Conserv 5:135–168

Golley FB (1993) A history of the ecosystem concept in ecology. Yale University Press, New Haven

Goodale CL, Apps MJ, Birdsey RA, Field CB, et al (2002) Forest carbon sinks in the northern hemisphere. Ecol Appl 12:891–899

Gosz JR (1992) Gradient analysis of ecological change in time and space: implications for forest management. Ecol Appl 2:248–261

Göttsche D (1972) Verteilung von Feinwurzeln und Mykorrhizen im Bodenprofil eines Buchen- und Fichtenbestandes im Solling. Bundesanst Forst-Holzwirtsch, Reinbek

Goulden ML, Munger JW, Fan S-M, Daube B, Wofsy SC (1996) Exchange of carbon dioxide by a deciduous forest: response to interannual climate variability. Science 271:1576–1578

Grabo J (1991) Ökologische Verteilung phytophager Arthropoda an Schilf (*Phragmites australis*) im Bereich der Bornhöveder Seenkette. Faun Oekol Mitt Suppl 12

Grajetzky B (1992) Nahrung und Brutverhalten von Rotkehlchen-Weibchen *Erithacus rubecula* einer schleswig–holsteinischen Knicklandschaft. Vogelwelt 113:282–288

Granier A, Ceschia E, Damesin C, Dufrène E, Epron D, Gross P, Lebaube S, Le Dantec V, Le Goff N, Lemoine D, Lucot E, Ottorini JM, Pontailler JY, Saugier B (2000) The carbon balance of young beech forests. Funct Ecol 14:312–325

Grayston SJ, Vaugham D, Jones D (1996) Rhizosphere carbon flow in trees, in comparison with annual plants: the importance of root exudation and its impact on microbial activity and nutrient availability. Appl Soil Ecol 5:29–56

Griffiths RP, Bradshaw GA, Marks B, Lienkaemper GW (1996) Spatial distribution of ectomycorrhizal mats in coniferous forests of the Pacific Northwest, USA. Plant Soil 180:147–158

Grime JP (1979) Plant strategies and vegetation processes. Wiley, New York

Grime JP (2001) Plant strategies, vegetation processes and ecosystem properties. Wiley, Chichester

Groffman PM, Tiedje JM, Robertson GP, Christensen S (1988) Denitrification at different temporal and geographical scales: proximal and distal controls. In: Wilson JR (ed) Advances in nitrogen cycling in agricultural ecosystems. CAB International, Wallingford, pp 174–192

Groot RS de (1992) Functions of nature. Wolters-Noorhoff, Amsterdam

Groot RS de (1994) Evaluation of environmental functions as a tool in planning, management and decision-making. PhD thesis, University of Wageningen, Wageningen

Grootjans, AP, Schipper PC, Windt HJ van der (1985) Influence of drainage on N-mineralization and vegetation response in wet meadows: I. Calthion stands. In: Grootjans AP (ed) Changes of groundwater regime in wet meadows. Groningen, Amsterdam, pp 75–92

Guderian R (ed) (1985) Air pollution by photochemical oxidants. Springer, Berlin Heidelberg New York

Guiguer N, Franz T (1997) Visual MODFLOW user's manual – the most complete, fully integrated modelling environment for MODFLOW, MODPATH and MT3D. Waterloo Hydrologic Software, Waterloo

Håkansson H, Olsson S, Jiang H, Garbe-Schönberg C-D (1998) The sediment diatom association and chemistry of surface sediments of Lake Belau, Northern Germany. Diatom Res 13:63–91

Haas CA (1995) Dispersal and use of corridors by birds in wooded patches on an agricultural landscape. Conserv Biol 9:845–854

Haberlehner E (1988) Comparative analysis of feeding and schooling behaviour of the Cyprinidae *Alburnus alburnus* (L., 1758) *Rutilus rutilus* (L., 1758), and *Scrardinius erythrophthalmus* (L., 1758) in a backwater of the Danube near Vienna. Int Rev Ges Hydrobiol 73:537–546

Habermas J (1992) Erläuterungen zur Diskursethik. Suhrkamp, Frankfurt

Hacke U, Sauter JJ (1995) Vulnerability of xylem to embolism in relation to leaf water potential and stomatal conductance in *Fagus sylvatica* f. *purpurea* and *Populus balsamifera*. J Exp Bot 46:1177–1183

Härdtle W (1995) Vegetation und Standort der Laubwaldgesellschaften (*Querco–Fagetea*) im nördlichen Schleswig–Holstein. Mitt AG Geobot Schl- Holst Hmbg 48

Hannappel S, Voigt HJ (1997) Beschaffenheitsmuster des Grundwassers im Lockergestein. In: Matschullat M, et al (eds) Geochemie und Umwelt. Relevante Prozesse in Atmo-, Pedo- und Hydrosphäre. Springer, Berlin Heidelberg New York, pp 359–379

Hanssen U, Irmler U (1995) Einfluß des Klimas auf den Massenwechsel und die Populationsdynamik wirbelloser Tierarten. EcoSys 2:124–137

Hári S, Müller F (2000) Ecosystems as hierarchical systems. In: Jørgensen SE, Müller F (eds) Handbook of ecosystem theories and management. Lewis, Boca Raton, pp 265–280

Harmon ME, Franklin JF, Swanson FJ, Sollins P, Gregory SV, Lattin JD, Anderson NH, Cline SP, Aumen NG, Sedell JR, Lienkaemper GW, Cromack K Jr, Cummins KW (1986) Ecology of coarse woody debris in temperate ecosystems. Adv Ecol Res 15:133–277

Hartje V, Klaphake A, Schliep R (2002) Consideration of the ecosystem approach on the conservation of biological diversity in Germany. BfN Skripten 69

Hartmann, J. (1992) Stoffeinträge in schleswig-holsteinische Böden während des Holozäns. Univ Kiel Schrift Inst Pflanz Bodenkde 17

Hasegawa M, Takeda H (1996) Carbon and nutrient dynamics in decomposing pine needle litter in relation to fungal and faunal abundances. Pedobiologia 40:171–184

Hasler AD (1975) Coupling of land and water systems. Springer, Berlin Heidelberg New York

Haude H (1958) Über die Verwendung verschiedener Klimafaktoren zur Berechnung potentieller Evaporation und Evapotranspiration. Meteor Rdsch 11:96–99

Hayasaka K, Takasuri N, Yamagishi N (1995) Energy metabolism in lactating Holstein cows (in Japanese, with English abstract). Anim Sci Technol 66:374–382

Haycock NE, Burt TP, Goulding KWT, Pinay G (eds) (1997) Buffer zones: their processes and potential in water protection. Quest International, Harpenden

Haygarth PM, Sharpley A (2000) Terminology for phosphorus transfer. J Environ Qual 29:10–15

Hayne DW, Ball RC (1956) Benthic productivity as influenced by fish predation. Limnol Oceanogr 1:163–175

He X, Kitchell JF, Carpenter SR, Hodgson JR, Schindler DE, Cottingham KL (1993) Food web structure and long-term phosphorous recycling: a simulation model evaluation. Trans Am Fish Soc 122:773–783

Heichel GH, Turner NC (1983) CO_2 assimilation of primary and regrowth foliage of red maple (*Acer rubrum* L.) and red oak (*Quercus rubra* L.): response of defoliation. Oecologia 57:14–19

Heijden M van der, et al (1998a) Mycorrhizal fungal diversity determinates plant biodiversity, ecosystem variability and productivity. Nature 396:69–72

Heijden M van der, Boller, T, Wiemken A, Sanders IR (1998b) Different arbuscular mycorrhizal fungal species are potential determinants of plant community structure. Ecology 79:2082–2091

Heiland S (1999) Voraussetzungen erfolgreichen Naturschutzes. Angew Umweltschutz. Ecomed, Landsberg

Heller K (1996) Vergleichende biozönotische und produktionsbiologische Untersuchungen an terricol-detritophagen Nematocera in einem Wald-Agrar-Ökosystemkomplex. Faun Oekol Mitt Suppl 22:41–86

Heller K, Irmler U, Ritter D (1991) Faktorenanalyse zur Dynamik der Bodenfauna. Ber Forschungszentr Waldoekosysteme B 22:380–382

Herbst M (1995) Stomatal behaviour in a beech canopy – an analysis of Bowen ratio measurements compared with porometer data. Plant Cell Environ 18:1010–1018

Herbst M (1997) Die Bedeutung der Vegetation für den Wasserhaushalt ausgewählter Ökosysteme. PhD thesis, Kiel University, Kiel

Herbst M, Hörmann G (1998) Predicting effects of temperature increase on the water balance of beech forest – an application of the KAUSHA model. Clim Change 40:683–698

Herbst M, Kappen L (1993) Die Rolle des Schilfs im standörtlichen Wasserhaushalt eines norddeutschen Sees. Phytocoenol 23:51–64

Herbst M, Kappen L (1999) The ratio of transpiration versus evaporation in a reed belt as influenced by weather conditions. Aquat Bot 63:113–125

Herbst M, Thamm F (1994) Kronendachinterzeption eines norddeutschen Buchenwaldes. Eine Anwendung des Interzeptionsmodells von Gash. Z Kulturtechn Landentwickl 35:311–319

Herbst M, Vanselow R (1997) Transpiration, Bodenverdunstung und Gesamtverdunstung in einem Maisfeld – gleichzeitige Messungen auf verschiedenen Maßstabsebenen. EcoSys Suppl 20:71–77

Herbst M, Eschenbach C, Kappen L (1999) Water use in neighbouring stands of beech (*Fagus sylvatica* L.) and black alder (*Alnus glutinosa* (L.) Gaertn.). Ann Sci Forest 56:107–120

Herzog C (2002) Das Methodenpaket IeMAX mit dem Fuzzy-Simulationsmodell FLUCS – Entwicklung und Anwendung eines Entscheidungsunterstützungs systems für die integrative Raumplanung. PhD thesis, University of Kiel, Kiel

Heyer J (1990) Der Kreislauf des Methans. Mikrobiologie – Ökologie – Nutzung. Akademischer Verlag, Berlin

Hiebner T (1985) Geomorphologische Detailaufnahme des TK 25 Blattes 1827 Stolpe. BSc thesis, University of Kiel, Kiel

Hills JM, Murphy KJ (1996) Evidence for consistent functional groups of wetland vegetation across a broad geographical range of Europe. Wetlands Ecol Management 4:51–63

Hingst H (1985) Großsteingräber in Schleswig–Holstein. Offa 42:57–112

Ho SC (1979) Structure, species diversity and primary production of epiphytic algal communities in the Schöhsee (Holstein), West Germany. PhD thesis, University of Kiel, Kiel

Hodgson JG (1986a) Commonness and rarity in plants with special reference to the Sheffield flora. Part I. The identity, distribution and habitat characteristics of the common and rare species. Biol Conserv 36:199–252

Hodgson JG (1986b) Commonness and rarity in plants with special reference to the Sheffield flora. Part II. The relative importance of climate, soils and land use. Biol Conserv 36:253–274

Hodgson JG, Thompson K, Wilson PJ (1998) Does biodiversity determine ecosystem function? The Ecotron experiment reconsidered. Funct Ecol 12:843–856

Hofmann W (1997) Seen-Beobachtungsprogramm 1991–1995. Die Seen im Vergleich. Max Planck Inst Limnol Ploen:18–33

Hoffmann F (1996) Die CERES-Modelle – Übersicht, Weiterentwicklungen, Erfahrungen. Agrarinformatik 24:139–150

Högberg P, Nordgren A, Buchmann N, Taylor AFS, Ekblad A, Högberg MN, Nyberg G, Ottosson-Löfvenius M, Read DJ (2001) Large-scale forest girdling shows that current photo-synthesis drives soil respiration. Nature 411:789–792

Hölker F (2000) Bioenergetik dominanter Fischarten *Abramis brama* (Linnaeus, 1758) und *Rutilus rutilus* (Linnaeus, 1758) in einem eutrophen See Schleswig-Holsteins – Ökophysiologie und individuenbasierte Modellierung. EcoSys Suppl 32

Hölker F, Breckling B (1998) Object orientation in fish modelling – simulation of roach activity (*Rutilus rutilus*) in Lake Belau (Germany). ASU Newsl 24[Suppl]:41–52

Hölker F, Breckling B (2001) An individual-based approach to depict the influence of the feeding strategy on the population structure of roach (*Rutilus rutilus* L.). Limnologia 31:69–78

Hölker F, Breckling B (2002) Influence of activity in a heterogeneous environment on the dynamics of fish growth: an individual-based approach of roach. J Fish Biol 60:1170–1189

Hölker F, Breckling B (2005) A spatiotemporal individual-based fish model to investigate emergent properties at the organismal and the population level. Ecol Model 184:406–426

Holling CS (1976) Resilience and stability of ecosystems. In: Jantsch E, Waddington CH (eds) Evolution and consciousness. Addison–Wesley, Reading, Mass., pp 73–92

Holling CS (1986) The resilience of terrestrial ecosystems: local surprise and global change. In: Clark WM, Munn RE (eds) Sustainable development of the biosphere. Oxford University Press, Oxford, pp 292–320

Hooper DU, Vitousek PM (1997) The effect of plant composition and diversity on ecosystem processes. Science 277:1302–1305

Horlitz T (1998) Naturschutzszenarien und Leitbilder – eine Grundlage für die Zielbestimmung im Naturschutz. NatSchutz LandschPlan 30:327–330

Hörmann G (1997) SIMPEL – Ein einfaches, benutzerfreundliches Bodenwassermodell zum Einsatz in der Ausbildung. Dtsch Gewässerkdl Mitt 41:67–72

Hörmann G, Branding A, Clemen T, Herbst M, Hinrichs A, Thamm F (1996) Calculation and simulation of wind controlled canopy interception of a beech forest in northern Germany. Agric For Meteorol 79:131–148

Hughes JB, Daily GC, Ehrlich PR (1997) Population diversity: its extent and extinction. Science 276:689–692

Hughes JB, Petchey OL (2001) Merging perspectives on biodiversity and ecosystem functioning. Trends Ecol Evol 16:222–223

Hunger H (2002) Anwendungsorientiertes Habitatmodell für die Helm-Azurjungfer (*Coenagrion mercuriale*, Odonata) aus amtlichen GIS-Grundlagendaten. Nat Landsch 77:261–265

Hupfer M (2001) Bindungsformen und Mobilität des Phosphors in Gewässersedimenten., In: Steinberg CH, Calmano W, Klapper H, Wilken R-D, Bernhardt H (eds) Handbuch der angewandten Limnologie, Kap IV-3.2. Ecomed, Landsberg

Hupfer M, Scharf B (2002) Seentherapie: Interne Maßnahmen zur Verminderung der Phosphorkonzentration. In: Steinberg CH, Calmano W, Klapper H, Wilken R-D, Bernhardt H (eds) Handbuch der angewandten Limnologie, Kap VI-2.1. Ecomed, Landsberg

Hurley Octavio KA, Jirka GH, Harleman DRF (1977) Vertical heat transport mechanisms in lakes and reservoirs. (Technical report 227) Massachusetts Institute of Technology, Cambridge, Mass.

Huston MA (1997) Hidden treatments in ecological experiments: re-evaluating the ecosystem function of biodiversity. Oecologia 10:449–460

Huston MA, Smith TM (1987) Plant succession: life history and competition. Am Nat 130:168–198

Huttula T (1992) Modelling resuspension and settling in lakes using a one-dimensional vertical model. Aqua Fenn 22:23–34

Idso SB (1973) On the concept of lake stability. Limnol Oceanogr 18:681–683

Irmler U (1995) Die Stellung der Bodenfauna im Stoffhaushalt schleswig-holsteinischer Wälder. Faun Oekol Mitt Suppl 18

Irmler U (1996) Sukzession der Streubesiedlung durch Bodentiere (Oribatida, Collembola) in verschiedenen Waldtypen. Verh Ges Oekol 26:275–282

Irmler U (1998a) Die vertikale Verteilung flugaktiver Käfer (Coleoptera) in drei Wäldern Norddeutschlands. Faun Oekol Mitt 7:387–404

Irmler U (1998b) Spatial heterogeneity of biotic activity in the soil of a beech wood and consequences for the application of the bait-lamina-test. Pedobiologia 42:102–108

Irmler U (1999) Environmental characteristics of ground beetle assemblages in northern German forests as a basis for an expert system. Z Oekol Naturschutz 8:227–237

Irmler U (2000) Changes in the fauna and its contribution to mass loss and N release during leaf litter decomposition in two deciduous forests. Pedobiologia 44:105–118

Irmler U, Bock W, Daunicht W, Hanssen U, Hingst R (1996a) Knicks als ökologische Verbundelemente in der Agrarlandschaft. EcoSys 5:193–203

Irmler U, Heller K, Warning J (1996b) Age and tree species as factors influencing the populations of insects living in dead wood (Coleoptera, Diptera: Sciaridae, Mycetophilidae). Pedobiologia 40:134–148

Irmler U, Heller K, Warning J (1997) Kurzflügelkäfer (Col.; Staphylinidae) an Totholz schleswig-holsteinischer Wälder. Faun Oekol Mitt 7:307–318

Irmler U, Hanssen U, Nötzold R, Schröter L (2000) Biodiversität in der Agrarlandschaft. Bedeutung von Landschaftsstrukturen und Nutzungsänderungen. Mitt Dtsch Ges Allg Angew Entomol 12:311–322

ISO 10693 Soil quality – determination of carbonate content – volumetric method. ISO, London

ISO 11260 Soil quality – determination of effective cation exchange capacity and base saturation using barium chloride solution. ISO, London

ISO 11261 Soil quality – determination of total nitrogen – modified Kjeldahl method. ISO, London

ISO 11272 Soil quality – determination of dry bulk density. ISO, London

ISO 11276 Soil quality – determination of the water retention characteristics – laboratory methods. ISO, London

ISO 11277 Soil quality – determination of particle size distribution in mineral soil material; method by sieving and sedimentation following removal of soluble salts, organic matter and carbonates. ISO, London

IUSS Working Group WRB (2006) World reference base for soil resources. FAO, Rome

Janecek A, Benderoth G, Lüdecke MKB, Kindermann J, Kohlmaier GH (1989) Model of seasonal and perennial carbon dynamics in deciduous type forests controlled by climatic variables. Ecol Model 49:101–124

Janiesch P (1981) Ökophysiologische Untersuchungen an Carex-Arten aus Erlenbruchwäldern. Habilitationsschrift, University of Münster, Münster

Janiesch P (2003) Vegetationsökologische Untersuchungen in einem Erlenbruchwald im nördlichen Münsterland – 25 Jahre im Vergleich. Abh Westf Mus Nat Kde 65:71–80

Janssens IA, Lankreijer H, Matteucci G, Kowalski AS, Buchmann N, Epron D, Pilegaard K, Kutsch W, Longdoz B, Grunwald T, Montagnani L, Dore S, Rebmann C, Moors EJ, Grelle A, Rannik U, Morgenstern K, Oltchev S, Clement R, Gudmundsson J, Minerbi S, Berbigier P, Ibrom A, Moncrieff J, Aubinet M, Bernhofer C, Jensen NO, Vesala T, Granier A, Schulze E-D, Lindroth A, Dolman AJ, Jarvis PG, Ceulemans R, Valentini R (2001) Productivity overshadows temperature in determining soil and ecosystem respiration across European forests. Global Change Biol 7:269–278

Jansson AM, Hammer M, Folke C, Costanza R (eds) (1994) Investing in natural capital: the ecological economics approach to sustainability. Island Press, Washington, D.C.

Jansson P-E, Karlberg L (2001) Coupled heat and mass transfer model for soil-plant-atmosphere systems. Royal Institute of Technology, Stockholm

Jantsch E, Waddington CH (eds) (1976) Evolution and consciousness. Addison–Wesley, Reading, Mass.

Jarvis PG (1976) The interpretation of the variation in leaf water potential and stomatal conductance found in canopies in the field. Philos Trans R Soc Lond 273:593–610

Jean GE, Bancroft GM (1986). Heavy metal adsoption by sulphide mineral surfaces. Geochim Cosmochim Acta 50:1455–1463

Jelinek S (1995) Einsatz hydrologischer Modelle zur Bewertung des Einflusses von Seeuferzonen auf diffuse Stoffeinträge. MSc thesis, University of Kiel, Kiel

Jelinek S (2000) Assessing the impact of lake shore zones on erosional sediment input using the EROSION-2D erosion model. In: Schmidt J (ed) Soil erosion. Application of physically based models. Springer, Berlin Heidelberg New York, pp 79–92

Jensen K, Schrautzer J (1999) Consequences of abandonment for a regional fen flora and mechanisms of succesional change. Appl Veg Sci 2:79–88

Jensen-Huß K (1990) Raum-zeitliche Analyse atmosphärischer Stoffeinträge in Schleswig–Holstein und deren ökologische Bewertung. PhD thesis, University of Kiel, Kiel

Jensen-Huß K (1992) Atmosphärische Stoffeinträge in Schleswig–Holstein: Herkunft und ökologische Bedeutung. Kieler Geogr Schrift 85:22–41

Jeschke WD (1977) K$^+$–Na$^+$ exchange and selectivity in barley root cells: effect of Na$^+$ on the K$^+$ fluxes. J Exp Bot 28:1289–1305

Jessel B, Reck H (1999) Umweltplanung. In: Fränzle O, Müller F, Schroeder W (eds) Handbuch der Umweltwissenschaften, Kap VI-3.6–3.5. Ecomed, Landsberg

Jödicke K, Trautz D (1994) Veränderungen der Samenbank im Boden von Ackerbrachen. Nat Landsch 69:258–264

Johnes PJ, Hodgkinson RA (1998) Phosphorus loss from agricultural catchments: pathways and implications for management. Soil Use Manage 14:175–183

Johnson CG (1957) The distribution of insects in the air and the empirical relation of density to height. J Anim Ecol 26:479–494

Jones RI (1993) Phytoplankton migrations: pattern, processes and profits. Arch Hydrobiol Beih Ergebn Limnol 39:67–77

Jones RI, Francis RC (1982) Dispersion patterns of phytoplankton in lakes. Hydrobiologia 86:21–28

Jones RI, Fulcher AS, Jayakody JKU, Laybourn-Parry J, Shine AJ, Walton MC, Young JM (1995) The horizontal distribution of plankton in a deep, oligotrophic lake – Loch Ness, Scotland. Freshwater Biol 33:161–170

Jopp F, Weigmann G, Reuter H (1998) Modelling movement and migration patterns of ground-dwelling invertebrates. ASU Newsl 24[Suppl]:53–63

Jules E, Shahani P (2003) A broader ecological context to habitat fragmentation: why matrix habitat is more important than we thought. J Veg Sci 14:459–464

Jungbluth T, Hartung E, Brose G (2001) Greenhouse gas emissions from animal houses and manure stores. Nutr Cycling Agroecosyst 60:122–145

Jørgensen SE (1983) Eutrophication models of lakes. In: Jørgensen SE (ed) Application of ecological modelling in environmental management. (Developments in environmental modelling 4A) Elsevier, Amsterdam, pp 227–282

Jørgensen SE (1988) Fundamentals of ecological modelling. Elsevier, Amsterdam

Jørgensen SE (1996) Integration of ecosystem theories – a pattern. Kluwer, Dortrecht

Jørgensen SE (2000) The tentative fourth law of thermodynamics. In: Jørgensen SE, Müller F (eds) Handbook of ecosystem theories and management. Lewis, Boca Raton, pp 161–176

Kahmen S, Poschlod P (2004) Plant functional trait responses to grassland succession over 25 years. J Veg Sci 15: 21–32

Kaila L, Martikainen P, Punttila P, Yakovlev E (1994) Saproxylic beetles (Coleoptera) on dead birch trunks decayed by different polypore species. Ann Zool Fenn 31:97–100

Kajak Z (1968) Experimental analysis of factors decisive for benthos abundance. Zesz Nauk Inst Ekol PAN 1:1–22

Kajak Z (1988) Considerations on benthos abundance in freshwaters, its factors and mechanisms. Int Rev Ges Hydrobiol 73:5–19

Kasule FK (1970) Field studies on the life histories of Othius (Gyrohypnus auct.) punctulatus (Goeze) and O. myrmecophilus (Kiesenwetter) (Coleoptera Staphylinidae). Proc R Entomol Soc Lond A 45:55–67

Kay JJ (2000) Ecosystems as self-organised holarchic open systems: narratives and the second law of thermodynamics. In: Jørgensen SE, Müller F (eds) Handbook of ecosystem theories and management. Lewis, Boca Raton, pp 135–160

Keitz S von, Schmalholz M (2002) Handbuch der EU-Wasserrahmenrichtlinie. Schmidt, Berlin

Kent M, Coker P (1992) Vegetation description and analysis – a practical approach. Belhaven, London

Kerger KE (1992) Modellierung der thermischen Struktur des Belauer Sees. MSc thesis, University of Kiel, Kiel

Kieckbusch JJ (2003) Ökohydrologische Untersuchungen zur Wiedervernässung von Niedermooren am Beispiel der Pohnsdorfer Stauung. PhD thesis, University of Kiel, Kiel

Kim J, Verma SB (1992) Soil surface CO_2 flux in a Minnesota peatland. Biogeochemistry 18:37–51

Kirschbaum MUF (1995) The temperature dependence of soil organic matter decomposition, and the effect of global warming on soil organic C storage. Soil Biol Biochem 27:753–760

Kleyer M (1995) Biological traits of vascular plants. A database. Arb Inst Landschplan Oekol Univ Stuttgart NF 2

Kleyer M (1999) The distribution of plant functional types on gradients of disturbance intensity and resource supply in an agricultural landscape. J Veg Sci 10:697–708

Kleyer M, Kratz R, Lutze G, Schröder B (2000) Habitatmodelle für Tierarten: Entwicklung, Methoden und Perspektiven für die Anwendung. Z Oekol Natschutz 8:177–194

Kluge W (2004) Einfluss von Uferzonen auf die diffusen Einträge von N und P in den Belauer See (Schleswig–Holstein). Archiv Naturschutz Landschaftsforsch 43:31–52

Kluge W, Jelinek S (1999) Anforderungen an die Modellierung des Wasser- und Stofftransportes in Tieflandeinzugsgebieten Schleswig-Holsteins – Erfahrungen aus dem Bornhöved- und Störprojekt. In: Fohrer N, Döll P (eds) Modellierung des Wasser- und Stofftransportes in großen Einzugsgebieten. Kassel University Press, Kassel, pp 135–142

Kluge W, Theesen L (1996) Wasserstands- und Wasserhaushaltsmodelle für Uferökotone. In: Breckling B, Asshoff M (eds) Modellbildung und Simulation im Projektzentrum Ökosystemforschung. EcoSys 4:179–194

Kluge W, Müller-Buschbaum P, Theesen L (1994) Parameter acquisition for modelling exchange processes between terrestrial and aquatic ecosystems. Ecol Model 75/76:399–408

Kluge W, Jelinek S, Martini M (2000) Einfluss von Talniederungen auf die diffusen Stoffeinträge in Kleingewässer über den Grundwasserpfad. In: Friese K, Witter B, Miehlich G, Rode M (eds) Stoffhaushalt von Auenökosystemen – Böden und Hydrologie, Schadstoffe, Bewertungen. Springer, Berlin Heidelberg New York, pp 129–138

Kluge W, Fränzle O, Müller F (2003a) Stoffliche und energetische Beziehungen zwischen Ökosystemen. In: Fränzle O, Müller F, Schroeder W (eds) Handbuch der Umweltwissenschaften, Kap IV-2.4. Ecomed, Landsberg

Kluge W, Martini M, Baumann R, Kersebaum K-C, Venohr M (2003b) Einfluss der Talniederungen auf die diffusen Stoffeinträge am Beispiel der oberen Stör (Schleswig-Holstein). In: Becker A, Lahmer W (eds) Wasser- und Nährstoffhaushalt im Elbgebiet und Möglichkeiten zur Stoffeintragsminderung. Konzepte für die nachhaltige Entwicklung einer Flusslandschaft (Band 1). Weißensee, Berlin

Kniess A (2001) Simulation des Bodenwasserhaushaltes eines Buchenwaldstandortes in Schleswig-Holstein – Kalibrierung und Sensitivitätsanalyse des Modells CoupModel. MSc thesis, Kiel University, Kiel

Knohl A, Schulze ED, Kolle O, Buchmann N (2003) Large carbon uptake by an unmanaged 250-year-old deciduous forest in Central Germany. Agric For Meteorol 118:151–167

Koerselman W, Verhoeven JTA (1995) Eutrophication of fen ecosystems: external and internal nutrient sources and restoration strategies. In: Wheeler BD, Shaw SC, Foijt WJ, Robertson RA (eds) Restoration of temperate wetlands. Wiley, Chichester, pp 91–112

Köhler F (2000) Totholzkäfer in Naturwaldparzellen des nördlichen Rheinlandes. Vergleichende Studien zur Totholzkäferfauna Deutschlands und deutschen Naturwaldforschung. Naturwaldparzellen in Nordrhein-Westfalen VII. LÖBF-SchrR 18

Koike T (1987) Photosynthesis and expansion of leaves of early, mid, and late successional tree species, birch, ash, and maple. Photosynthetica 21:503–508

Kolasa JA, Picket STA (eds) (1991) Ecological heterogeneity. (Ecological studies 86) Springer, Berlin Heidelberg New York

Kowarik I (1999) Natürlichkeit, Naturnähe und Hemerobie als Bewertungskriterien. In: Konold W, Böcker R, Hampicke U (1999) Handbuch Naturschutz und Landschaftspflege, Kap V-2.1. Ecomed, Landsberg

Kraft CE (1992) Estimates of phosphorus and nitrogen cycling by fish using a bioenergetics approach. Can J Fish Aquat Sci 49:2596–2604

Kraft CE (1993) Phosphorus regeneration by Lake Michigan alewives in the mid-1970s. Trans Am Fish Soc 122:749–755

Krinitz J, Garbe-Schönberg CD, Schleuß U (1996) Filterwirkung von Knicks für atmosphärische Schadstoffe am Beispiel des Schwermetalls Blei. EcoSys 5:205–216

Krohn W, Küppers G (eds) (1990) Selbstorganisation: Aspekte einer wissenschaftlichen Revolution. Vieweg, Braunschweig

Kühl H (1989) Produktivität und Vitalität von Röhrichtbeständen (Phragmites australis (Cav.) Trin. ex Steudel) verschiedener Seen in der Uckermark und in Ostbrandenburg. PhD thesis, Humboldt University, Berlin

Kuhn W (1997) Die Ableitung artspezifischer Habitateignungskarten aus vegetationskundlichen und topographischen Karten. In: Kratz R, Suhling F (eds) Geographische Informationssysteme im Naturschutz: Forschung, Planung, Praxis. Westarp Wissenschaften, Magdeburg, pp 95–193

Kuhn W, Kleyer M (2000) A statistical habitat model for the blue winged grasshopper (Oedipoda caerulescens) considering the habitat connectivity. Z Oekol Naturschutz 8:207–218

Kutsch W (1996) Untersuchungen zur Bodenatmung zweier Ackerstandorte im Bereich der Bornhöveder Seenkette. EcoSys Suppl 16

Kutsch W, Dilly O (1999) Ecophysiology of plant-microbial interactions in terrestrial ecosystems. Bielefelder Oekol Beitr 14:74–84

Kutsch W, Kappen L (1997) Aspects of carbon and nitrogen cycling in soils of the Bornhöved Lake District II. Modelling the influence of temperature increase on soil respiration and organic carbon content in soils under different managements. Biogeochemistry 39:207–224

Kutsch W, Dilly O, Steinborn W, Müller F (1998) Quantifying ecosystem maturity – a case study. In: Müller F, Leupelt M (eds) (1998) Eco targets, goal functions and orientors. Springer Berlin Heidelberg New York, pp 209–231

Kutsch W, Hörmann G, Barkmann J (2001a) Die Bedeutung von Wäldern für die Integrität von divers strukturierten Agrarlandschaften. Forstarch 72:138–145

Kutsch W, Herbst M, Vanselow R, Hummelshøj P, Jensen NO, Kappen L (2001b) Stomatal acclimation influences water and carbon fluxes of a beech canopy in northern Germany. Basic Appl Ecol 2:265–281

Kutsch W, Eschenbach C, Dilly O, Middelhoff U, Steinborn W, Vanselow R, Weisheit K, Wötzel J, Kappen L (2001c) The carbon cycle of contrasting landscape elements of the Bornhöved Lake District. In: Lenz R, Hantschel R, Tenhunen JD (eds) Ecosystem properties and landscape function in Central Europe. (Ecological Studies 147) Springer, Berlin Heidelberg New York, pp 75–95

Kuzyakov Y, Friedel JK, Stahr K (2000) Review of mechanisms and quantification of priming effects. Soil Biol Biochem 32:1485–1498

Lachavanne JB, Juge R (eds) (1997) Biodiversity in land–inland water ecotones. UNESCO, Paris

Lalubie C (1991) Stoffeinträge durch Streufall in verschiedenen Waldökosystemen des Untersuchungsgebietes "Bornhöveder Seenkette". MSc thesis, University of Kiel, Kiel

Laminger H (1980) Bodenprotozoologie. Mikrobios 1:1–142

Lampert W, Sommer U (1993) Limnoökologie. Thieme, Stuttgart

Landmesser B (1993) Untersuchungen zur Struktur und zur Primärproduktion des Phytoplanktons im Belauer See. PhD thesis, University of Hamburg, Hamburg

Langford AO, Fehsenfeld FC, Zachariassen J, Schimmel DS (1990) Gaseous ammonia fluxes and background concentrations in terrestrial ecosystems of the United States. Global Biochem Cycles 6:459–483

LANU (2000) Seenbewertung in Schleswig–Holstein. Landesamt für Natur und Umwelt Schleswig–Holstein, Flintbek

Larcher W (1994) Ökophysiologie der Pflanzen. 5th ed Ulmer, Stuttgart

Läuchli A, Pflüger R (1978) Potassium transport through plant cell membranes and metabolic role of potassium in plants. Proc Congr Int Potash Inst Bern 11:111–163

Lauer W, Rafiqpoor MD (2002) Die Klimate der Erde: eine Klassifikation auf der Grundlage der ökophysiologischen Merkmale der realen Vegetation. Steiner, Stuttgart

LAWA (1999) Gewässerbewertung – stehende Gewässer: Vorläufige Richtlinie für eine Erstbewertung von natürlich entstandenen Seen nach trophischen Kriterien. Länderarbeitsgemeinschaft Wasser/Kulturbuch, Berlin

LAWAKÜ (1995) Ein Jahrzehnt Beobachtung der Niederschlagsbeschaffenheit in Schleswig-Holstein 1985–1994. Landesamt für Wasserhaushalt und Küsten des Landes Schleswig–Holstein, Kiel

Lawton JH, Brown VK (1993) Redundancy in ecosystems. In: Schulze ED, Mooney HA Biodiversity and ecosystem function. Springer, Berlin Heidelberg New York, pp 255–270

Leitungsgremium des Bornhöved-Projekts (eds) (1991) Ökosystemforschung im Bereich der Bornhöveder Seenkette. Interne Mitteilungen 3

Lenfers U (1994) Stoffeintrag durch Streufall in verschiedenen Waldökosystemen im Bereich der Bornhöveder Seenkette. MSc thesis, University of Kiel, Kiel

Lenz U (1992) Die Auswirkungen der Frühjahrsblüte auf den Chemismus des Belauer Sees unter besonderer Berücksichtigung der Calcitfällung und ihrer Modellierung. MSc thesis, University of Kiel, Kiel

Levin SA (1974) Dispersion and population interactions. Am Nat 108:207–228

Levitt J (1980) Responses of plants to environmental stresses. Academic Press, New York

Lewis T (1966) An analysis of components of wind affecting the accumulation of flying insects near artificial windbreaks. Ann Appl Biol 58:365–370

Li YH (1973) Vertical eddy diffusion coefficient in Lake Zürich. Schweiz Zeitschr Hydrol 35:1–7

Lichtenthaler HK (1998) The stress concept in plants. In: Csermely P (ed) Stress of life from molecules to man. Ann NY Acad Sci 851:187–198

Likens GE, Bormann FH, Pierce RS, Eaton JS, Johnson NM (1977) Biogeochemistry of a forested ecosystem. Springer, Berlin Heidelberg New York

Lille R (1996) Zur Bedeutung von Bracheflächen für die Avifauna der Agrarlandschaft: Eine nahrungsbiologische Studie an der Goldammer *Emberiza citrinella*. Agrarökologie 21:1–150

Lindsay WL (1979) Chemical equilibria in soils. Wiley, New York

Lock MA (1993) Attached microbial communitites in rivers. In: Ford TE (ed) Aquatic microbiology. Blackwell, Boston, Mass., pp 113–138

Löhlein B (1998) Nematoda und Oligochaeta im Aufwuchs auf Schilf eines eutrophen Sees: Ökologie, Populationsdynamik und Rolle im trophischen Gefüge. EcoSys Suppl 24

Loreau M, Nolf CL (1993) Occupation of space by the carabid beetle *Abax ater*. Acta Oecol 14:247–258

Loreau M, Naeem S, Inchausti P, Bengtsson J, Grime JP, Hector A, Hooper DU, Huston MA, Raffaelli D (2001) Biodiversity and ecosystem functioning: current knowledge and future challenges. Science 294:804–808

Loreau M, Naeem S, Inchausti P (eds) (2002) Biodiversity and ecosystem functioning: synthesis and perspectives. Oxford University Press, Oxford

Lugo AE (1996) Monitoring biodiversity at global scales. In: Castri F di, Younès T (eds) Biodiversity, science and development. IUBS, Paris, pp 189–196

Lyr H (1992) Symbiontische Ernährungsweisen. In: Lyr H, Fiedler JH, Tranquillini W (eds) Physiologie und Ökologie der Gehölze. Fischer, Jena, pp 117–142

Maarel E van der, Dauvellier PL (1978) Naar een globaal ecologisch model (GEM) voor de ruimtelijke ontwikkeling van Nederland. Min Volkshuisv Ruimt Ord, Den Haag

Maas MP van der, Breemen N van, Langenvelde I van (1990) Estimation of atmospheric deposition and canopy exchange in two Douglas fir stands in the Netherlands. Agricultural University of Wageningen, Wageningen

Macfadyen A (1962) Soil arthropod sampling. In: Gragg JB (ed) Advances in ecological research. Academic Press, London

Mamilov AS, Dilly OM (2002) Soil organic carbon as affected by microbial physiological state. In: Weber J, Jamroz E, Drozd J, Karczewska A (eds) Biogeochemical processes and cycling of elements in the environment. Polish Society of Humic Substances, Wrocław, pp 57–58

Mann RHK (1973) Observations on the age, growth, reproduction and food of the roach (*Rutilus rutilus* L.) in two rivers in southern England. J Fish Biol 5:707–736

Mansfeldt T (1994) Schwefeldynamik von Böden des Dithmarscher Speicherkoogs und der Bornhöveder Seenkette in Schleswig–Holstein. Univ Kiel SchriftenR Inst Pflanzenern Bodenkde 38

Manteifel BP, Girsa II, Pavlov DS (1978) On rhythms of fish behaviour. In: Thorpe J (ed) Rhythmic activity of fishes. Academic, London, pp 215–224

Margalef R (1995) Information theory and complex ecology. In: Patten BC, Jørgensen SE (eds) Complex ecology: the part-whole relation in ecosystems. Prentice–Hall, Englewood Cliffs, pp 40–50

Marschner H (1995) Mineral nutrition of higher plants. Academic Press, London

Marxen-Drewes H (1987) Kulturpflanzenentwicklung, Ertragsstruktur, Segetalflora und Arthropodenbesiedlung intensiv bewirtschafteter Äcker im Einflussbereich von Wallhecken. SchrR Inst Wasserwirtsch Landschaftsökol 6

Mason NWH, MacGillivray K, Steel JB, Wilson JB (2003) An index of functional diversity. J Veg Sci 14:571–578

Mather PM (1972) Areal classification in geomorphology. In: Chorley RJ (ed) Spatial analysis in geomorphology. Methuen, London, pp 305–322

Mathes K, Breckling B, Ekschmitt K (eds) (1996) Systemtheorie in der Ökologie. Ecomed, Landsberg

Matschullat J, Kritzer P (1997) Atmosphärische Deposition von Spurenelementen in "Reinluftgebieten". In Matschullat M, et al (eds) Geochemie und Umwelt. Relevante Prozesse in Atmo-, Pedo- und Hydrosphäre. Springer, Berlin Heidelberg New York, pp 3–23

Matsunga K, Igarashi K, Fukase S (1982) Behaviour of organically bound iron in Lake Ohnuma. Jpn J Limnol 43:182–188

Matthiesen D (1995) Die terricole Testacea-Fauna in einem Wald-Agrar-Ökosystemkomplex. Faun Oekol Mitt Suppl 22:11–39

Maturana HR, Varela FG (1980) Autopoiesis and cognition. Kluwer, Dordrecht

Matzner E, Alewell C, Bittersohl J, Liecheid G, Kammerer G, Manderscheid B, Matschonat G, Moritz K, Tenhunen JD, Totsche K (2001) Biogeochemistry of a spruce forest catcment of the Fichtelgebirge in response to changing atmospheric deposition. In: Tenhunen JD, Lenz R, Hantschel R (eds) Ecosystem approaches to landscape management in central Europe. (Ecological studies 147) Springer, Berlin Heidelberg New York, pp 463–503

Mayer R, Ulrich B (1982) Calculation of deposition rates from the flux balance and ecological effects of atmospheric deposition upon forest ecosystems. In: Georgii HW, Pankrath J (eds) Deposition of atmospheric pollutants. Reidel, Dordrecht, pp 195–200

Mazumder A, Taylor WD, McQueen DJ, Lean DRS (1989) Effects of nutrients and grazers on periphyton phosphorus in lake enclosures. Freshwater Biol 22:405–415

McCann KS (2000) The diversity–stability debate. Nature 404:228–233

McDonald MG, Harbaugh AW (1988) A modular three-dimensional finite-difference groundwater flow model. (Techniques of water resources investigation 06-A1) United States Geological Survey, Washington, D.C.

McMurtrie RE (1975) Determinants of stability of large randomly connected systems. J Theor Biol 50:1–11

Meffert M-E, Wulff W-R (1987) Morphometrie und Chlorophyllproduktion von ostholsteinischen Seen. Z Wasser Abwasser Forsch 20:13–15

Mehner T, Schultz H, Herbst R (1995) Interaction of zooplankton dynamics and diet of 0+ perch (*Perca fluviatilis* L.) in the top-down manipulated Bautzen Reservoir (Saxony, Germany) during summer. Limnologica 25:1–9

Mehner T, Plewa M, Hülsmann S, Voigt H, Benndorf J (1997) Age-0 fish predation on daphnids – spatial and temporal variability in the top-down manipulated Bautzen Reservoir, Germany. Arch Hydrobiol Spec Iss Adv Limnol 49:13–25

Meissner P, Ostendorp W (1988) Ein Strömungsmodell der temperaturinduzierten Dichteströmung in geschlossenen Uferröhrichten des Bodensee-Untersees. Arch Hydrobiol 112:433–448

Meister HP, Caldwell MM, Tenhunen JD, Lange OL (1987) Ecological implications of sun/shade-leaf differentiation in sclerophyllous canopies: Assesment by canopy modeling. In: Tenhunen JD (ed) Plant response to stress. NATO ASI Series, vol G15. Springer, Berlin Heidelberg New York, pp 339–354

Mette R (1994) Ertragsstruktur und Mineralstoffaufnahme von Mais und Hafer im Einflußbereich von Wallhecken. PhD thesis, University of Kiel, Kiel

Meulemans JT, Roos PJ (1985) Structure and architecture of the periphytic community on dead reed stems in Lake Maarsseveen. Arch Hydrobiol 102:487–502

Meyer M (1996) Erprobung und Anwendung von Methoden zur einzugsgebietsbezogenen Modellierung der Phosphatdynamik terrestrischer Ökosysteme. MSc thesis, University of Kiel, Kiel

Meyer M (2000) Entwicklung und Modellierung von Planungsszenarien für die Landnutzung im Gebiet der Bornhöveder Seenkette. PhD thesis, University of Kiel, Kiel

Meyer M, Reiche E-W, Heinrich U, Filipinski M (1998) Einzugsgebietsbezogene Erosionsmodellierung und Erstellung von Gewässerbelastungskarten. Verh Ges Oekol 28:91–98

Middelhoff U (1998) An object-oriented model developing competing root systems of black alder trees. ASU Newsl 24[Suppl]:65–74

Middelhoff U (2000) Simulationsgestützte Analyse der raum-zeitlichen Verteilung der Biomasse in einem Erlenbruch unter besonderer Beachtung der Feinwurzeldynamik. Shaker, Aachen

Miller A (1995) Technical thinking: its impact on environmental management. Environm managem 9:179–190

Mills EL, Forney JS (1981) Energetics, food consumption and growth of young yellow perch in Oneida Lake, New York. Trans Am Fish Soc 110:479–488

Mogge B (1995) N_2O-Emissionen und Denitrifikationsabgaben von Böden einer Jungmoränenlandschaft in Schleswig–Holstein. EcoSys Suppl 9

Möller CM, Müller D, Nielsen J (1954) Respiration in stem and branches of beech. Det Forstl Forsøgsv Denmark 21:273–301

Monteith JL, Unsworth MH(1990) Principles of environmental physics. Arnold, London

Mooney HA, Cushman JH, Medina E, Sala OE, Schulze ED (1996) What have we learned about ecosystem functioning of biodiversity? In: Mooney HA, Cushman JH, Medina E, Sala OE, Schulze ED (eds) Functional roles of biodiversity – a global perspective. Wiley, Chichester

Morel FMM, Hering J (1993) Principles and applications of aquatic chemistry. Wiley, New York

Morgan NC, Backiel T, Bretscko G, Duncan A, Hillbricht-Ilkowska A, Kajak Z, Kitchell JF, Larsson P, Lévèque C, Nauwerck A, Schiemer F, Thorpe JE (1980) Secondary production. In: LeCren ED, Lowe-McConnell RH (eds) The functioning of freshwater ecosystems. (International biological programme 22) Cambridge University Press, Cambridge

Mühle H, Eichler S (1997) Tern-Tagung. Terrestrische und ökosystemare Forschung in Deutschland – Stand und Ausblick. UFZ-Bericht 5

Müller F (1992) Hierarchical approaches to ecosystem theory. Ecol Model 63:215–242

Müller F (1997) State-of-the-art in ecosystem theory. Ecol Model 100:135–161

Müller F (1998) Gradients in ecological systems. Ecol Model 108:3–21

Müller F, Fath B (1998) The physical basis of ecological goal functions. In: Müller F, Leupelt M (eds) Eco targets, goal functions and orientors. Springer, Berlin Heidelberg New York, pp 269–288

Müller F, Jørgensen SE (2000) Ecological orientors – a path to environmental applications of ecosystem theories. In: Jøgernsen SE, Müller F (eds) Handbook of ecosystem theories and management. Lewis, Boca Raton, pp 561–576

Müller F, Nielsen S (2000) Ecosystems as subjects of self-organising processes. In: Jørgensen SE, Müller F (eds) Handbook of ecosystem theory and management. Lewis, Boca Raton, pp 177–194

Müller F, Windhorst W (2000) Ecosystems as functional entities. In: Jørgensen SE, Müller F (eds) Handbook of ecosystem theories and management. Lewis,Boca Raton, pp 33–50

Müller F, Breckling B, Bredemeyer M, Grimm V, Malchow H, Nielsen SN, Reiche EW (1997a) Ökosystemare Selbstorganisation. In: Fränzle O, Müller F, Schröder W (eds) Handbuch der Umweltwissenschaften, Kap III-2.4. Ecomed, Landsberg

Müller F, Breckling B, Bredemeyer M, Grimm V, Malchow H, Nielsen SN, Reiche EW (1997b) Emergente Ökosystemeigenschaften. In: Fränzle O, Müller F, Schröder W (eds) Handbuch der Umweltwissenschaften, III-2.5. Ecomed, Landsberg

Müller F, Hoffmann-Kroll R, Wiggering H (2000) Indicating ecosystem integrity – from ecosystem theories to eco targets, models, indicators and variables. Ecol Model 130:13–23

Müller HE (1981) Vergleichende Untersuchungen zur hydrochemischen Dynamik von Seen im schleswig-holsteinischen Jungmoränengebiet. Kieler Geogr Schr 53

Müller HJ (1980) Die Bedeutung abiotischer Faktoren für die Einnischung der Organismen in Raum und Zeit. Biol Rundsch 18:373–388

Münchmeyer U, Koppisch D, Augustin J, Merbach W, Succow M (1998) Untersuchungen zur Stickstoff-Netto-Mineralisierung unter Wald- und Wiesenstandorten des Niedermoores "Friedländer Große Wiese" in Mecklenburg-Vorpommern. In: Merbach W (ed) Pflanzenernährung, Wurzelleistung und Exsudation. Teubner, Stuttgart, pp 13–20

Murray JW (1987) Mechanisms controlling the distribution of trace elements in ocean and lakes. In: Hites RA, Eisenreich SJ (eds) Sources and fates of aquatic pollutants. Am Chem Soc 1987:153–184

Nabuurs GJ, Schelhaas MJ, Mohren GMJ, Field CB (2003) Temporal evolution of the European forest sector carbon sink 1950–1999. Global Change Biol 9:152–160

Naiman RJ, Décamps H (1990) The ecology and management of aquatic-terrestrial ecotones. (Man and the biosphere, series 4) Parthenon Casterton Hall, London

Nakashima BS, Leggett WC (1975) Yellow perch (*Perca flavescens*) biomass responses to different levels of phytoplankton and benthic biomass in Lake Memphremagog, Quebec–Vermont. J Fish Res Bd Can 32:1785–1797

Naujokat D (1991) Modellierung von Oberflächenwiderständen der trockenen Deposition von SO_2 im Bereich der Bornhöveder Seenkette. MSc thesis, Kiel University, Kiel

Naujokat D (1996) Nährstoffbelastung und Eutrophierung stehender Gewässer. Möglichkeiten und Grenzen ökosystemarer Entlastungsstrategien am Beispiel der Bornhöveder Seenkette. PhD thesis, University of Kiel, Kiel

Naujokat D (1997) Nährstoffbelastung und Eutrophierung stehender Gewässer. Diss Druck Darmstadt Oekol Reihe Bd 2

Neumann F (1998): Auswirkungen verschiedener Bewirtschaftungsweisen im Feuchtgrünland auf die Gastropoden-Fauna. Faun Oekol Mitt Suppl 24:5–43

Neumann U (1971) Die Ausbreitungsfähigkeit von Carabiden in den forstlichen Rekultivierungen des Rheinischen Braunkohlenreviers. In: Den Boer PJ (ed) Dispersal and dispersal power of carabid beetles. (Miscellaneous papers 8) Landbouwhogeschool Wageningen, pp 89–103

Nicolis G, Prigogine I (1977) Self-organization in nonequilibrium systems: from dissipative structures to order through fluctuations. Wiley, New York

Niedermeier-Lange R (2000) Hydrochemische Untersuchungen von Porenlösungen der Wasserungesättigten und -gesättigten Zonen im Bereich der Bornhöveder Seenkette. EcoSys Suppl 31

Nielsen SN, Müller F (2000) Emergent properties of ecosystems. In: Jørgensen SE, Müller F (eds.) Handbook of ecosystem theories and management. Lewis, Boca Raton, pp 195–216

Nishio T, Ishida Y (1990) Organically bound iron in lake sediments and its availability to Uroglena americana, a freshwater red tide Chrysophyceae. Jpn J Limnol 51:281–291

Nonami H, Schulze E-D, Ziegler H (1990) Mechanisms of stomatal movement in response to air humidity, irradiance and xylem water potential. Planta 183:57–64

Noriki S, Ishimori N, Harada K, Tsungai S (1985) Removal of trace metals from seawater during a phytoplankton bloom as studied with sediment traps in Funka Bay, Japan. Mar Chem 17:75–89

Nötzold R (1996) Die Kurzflügel- und Laufkäfergemeinschaften (Staphylinidae und Carabidae) des Bodens verschiedener Erlenbruchbiotope und ihre Abhängigkeit von Standortfaktoren. Faun Oekol Mitt Suppl 20:9–46

Nowok C (1994) Räumliche Struktur der Gewässerbelastung im Quellgebiet der Alten Schwentine. MSc thesis, University of Kiel, Kiel

Odum EP (1969) The strategy of ecosystem development. Science 104:262–270

Odum HT (1983) Maximum power and efficiency: a rebuttal. Ecol Model 20:71–82

OECD (1982) Eutrophication of waters – monitoring assessment and control. Organisation for Economic Co-operation and Development, Paris

Oheimb G von (2003) Einfluß forstlicher Nutzung auf die Artenvielfalt und Artenzusammensetzung der Gefäßpflanzen in norddeutschen Laubwäldern. Naturwiss Forschung 70

Oles B (2001) Ökopunkt ist nicht gleich Ökopunkt – Ergebnisse eines quantitativen Vergleichs von Biotopwertverfahren. NatSchutz LandschPlan 33:213–217

Olsson L, Ardö J (2002) Soil carbon sequestration in degraded semiarid agro-ecosystems – perils and potentials. Ambio 31:471–477

O'Neill RV (1988) Hierarchy theory and global change. In: Rosswall T, Woodmansee RG, Risser PG (eds) Scales and global change. Wiley, Chichester, pp 29–46

O'Neill RV, DeAngelis DL, Waide JB, Allen TFH (1986) A hierarchical concept of ecosystems. Princeton University Press, Princeton

O'Neill RV, Johnson AR, King AW (1989) A hierarchical framework for the analysis of scale. Landscape Ecol 3:193–206

Opitz S, Barkmann S, Bertram C, Dienemann P, Gessner M, Hölker F, Löhlein B, Müller U, Newzella R, Schieferstein B, Zimmermann H, Pöpperl R (1997) Ein quantitatives Nahrungsnetz-Modell für das Litoral des Belauer Sees (Schleswig–Holstein). Erweiterte Zusammenfassung Jahrestagung DGL, pp 186–190

Orth CB (1996) Die Bedeutung des troposphärischen Ozons für die Luftqualität Schleswig-Holsteins. MSc thesis, University of Kiel, Kiel

Ostendorp W (1992) Sedimente und Sedimentbildung in Seeuferröhrichten des Bodensee-Untersees. Limnologica 22:16–33

Ostendorp W (1995) Seeuferrenaturierung als Teil einer Seesanierung. Limnol Aktuell 8:53–68

Ott K (1994) Ökologie und Ethik: ein Versuch praktischer Philosophie. Attempto, Tübingen

Ott K (1996) Vom Handeln zum Begründen. Aufsätze zur angewandten Ethik. Attempto, Tübingen

Otto C-J (1991) Benthosuntersuchungen am Belauer See: Eine ökologisch-phänologische und produktionsbiologische Studie unter besonderer Berücksichtigung der merolimnischen Insekten. PhD thesis, University of Kiel, Kiel

Overgaard Nielsen B (1987) Vertical distribution of insect populations in the free air space of beech woodland. Entomol Meddr 54:169–178

Palmer MW (1992) The coexistence of species in fractal landscapes. Am Nat 139:375–397

Palmer MW (1994) Variation in species richness: towards a unification of hypotheses. Folia Geobot Phytotax 29:511–530

Parkhurst DL, Thorstenson DC, Plummer LN (1980) PHREEQUE – a computer program for geochemical calculations. US Geol Surv Water Resour Invest Rep 1980:80–96

Patrick WHJ, Gotoh S, Wiliams BG (1973) Strengite dissolution in flooded soils and sediments. Science 179:564–565

Patten BC (1992) Energy, emergy, and environs. Ecol Model 62:29–70

Paulson LJ (1977) The significance of ammonia regeneration by trout on phytoplankton productivity in Castle Lake, California. PhD thesis, University of California, Davis

Perrings C (1995) Ecology, economics and ecological economics. Ambio 24:60–64

Peters M (1990) Nutzungseinfluß auf die Stoffdynamik schleswig-holsteinischer Böden.Wasser-, Luft-, Nähr- und Schadstoffdynamik. SchriftenR Inst Pflanzenern Bodenkde Univ Kiel 8

Peters RH (1991) A critique for ecology. Cambridge University Press, Cambridge

Peterson AG, Ball JT, Luo YQ, Field CB, Reich PB, Curtis PS, Griffin KL, Gunderson CA, Norby RJ, Tissue DT, Forstreuter M, Rey A, Vogel CS (1999) The photosynthesis leaf nitrogen relationship at ambient and elevated atmospheric carbon dioxide: a meta-analysis. Global Change Biol 5:331–346

Peterson H, Luxton M (1982) A comparative analysis of soil fauna populations and their role in decomposition processes. Oikos 39:288–422

Petit S, Burel F (1997) The effects of spatial isolation on the distribution of forest carabid beetles in a hedgerow network landscape. Ecol Mediterr 23:27–36

Pfeiffer HW (2000) Die Bedeutung von Brassen (*Abramis brama* (L.)) und Plötzen (*Rutilus rutilus* (L.)) für die Stoff- und Energiekreisläufe im Belauer See, Schleswig–Holstein. PhD thesis, University of Hamburg, Hamburg

Pilegaard K, Hummelshoj P, Jensen NO, Chen Z (2002) Two years of continuous CO_2 eddy flux measurements over a Danish beech forest. Agric For Meteorol 107:29–41

Pimm SL (1982) Food webs. Chapman and Hall, London

Pinel-Alloul B (1995) Spatial heterogeneity as a multiscale characteristic of zooplankton community. Hydrobiologia 301:17–42

Piotrowski JA (1991) Quartär- und hydrogeologische Untersuchungen im Bereich der Bornhöveder Seenkette, Schleswig–Holstein. Ber Rep Geol Palaeont Inst Univ Kiel 43

Piotrowski JA, Kluge W (1994) Die Uferzone als hydrogeologische Schnittstelle zwischen Aquifer und See: Sedimentfacies und Grundwasserdynamik am Belauer See, Schleswig-Holstein. Z Geol Ges 145:131–142

Podsetchine V, Schernewski G (1999) The influence of spatial wind inhomogeneity on flow patterns in a small lake. Water Res 33:3348–3356

Pollet M, Grootaert P (1991) Horizontal and vertical distribution of Dolichopodidae (Diptera) in a woodland ecosystem. J Nat Hist 25:1297–1312

Pollock MM, Naiman RJ, Hanley TA (1998) Plant species richness in riparian wetlands – a test of biodiversity theory. Ecology 79:94–105

Popper K (1959) The logic of scientific discovery. Harper and Row, New York

Pöpperl R, Kluge W, Schernewski G, Garbe-Schönberg CD, Nellen W (2001) Spatial and temporal variability of limnological processes. In: Tenhunen JD, Lenz R, Hantschel R (eds) Ecosystem approaches to landscape management in Central Europe. (Ecological studies 147) Springer, Berlin Heidelberg New York, pp 117–161

Prigogine I (1967) Thermodynamics of irreversible processes. Wiley, New York

Prigogine I (1976) Order through fluctuation: self-organization and social system. In: Jantsch E, Waddington CE (eds) Evolution and consciousness. Addison–Wesley, Reading, Mass., pp 93–133

Prigogine I (1985) Vom Sein zum Werden – Zeit und Komplexität in den Naturwissenschaften. Piper, Munich

Purvis A, Hector A (2000) Getting the measure of biodiversity. Nature 405:212–219

Raabe M, Irmler U, Meyer H (1996) Vertikalzonierung flugaktiver Empidoidea (Diptera: Empididae, Hybotidae, Dolichopodidae) in Waldökosystemen. Faun Oekol Mitt 7:93–108

Rambow K (1996) Untersuchungen zum Stoffverhalten in einer forstlich und einer landwirtschaftlich genutzten Braunerde im Bereich der Bornhöveder Seenkette – Messung und Simulation. SchriftenR Inst Pflanzenern Bodenkde Univ Kiel 34

Raschke K (1979) Movements of stomata. In: Haupt W, Feinlieb ME (eds) Encyclopedia of plant physiology, N.S. 7. Springer, Berlin Heidelberg New York, pp 383–441

Reck H (2003) Tierökologie und räumliche Planung: Die Eignung arten- und populationsorientierter Ansätze für die Umweltplanung, untersucht am Beispiel des Überlebens des Feldgrashüpfers (*Chorthippus apricarius* L.) in Agrarlandschaften. Habilitationsschrift Agrarwiss Fak Univ Kiel

Reddy KA, Rao PSC (1983) Nitrogen and phosphorus fluxes from a flooded organic soil. Soil Sci 136:300–307

Reddy KR, Kadlec RH, Laig E, Gale PM (1999) Phosphorus retention in streams and wetlands: a review. Crit Rev Environ Sci Technol 29:83–146

Reiche E-W (1991) Entwicklung, Validierung und Anwendung eines Modellsystems zur Beschreibung und flächenhaften Bilanzierung der Wasser- und Stickstoffdynamik in Böden. Kieler Geogr Schr 79

Reiche E-W (1994) Modelling water and nitrogen dynamics on catchment scale. Ecol Model 75/76:372–384

Reiche E-W (1996) WASMOD – Ein Modellsystem zur gebietsbezogenen Simulation von Wasser- und Stoffflüssen. In: Breckling B, Asshoff A (eds) Modellbildung und Simulation im Projektzentrum Ökosystemforschung. EcoSys 4:143–163

Reiche E-W (1998) Bodenschätzungs-Standardauswertung Schleswig-Holstein: Eine Methode zur computergestützten Übersetzung, Parameterisierung und planungsbezogenen Bodenbewertung. Mitteil Dtsch Bodenkde Ges 87:42–93

Reiche E-W, Dibbern I (1996) Analyse räumlicher Musterbildungsprozesse am Beispiel des Stoffbestandes und der floristischen Ausstattung in einem Buchenwaldökosystem. Verh Ges Oekol 26:471–478

Reiche E-W, Meyer M, Dibbern I (1999) Modelle als Bestandteile von Umweltinformationssystemen dargestellt am Beispiel des Methodenpaketes "DILAMO". In: Blaschke T (ed) Umweltmonitoring und Umweltmodellierung – GIS und Fernerkundung als Werkzeuge einer nachhaltigen Entwicklung. Wichmann, Heidelberg, pp 131–141

Reiche E-W, Müller F, Dibbern I, Kerrines A (2001) Spatial heterogeneity in forest soils and understory communities of the Bornhöved Lake District. In: Tenhunen JD, Lenz R, Hantschel R (eds) Ecosystem approaches to landscape management in central Europe. Springer, Berlin Heidelberg New York, pp 49–72

Renkonen O (1938) Statistisch-ökologische Untersuchungen über die terrestrische Käferwelt der finnischen Bruchmoore. Ann Zool Soc Zool Bot Fenn 6:1–231

Rennenberg H, Kreutzer K, Papen H, Weber P (1998) Consequences of high loads of nitrogen for spruce (*Picea abies* L.) and beech (*Fagus sylvatica* L.) forests. New Phytol 139:71–86

Reuter H (1996) An individual-based model on the reproductive success of the European robin (*Erithacus rubecula*) in a complex environment. EcoSys 4:223–239

Reuter H (2001) Individuum und Umwelt. Wechselwirkungen und Rückkopplungsprozesse in individuenbasierten tierökologischen Modellen. PhD thesis, University of Bremen, Bremen

Reuter H, Breckling B (1999) Emerging properties on the individual level: modelling the reproduction phase of the European robin *Erithacus rubecula*. Ecol Model 121:199–219

Richards BN (1987) The microbiology of terrestrial ecosystems. Longman, London

Richter D (1995) Ergebnisse methodischer Untersuchungen zur Korrektur des systematischen Meßfehlers des Hellmann Niederschlagsmessers. Ber Dtch Wetterdienst 194

Riecken U, Ries U (1992) Untersuchung zur Raumnutzung von Laufkäfern (Col: Carabidae) mittels Radiotelemetrie – Methodenentwicklung und erste Freilandversuche. Z Oekol Natursch 1:147–149

Rief S (1996) Einfluß der Bewirtschaftung auf ausgewählte Diptera (Nematocera: Limoniidae; Tipulidae; Trichoceridae; Brachycera: Empididae; Hybotidae; Dolichopodidae) verschiedener Ökosysteme auf Niedermoor. Faun Oekol Mitt Suppl 20:47–76

Roden JS, Pearcy RW (1993) The effect of flutter on the temperature of poplar leaves and its implications for carbon gain. Plant Cell Environ 16:571–577

Rodewald-Rudescu L (1974) Das Schilfrohr Phragmites communis TRIN. Binnengewässer 27

Romahn KS (2003) Rationalität von Werturteilen im Naturschutz. (Theorie in der Ökologie 8), Frankfurt

Rossknecht H (1977) Zur autochthonen Calcitfällung im Bodensee-Obersee. Arch Hydrobiol 81:35–64

Rossknecht H (1980) Phosphatelimination durch autochthone Calcitfällung im Bodensee-Obersee. Arch Hydrobiol 88:328–344

Roth CH (1999) Physikalische Ursachen der Wassererosion. In: Blume H-P, Felix-Henningsen P, Fischer WR, Frede H-G, Horn R, Stahr K (eds) Handb Bodenkde. Ecomed, Landsberg, pp 1–33

Rudd JWM, Hamilton RD (1978) Methane cycling in a eutrophic shield lake and its effects on whole metabolism. Limnol Oceanogr 23:337–348

Ruhmohr S (1996) Die Grundwasserdynamik zwischen Bornhöveder Seenkette und Großem Plöner See. Ber Rep Geol Paläont Inst Univ Kiel 77

Ryan PJ, Harleman DRF (1971) Prediction of the annual cycle of temperature changes in a stratified lake or reservoir: mathematical model and user's manual. (Technical report 137) Massachusetts Institute of Technology, Cambridge, Mass.

Ryden JC, Syers JK (1976) Calcium retention in response to phosphate sorption by soils. Soil Sci Soc Am J 40:845–846

Sach W (1999) Vegetation und Nährstoffdynamik unterschiedlich genutzten Grünlandes in Schleswig–Holstein. Diss Bot 308

Sala ES, Chapin FS III, Armesto JJ, Berlow E, Bloomfield J, Dirzo R, Huber-Sanwald E, Huenneke LF, Jackson RB, Kinzig A, Leemans R, Lodge DM, Mooney HA, Oesterheld M, Poff NL, Sykes MT, Walker BH, Walker M, Wall DH (2000) Global change scenarios for the year 2100. Science 287:1770–1774

Salomons W, Förstner U (1984) Metals in the hydrocycle. Springer, Berlin Heidelberg New York

Salski A, Fränzle O, Kandzia P (1996) Fuzzy logic in ecological modelling. Ecol Model Spec Iss 85

Schaefer M, Schauermann J (1990) The soil fauna of beech forests: comparison between a mull and a moder soil. Pedobiologia 34:299–314

Schenk D, Kaupe M (1998) Grundwassererfassungssysteme in Deutschland. Materialien zur Umweltforschung. Metzel–Poeschel, Stuttgart

Schernewski G (1992) Raumzeitliche Prozesse und Strukturen im Wasserkörper des Belauer Sees. EcoSys Suppl 1

Schernewski G (1999) Der Stoffhaushalt von Seen: Bedeutung zeitlicher Variabilität und räumlicher Heterogenität von Prozessen sowie des Betrachtungsmaßstabs. Meereswiss Ber 36

Schernewski G (2003) Nutrient budgets, dynamics and storm effects in a eutrophic, stratified Baltic Lake. Acta Hydrochim Hydrobiol 31:152–161

Schernewski G, Schulz U (1999) Zustandsentwicklung schleswig-holsteinischer Seen zwischen 1983 und 1993: Eine Betrachtung unter Anwendung der Clusteranalyse. Limnologica 29:146–159

Schernewski G, Wetzel H (1997) Phosphorhaushalt. In: Fränzle O, Müller F, Schröder W (eds) Handbuch der Umweltwissenschaften, Kap IV – 2.2.4. Ecomed, Landsberg

Schernewski G, Theesen L, Kerger KE (1994) Modelling thermal stratification and calcite precipitation of Lake Belau (Northern Germany). Ecol Model 75/76:421–433

Schernewski G, Schleuß U, Wetzel H (1996) Bedeutung von Wallhecken für den Gewässerschutz. EcoSys 5:217–232

Schernewski G, Podsetchine V, Huttula T (2005) Effects of the flow field on small-scale phytoplankton patchiness. Nord Hydrol 36:85–98

Scheytt T (1994) Örtliche und zeitliche Veränderungen der Grund wasserbeschaffenheit im Bereich der Bornhöveder Seenkette. EcoSys Suppl 7

Schieferstein B (1997) Ökologische und molekularbiologische Untersuchungen an Schilf (*Phragmites australis* (Cav.) Trin. ex Steud.) im Bereich der Bornhöveder Seen. EcoSys Suppl 22

Schimitschek E (1952) Forstentomologische Studien im Urwald Rotwald. Teil I. Z Angew Entomol 34:178–215

Schimming C-G (1991) Wasser-, Luft-, Nähr- und Schadstoffdynamik charakteristischer Böden Schleswig–Holsteins. Schriftenr Inst Pflanzenern Bodenkde Univ Kiel 13

Schimming C-G, Schrautzer J, Reiche E-W, Munch JC (2001) Nitrogen retention and loss from ecosystems of the Bornhöved Lake District. In: Tenhunen JD, Lenz R, Hantschel R (eds) Ecosystem approaches to landscape management in Central Europe. Ecol Stud 147:97–115

Schindler DE, Kitchell JF, He X, Carpenter SR, Hodgson JR, Cottingham KL (1993) Food web structure and phosphorus cycling in lakes. Trans Am Fish Soc 122:756–772

Schjørring JK (1991) Ammonia emission from the foliage of growing plants. In: Sharkey TD, Holland EA, Mooney HA (eds) Trace gas emissions by plants. Academic, San Diego, pp 267–292

Schläpfer F, Schmid B (1999) Ecosystem effects of biodiversity: a classification of hypotheses and exploration of empirical results. Ecol Appl 9:893–912

Schleuß U (1992) Böden und Bodenschaften einer Norddeutschen Moränenlandschaft. EcoSys Suppl 2

Schleuß U, Blume H-P (1996) Bodengesellschaften einer Jungmoränenlandschaft in Nordwestdeutschland (Bornhöveder Seenkette, Schleswig–Holstein). Petermanns Geogr Mitt 140:3–13

Schleuß U, Trepel M, Wetzel H, Schimming C-G, Kluge W (2001) Interactions between hydrologic parameters, soils, and vegetation at three minerothrophic peat ecosystems. In: Broll G, Merbach W, Pfeiffer EM (eds) Wetlands in central Europe. Soil organisms, soil ecological processes and trace gas emissions. Springer, Berlin Heidelberg New York, pp 117–132

Schlichting A, Leinweber P, Meissner R, Altermann M (2002) Sequentially extracted phosphorus fractions in peat-derived soils. J Plant Nutr Sci 165:190–198

Schlichting E (1975) Bedingungen und Bedeutung landwirtschaftlicher Umsatz- und Bilanzuntersuchungen. Forstw Cbl 94:273–280

Schlichting E, Blume H-P, Stahr K (1995) Bodenkundliches Praktikum. Parey–Blackwell, Berlin

Schmidt J (1991) A mathematical model to simulate rainfall erosion. In: Bork HR, De Ploey J, Schick AP (eds) Erosion, transport and deposition processes – theories and models. Catena Suppl 19:145–165

Schmidt J (ed) (2000) Soil erosion – application of physically based models. (Environmental science) Springer, Berlin Heidelberg New York

Schmidt J, Werner M von, Michael A (1996) EROSION 2D – Ein Computermodell zur Simulation der Bodenerosion durch Wasser. Sächsische Landesanstalt Landwirtschaft, Sächsiches Landesamt für Umwelt und Geologie, Freiberg

Schmidt R (1997) Grundsätze der Bodenvergesellschaftung; Schmidt (1999) Klassifikation von Bodengesellschaften. In: Blume H-P, Felix-Henningsen P, Fischer W, Frede H-G, Horn R, Stahr K (eds) Handbuch der Bodenkunde, Chapters 3.4.1 and 3.4.3. Ecomed, Landsberg

Schmitt R (1997) Untersuchungen zur Kohlenstoffbilanz im Einzugsgebiet des Belauer Sees. MSc thesis, University of Kiel, Kiel

Schmitz OJ (2000) Combining field experiments and individual-based modelling to identify the dynamically relevant organizational scale in a field system. Oikos 89:471–484

Schneider ED, Kay J (1994a) Life as a manifestation of the second law of thermodynamics. Math Comput Model 19:25–48

Schneider ED, Kay J (1994b) Complexity and thermodynamics: towards a new ecology. Futures 26:626–647

Scholle D (1997) GIS-gestützte Zusammenführung vegetationskundlicher, bodenkundlicher und nutzungsbezogener Daten zu einem landschaftsökologischen Indikationsverfahren. EcoSys Suppl 21

Scholle D, Schrautzer J (1993) Zur Grundwasserdynamik unterschiedlicher Niedermoor-Gesellschaften Schleswig–Holsteins. Z Oekol Natursch 2:87–98

Scholtissek B (2000) Naturschutzziele in der Landschaftsplanung. C.D.C. Heydorn, Uetersen

Schönthaler K, Meyer U, Pokorny D, Reichenbach M, Schuller D, Windhorst W (2003) Ökosystemare Umweltbeobachtung – Vom Konzept zur Umsetzung. Bayerisches Staatsministerium für Landesentwicklung und Umweltfragen, Umweltbundesamt. Schmidt, Berlin

Schrautzer J (1988) Pflanzensoziologische und standörtliche Charakteristik von Seggenriedern und Feuchtwiesen in Schleswig-Holstein. Mitt AG Geobot Schl Holst Hambg 38

Schrautzer J (2004) Niedermoore Schleswig-Holsteins: Charakterisierung und Beurteilung ihrer Funktion im Landschaftshaushalt. Mitt AG Geobot Schl Holst Hambg 63

Schrautzer J, Trepel M (1997) Wechselwirkungen zwischen bodenphysikalischen Parametern, Grundwasserdynamik und der Vegetationszusammensetzung in unterschiedlich stark genutzten Niedermoor-Ökosystemen. Feddes Repert 108:119–137

Schrautzer J, Wiebe C (1993) Geobotanische Characterisierung des Grünlands in Schleswig–Holstein. Phytocoenologia 22:105–144

Schrautzer J, Härdtle W, Hemprich G, Wiebe C (1991) Zur Synökologie und Synsystematik gestörter Erlenwälder im Gebiet der Bornhöveder Seenkette (Schleswig–Holstein). Tuexenia 11:293–307

Schröder B, Richter O (2000) Are habitat models transferable in space and time? Z Oekol NatSchutz 8:195–205

Schröder R (1975) Release of plant nutrients from reed borders and their transport into the open waters of the Bodensee-Untersee. Symp Biol Hung 15:27–37

Schröder W (1998) Ökologie und Umweltrecht als Herausforderung natur- und sozialwissenschaftlicher Forschung und Lehre. In: Daschkeit A, Schröder W (eds) Umweltforschung quergedacht: Perspektiven integrativer Unweltforschung und -lehre. Springer, Berlin Heidelberg New York, pp 329–358

Schuller D, Brunken-Winkler H, Busch P, Förster M, Janiesch P, Lemm R von, Niedrighaus R, Strasser H (2000) Sustainable land use in an agricultural misused landscape in northwest Germany through ecotechnical restoration by a 'Patch-Network-Concept'. Ecol Eng 16:99–117

Schulte M (1993) Saisonale und interannuelle Variabilität des CO_2-Gaswechsels von Buchen (*Fagus sylvatica* L) – Bestimmung von C-Bilanzen mit Hilfe eines empirischen Modells. Shaker, Aachen

Schulten HR (1993) A state of the art structural concept for humic substances. Naturwissenschaften 80:29–30

Schulz U (1996) Zustand, Entwicklung und Bewertung ausgewählter Seen Schleswig-Holsteins unter Anwendung der Clusteranalyse. MSc thesis, University of Kiel, Kiel

Schulze E-D (1970) Der CO_2-Gaswechsel der Buche (*Fagus sylvatica* L.) in Abhängigkeit von den Klimafaktoren im Freiland. Flora 159:177–232

Schulze E-D (2000) The carbon and nitrogen cycle of forest ecosystems. In: Schulze E-D (ed) Carbon and nitrogen cycling in European forest ecosystems. (Ecological studies 142) Spinger, Berlin Heidelberg New York, pp 3–13

Schulze E-D, Mooney HA (eds) (1993) Biodiversity and ecosystem function. Springer, Berlin Heidelberg NewYork

Schulze E-D, Lange OL, Oren R (eds) (1989) Forest decline and air pollution a study of spruce (*Picea abies*) on acid soils. Springer, Berlin Heidelberg NewYork

Schwartze, P (1992) Nordwestdeutsche Feuchtgrünlandgesellschaften unter kontrollierten Nutzungsbedingungen. Diss Bot 183

Schwerin v Krosigk H (1976) Untersuchungen zum vor- und frühgeschichtlichen Siedlungsablauf am Fundbild der Gemarkungen Bornhöved – Gönnebek – Groß-Kummerfeld – Schmalensee, Kr Segeberg/Holstein. Offa-Erg-R 1

Schwertmann U, Huth M (1975) Erosionsbedingte Stoffverteilung in zwei hopfengenutzten Kleinlandschaften der Hallertau (Bayern). Z Pflanzenern Bodenkde 138:397–405

Schwertmann U, Vogl W, Kainz M (1987) Bodenerosion durch Wasser: Vorhersage des Abtrags und Bewertung von Gegenmaßnahmen. Ulmer, Stuttgart

Seitzinger SP (1988) Denitrification in freshwater and coastal marine ecosystems: ecological and geochemical significance. Limnol Oceanogr 33:702–724

Shannon CE, Weaver W (1949) The mathematical theory of communication. Urbana, New York

Shapiro J, Carlson R (1982) Comment on the role of fish in the regulation of phosphorus availability in lakes. Can J Fish Aquat Sci 39:364

Shugart HH, Urban DL (1998) Scale, synthesis and ecosystem dynamics. In: Pomeroy LR, Alberts JJ (eds) Concepts of ecosystem ecology. (Ecological studies 67) Springer, Berlin Heidelberg New York, pp 279–290

Shuttleworth WJ, Wallace JS (1985) Evaporation from sparse crops – an energy combination theory. Q J R Meteorol Soc 111:839–855

Sigg L, Stumm W (1991) Aquatische Chemie. Eine Einführung in die Chemie wässriger Lösungen und in die Chemie natürlicher Gewässer. Teubner, Stuttgart

Siitonen J (1994) Decaying wood and saproxylic coleoptera in two old spruce forests: a comparison based on two sampling methods. Ann Zool Fenn 31:89–95

Simandl J (1993) The spatial pattern, diversity and niche partitioning in xylophagous beetles (Coleoptera) associated with *Frangula alnus* Mill. Acta Oecol 14:161–171

Simmering D, Waldhardt R, Otte A (2001) Zur vegetationsökologischen Bedeutung von scharfen Grenzlinien in Agrarlandschaften – Beispiele aus einer kleinstrukturierten Mittelgebirgslandschaft. Peckiana 1:79–87

Sioli E (1996) Die Phytophagenfauna der Krautschicht (Cicadina, Heteroptera und Symphyta) verschiedener Waldtypen Schleswig-Holsteins. Faun Oekol Mitt Suppl 21:1–94

Slanina J (1983) Collection and analysis of precipitation. Methods, data evaluation and interpretation. VDI Ber 500:117–124

Smith P (2004) Carbon sequestration in European croplands. Eur J Agronomy 20:229–236

Smith P, Goulding KW, Smith KA, Powlson DS, Smith JU, Falloon PD, Coleman K (2001) Enhancing the carbon sink in European agricultural soils: Including trace gas fluxes in estimates of carbon mitigation potential. Nutr Cycling Agroecosyst 60:237–252

Sodtke RM (2003) Ein Entscheidungsunterstützungssystem für den Zwischenfruchtanbau – Konzeption, Entwicklung, Validierung. PhD thesis, University of Kiel, Kiel

Solbrig OT, Medina E, Silva JF (eds) (1996) Biodiversity and savanna ecosystem processes – a global perspective. Springer, Berlin Heidelberg New York

Soussana JF, Salètes S, Smith P, Schils R, Ogle S, et al (2004) Greenhouse gas emissions from European grasslands. (CarboEurope-GHG discussion paper) Clermont–Ferrand, Paris

Sparling GP (1992) Ratio between microbial biomass carbon to soil organic carbon as a sensitive indicator of changes in soil organic matter. Austr J Soil Res 30:195–207

Spranger T (1992) Erfassung und ökosystemare Bewertung der atmosphärischen Deposition und weiterer oberirdischer Stoffflüsse im Bereich der Bornhöveder Seenkette. EcoSys Suppl 2

Srinivasan R, Arnold JG, Muttiah RS, Dyke PT (1995) Plant and hydrologic simulation for the conterminous U.S. using SWAT and GIS. Hydro Sci Technol 11:160–168

SRU (1987) Derzeitige Situation und Trends der Belastung der Nahrungsmittel durch Fremdstoffe. (Der Rat von Sachverständigen für Umweltfragen) Metzler–Poeschel, Stuttgart

SRU (1991) Allgemeine ökologische Umweltbeobachtung. Sondergutachten 1990. (Der Rat von Sachverständigen für Umweltfragen) Metzler–Poeschel, Stuttgart

SRU (1994) Umweltgutachten 1994. (Der Rat von Sachverständigen für Umweltfragen) Metzler–Poeschel, Stuttgart

SRU (1996) Umweltgutachten. Für eine dauerhaft umweltgerechte Entwicklung. (Der Rat von Sachverständigen für Umweltfragen) Metzler–Poeschel, Stuttgart

SRU (2000) Umweltgutachten 2000. Schritte ins nächste Jahrtausend. (Der Rat von Sachverständigen für Umweltfragen) Metzler–Poeschel, Stuttgart

SRU (2002) Sondergutachten. Für eine Stärkung und Neuorientierung des Naturschutzes. (Der Rat von Sachverständigen für Umweltfragen) Metzler–Poeschel, Stuttgart

Staack A (1996) Untersuchungen zum CO_2-Stoffwechsel von Wurzeln von *Alnus glutinosa* (L.) Gaertner in einem Erlenbruchwald. MSc thesis, University of Kiel, Kiel

Stamm S von (1992) Untersuchungen zur Primärproduktion von *Corylus avellana* an einem Knickstandort in Schleswig–Holstein und Erstellung eines Produktionsmodells. EcoSys Suppl 3

Stamm S von (1994) Linked stomata and photosynthesis model for *Corylus avellana* (hazel). Ecol Model 176:345–358

Stark J (1993) Zur rezenten Sedimentologie und Geochemie des Belauer Sees, Bornhöveder Seenkette (Schleswig–Holstein). MSc thesis, University of Kiel, Kiel

Stefan H, Ford DE (1975) Temperature dynamics in dimictic lakes. J Hydraul Div 101:97–114

Steffan-Dewenter I, Tscharnke T (1999) Effects of habitat isolation on pollinator communities and seed set. Oecologia 121:432–440

Steinborn W (2000) Quantifizierung von Ökosystemeigenschaften als Grundlage für die Umweltbewertung. PhD thesis, University of Kiel, Kiel

Steinborn W, Kutsch WL, Eschenbach C, Kappen L (1997) Photosynthese und Atmung bei Ästen von *Alnus glutinosa*. Schriftenreihe 'Landschaftsentwicklung und Umweltforschung' der TU Berlin 107:7–22

Steiner N (2002) Modellierung der Artendiversität auf verschiedenen Skalenebenen in Abhängigkeit von der Landschaftsstruktur. Treffpunkt Biol Vielfalt II BfN: 205–208

Stephan H-J, Menke B (1977) Untersuchungen über den Verlauf der Weichsel-Kaltzeit in Schleswig–Holstein. Z Geomorph NF Suppl 27:12–28

Sterner RW, Elser, JJ (2002) Ecological stoichiometry – the biology of elements from molecules to the biosphere. Princeton University Press, Princeton

Stickan W, Zhang X (1992) Seasonal changes in CO_2 and H_2O gas exchange of young European Beech (*Fagus sylvatica* L.). Trees 6:96–102

Stickan W, Gansert D, Neemann G, Rees U (1994) Modelling the influence of climatic variability on carbon and water budgets of beech saplings (*Fagus sylvatica* L.). Ecol Model 75/76:332–343

Stocker R, Korner C, Schmid B, Niklaus PA, Leadley PW (1999) A field study of the effects of elevated CO_2 and plant species diversity on ecosystem-level gas exchange in a planted calcareous grassland. Global Change Biol 5:95–105

Stork R, Dilly O (1998) Maßstabsabhängige räumliche Variabilität mikrobieller Bodenkenngrößen in einem Buchenwald. Z Pflanzenern Bodenkde 161:235–242

Stork R, Dilly O (1999) Einflußfaktoren auf die räumliche Heterogenität bodenmikrobiologischer Kenngrößen in einem Buchenwald. Mitt Bodenkdl Ges 89:149–152

Stottele T (1991) Ökologisch orientierte Grünpflege an Straßen. Beitrag zu einem flächendeckenden Naturschutzkonzept. Dtsch Gartenbau 45:3134–3137

Strahler AN (1957) Quantitative analysis of watershed geomorphology. Trans Am Geophys Union 38:6

Straškraba M, Gnauck A (1985) Freshwater ecosystems. Modelling and simulation. Elsevier, Amsterdam

Strenzke K (1952) Untersuchungen über die Tiergemeinschaften des Bodens: Die Oribatiden und ihre Synusien in den Böden Norddeutschlands. Zoologica 104

Strong DR, Kaya HK, Whipple AV, Child AL, Kraig S, Bondonno M, Dyer K, Maron JL (1996) Entomopathogenic nematodes: natural enemies of root-feeding caterpillars on bush lupine. Oecologica 108:167–173

Sugiyama M, Hori T, Kihara S, Matsui M (1992) A geochemical study on the specific distribution of barium in Lake Biwa, Japan. Geochim Cosmochim Acta 56:597–605

Sutton MA (1990) The surface/atmosphere exchange of ammonia. PhD thesis, University of Edinburgh, Edinburgh

Svirezhev YM, Steinborn W (2001) Exergy of solar radiation: thermodynamic approach. Ecol Model 145:101–110

Sykora KV, Kalvij JM, Keizer PJ (2002) Phytosociological and floristic evaluation of a 15-year ecological management of roadside verges in the Netherlands. Preslia 74:421–436

Symondson WOC (1994) The potential of *Abax parallelepipedus* (Col.: Carabidae) for mass breeding as a biological control agent against slugs. Entomophaga 39:323–333

Takamura N, Nojiri Y (1994) Picophytoplankton biomass in relation to lake trophic state and the TN:TP ratio of lake water in Japan. J Phycol 30:439–444

Takeda H (1995) Changes in the collembolan community during the decomposition of needle litter in a coniferous forest. Pedobiologia 39:304–317

Tardieu F (1988) Analysis of the spatial variability of maize root density. Plant Soil 107:259–266

Taylor G, Davies WJ (1988) The influence of photosynthetically active radiation and simulated shadelight on the control of leaf growth of *Betula* and *Acer*. New Phytol 108:393–398

Taylor LR (1958) Aphid dispersal and diurnal periodicity. Proc Linn Soc Lond 169:58

Tenhunen JD, Yocum CS, Gates DM (1976a) Development of a photosynthesis model with emphasis on ecological applications, I. Theory. Oecologia 26:89–100

Tenhunen JD, Weber JA, Yocum CS, Gates DM (1976b) Development of a photosynthesis model with emphasis on ecological applications, II. Analysis of a data set describing the PM-surface. Oecologia 26

Tenhunen JD, Harley PC, Beyschlag W, Lange OL (1987) A model of net photosynthesis for leaves of the sclerophyll *Quercus coccifera*. In: Tenhunen JD (ed) Plant response to stress. (NATO ASI series G15) Springer, Berlin Heidelberg New York, pp 339–354

Tenhunen JD, Lenz R, Hantschel R (eds) (2001) Ecosystem approaches to landscape management in Central Europe. Springer, Berlin Heidelberg New York

Tennant D (1975) A test of a modified line intersect method of estimating root length. J Ecol 63:995–1001

Thom R (1975) Structural stability and morphogenesis. Benjamin, Reading, Mass.

Thomasius W (1992) Dynamik von Waldökosystemen. In: Lyr H, Fiedler HJ, Tranquilini W (eds) Physiologie und Ökologie der Gehölze. Fischer, Jena, pp 575–600

Tilman D, Knops J, Wedin D, Reich P, Ritchie M, Siemann E (1997a) The influence of functional diversity and composition on ecosystem processes. Science 277:1300–1302

Tilman D, Lehmann CL, Thomson KT (1997b) Plant diversity and ecosystem productivity: theoretical considerations. Proc Natl Acad Sci USA 94:1857–1861

Tischler W (1948) Biozönologische Untersuchungen an Wallhecken. Zool Jb 77:283–400

Törne E von (1990) Assessing feeding activities of soil-living animals. Pedobiologia 34:89–101

Tóth L (1972) Reeds control eutrophication of Balaton Lake. Water Res 6:1533–1539

Tréhen P (1977) Recherches sur les Empidides à larves édaphiques. BSc thesis, University of Rennes, Rennes

Trepel M (1999) Spatiotemporal simulation of water and nitrogen dynamics as a tool in fen restoration. Int Peat J 9:45–52

Trepel M (2000) Quantifizierung der Stickstoffdynamik von Ökosystemen auf Niedermoorböden mit dem Modellsystem WASMOD. EcoSys Suppl 29

Trepel M, Kluge W (2002a) Analyse von Wasserpfaden und Stofftransformation in Feuchtgebieten zu Bewertung der diffusen Austräge. KA Wasserwirtsch Abwasser Abfall 49:807–815

Trepel M, Kluge W (2002b) Das Pfad-Transformations-Konzept als Grundlage für ein Wasser- und Stoffstrommanagement in Flusseinzugsgebieten. In: Stephan K, Bormann H, Diekkrüger B (eds) Tagungsbericht 5. Workshop zur hydrologischen Modellierung: Möglichkeiten und Grenzen für den Einsatz hydrologischer Modelle in Politik, Wirtschaft und Klimafolgenforschung. Kassel University Press, Kassel, pp 93–102

Trepel M, Kluge W (2004) WETTRANS: a flowpath oriented decision support system for the assesment of water and nitrogen exchange in riparian peatlands. Hydrol Process 18:357–371

Trepel M, Palmeri L (2002) Quantifying nitrogen retention in surface flow wetlands for environmental planning at the landscape-scale. Ecol Eng 19:127–140

Trewavas A (2001) Urban myths of organic farming. Nature 410:409–410

Trümpler D (1996) Räumlich hochauflösende Untersuchungen zum anthropogenen Energiehaushalt im Bereich der Bornhöveder Seenkette. MSc thesis, University of Kiel, Kiel

Tschirsich C (1994) Untersuchungen zur Quantifizierung von Denitri Fikationsverlusten aus Niedermoorböden. Dargestellt am Beispiel eines sauren Niedermoorbodens Nordwest-Deutschlands. PhD thesis, University of Göttingen, Göttingen

Tukey HB, Morgan JV (1963) Injury to foliage and its effect upon the leaching of nutrients from aboveground plant parts. Physiol Plant 16:557–564

Turner MG, Gardner RH (1990) Quantitative methods in landscape ecology. Springer, Berlin Heidelberg New York

Uhler AD, Helz GR (1984) Precipitation of PbS from solutions containing EDTA. J Crystal Growth 66:401–411

Ulanowicz RE (1986) Growth and development: ecosystem phenomenology. Springer, Berlin Heidelberg New York

Ulanowicz RE (1995) Ecosystem integrity: a causal necessity. In: Lemons J, Westra L (eds) Perspectives on implementing ecological integrity. Kluwer, Dordrecht, pp 77–87

Ulanowicz RE (2000) Ascendancy: a measure of ecosystem performance. In: Jørgensen SE, Müller F (eds) Handbook of ecosystem theories and management. Lewis, Boca Raton, pp 303–316

Ulrich B (1983) Interaction of forest canopies with atmospheric constituents: SO_2, alkali and earth alkali cations and chloride. In: Ulrich B, Pankrath J (eds) Effects of accumulation of air pollutants in forest ecosystems. Reidel, Dordrecht, pp 33–45

Ulrich B (1991) Beiträge zur Methodik der Waldschadensforschung. Göttingen Ber Forschungszentr Waldökosysteme/Waldsterben B 24

UNESCO (1988) Man belongs to the earth. International co-operation in environmental research. UNESCO, Paris

Vanni, MJ, Findlay, DL (1990) Trophic cascades and phytoplankton community structure. Ecology 71:921–937

Vanselow R (1997) Der stomatäre Widerstand als Regelgröße für die trockene Deposition gasförmiger Immissionen. PhD thesis, Kiel University, Kiel

Venebrügge G (1988) Analyse des ökologischen Hauptforschungsraumes "Bornhöveder Seengebiet" mittels eines Geographischen Informationssystems. Schriftl Hausarbeit Wiss Pruefung Lehramt Gymnasien Geogr Inst Univ Kiel 1988

Venebrügge G (1996) Die Ableitung der reliefabhängigen kurzwelligen Strahlungsbilanz im Bornhöveder Seengebiet mit Hilfe eines GIS. EcoSys Suppl 12

Verhagen JHG (1994) Modeling phytoplankton patchiness under the influence of wind-driven currents in lakes. Limnol Oceanogr 39:1551–1565

Verruijt A (1991) MULAT 1.0 – multi-layered aquifer transport. Delft University of Technology, Delft

Vleeshouwers LM (2002) Carbon emission and sequestration by agricultural land use: a model study for Europe. Global Change Biol 8:519

Vogel EF (1980) Seedlings of dicotyledons: structure, development, types: descriptions of 150 woody Malesian taxa. Centre for Publishing and Documentation, Wageningen

Vries HH de, Boer PJ de, Dijk TS van (1996) Ground beetle species in heathland fragments in relation to survival, dispersal, and habitat preference. Oecologia 107:332–342

Wachendorf C (1996) Eigenschaften und Dynamik der organischen Bodensubstanz ausgewählter Böden unterschiedlicher Nutzung einer norddeutschen Moränenlandschaft. EcoSys Suppl 13

Wachendorf C, Irmler U, Blume H-P (1997a) Der Einfluß einer ehemaligen Ackernutzung auf Morphologie, Biozönose und Chemismus des Humuskörpers. Mitt Dtsch Boden Ges 85:619–622

Wachendorf C, Irmler U, Blume H-P (1997b) Relationships between litter fauna and chemical changes of litter during decomposition under different moisture conditions. In: Cadisch G,

Giller KE (eds) Driven by nature: plant litter qualitity and decomposition. CAB International, Wallingford, pp 135–144

Waldhardt R, Fuhr-Boßdorf K, Otte A (2001) The significance of the seed bank as a potential for the reestablishment of arable-land vegetation in a marginal cultivated landscape. Web Ecol 2:83–87

WBGU(2000) Biosphere report 1999. Conservation and sustainable use of the biosphere. German Advisory Council on Global Change (WBGU)/Earthscan, London

WCED (1987) Our common future (Brundtland Report, World Commission on Environment and Development). Oxford University Press, Oxford

Weber HE (1967) Über die Vegetation der Knicks in Schleswig-Holstein. Mitt AG Floristik SH/HH 15

Weidemann G (1986) Zeitlicher Ablauf des Tierlebens. In: Ellenberg H, Mayer R, Schauermann J (eds) Ökosystemforschung. Ergebnisse des Sollings-Projektes. Ulmer, Stuttgart, pp 195–208

Weidemann G, Schauermann J (1986) Die Tierwelt, ihre Nahrungsbeziehungen und ihre Rolle. In: Ellenberg H, Mayer R, Schauermann J (eds) Ökosystemforschung. Ergebnisse des Solling-Projektes, Ulmer, Stuttgart, pp 179–266

Weisheit K (1995) Kohlenstoffdynamik am Grünlandstandort untersucht an 4 dominanten Grasarten. PhD thesis, University of Kiel, Kiel

Weiss U (2000) Eignung von praxisorientierten Modellansätzen zur Schätzung von Stickstoffausträgen aus landwirtschaftlich genutzten Böden. Schr Inst Pflanz Bodenkde 54

Wellbrock N (2002) Veränderungen und ökosystemare Bewertung der atmosphärischen Deposition eines Buchenwaldes und Übertragung des Bewertungskonzeptes auf ausgewählte Waldökosysteme in Schleswig–Holstein. EcoSys Suppl 35

Wendland F, Kunkel R (1999) Das Nitratabbauvermögen im Grundwasser des Elbeeinzugsgebietes. Umwelt Schr Forschungsz 13

Werner A (1994) Biomasseverteilung in einem Erlenbaum – Erfassung anhand der Jahrringanalyse. MSc thesis, University of Kiel, Kiel

Werner A, Müller K, Wenkel KO, Bork HR (1997) Partizipative und iterative Planung als Voraussetzung für die Integration ökologischer Ziele in die Landschaftsplanung des ländlichen Raumes. Z Kulturtechn Landes 38:209–217

Werwer W (2003) Regionalisierende Erfassung und Bewertung der ökologischen Bodenfunktionen als Fachbeitrag *Boden* zur Integration in die Landschaftsrahmenplanung/Regionalplanung. EcoSys Suppl 37

Wesely ML (1989) Parametrization of surface resistances to gaseous dry deposition in regional-scale numerical models. Atmos Environ 23:1293–1304

Westlake DF (1980) Primary production. In: Le Cren ED, Lowe-McConnell RH (eds) The functioning of freshwater ecosystems. Cambridge University Press, Cambridge, pp 141–246

Wetselaar R, Farquahr GD (1980) Nitrogen losses from tops of plants. Adv Agron 33:263–302

Wetzel H (1998) Prozessorientierte Deutung der Kationendynamik von Braunerden als Glieder von Acker- und Waldcatenen einer norddeutschen Jungmoränenlandschaft – Bornhöveder Seenkette. EcoSys Suppl 25

Wetzel RG (1972) The role of carbon in hardwater marl lakes. In: Likens GE (ed) Nutrients and eutrophication: the limiting-nutrient controversy. Spec Symp Am Soc Limnol Oceanogr 1:84–97

Wetzel RG (1983) Limnology. CBS College Publishing, Michigan

Wheeler BD, Shaw SC (1991) Above-ground crop mass and species richness of the principle types of herbaceous rich-fen vegetation of lowland England and Wales. J Ecol 79:285–301

Whittaker RH (1977) Evolution of species diversity in land communities. Evol Biol 10:1–67

Wiebe C (1998) Ökologische Charakterisierung von Erlenbruchwäldern und ihren Entwässerungsstadien: Vegetation und Standortverhältnisse. Kiel Mitt AG Geobot Schl-Holst Hmbg 56

Wiegleb G (1997) Leitbildmethode und naturschutzfachliche Bewertung. Z Oekol NatSchutz 6:43–62

Wiens JA, Crist TO, With KA, Milne BT (1995) Fractal patterns of insect movement in micro-landscap. Ecology 76:663–666

Wieser W (1991) Physiological energetics and ecophysiology. In: Winfield IJ, Nelson S (eds) Cyprinid fishes, systematics, biology and exploitation. Chapman & Hall, London, pp 426–455

Wiethold J (1998) Studien zur jüngeren postglazialen Vegetations- und Siedlungsgeschichte im östlichen Schleswig–Holstein. Univ Praehist Archaeol 45

Williamson MA, Parnall RA (1994) Partitioning of copper and zinc in the sediments and pore-waters of a high-elevation alkaline lake, east-central Arizona, USA. Appl Geochem 9:597–608

Wilson AG (1981) Catastrophe theory and bifurcation. Croom Helm, London

Windhorst W, Urff W von, Ahrens H, Becker H, Lippert C (2002) Erfüllung der gesellschaftlichen Funktionen in den Fallstudienregionen und Nachhaltigkeit – Vergleichende Analyse. In: Urff W von, Ahrens H, Neander H (eds) Landbewirtschaftung und nachhaltige Entwicklung ländlicher Räume. Forsch Sitzungsber ARL 124:225–255

Wingerden WKRE van, Musters JCM, Maaskamp FIM (1991) The influence of temperature on the duration of egg development in west European grasshoppers (Orthoptera: Acrididae). Oecologia 87:417–423

Wittman P (1997) Soil classification of the Federal Republic of Germany. Mitt Dtsch Bodenkde Ges 84:253–296

Woodwell GM (2002) On purpose of science, conservation and government. Ambio 31:432–436

Wootton RJ (1990) Ecology of teleost fishes. Chapman & Hall, London

Wulf AJ (2001) Die Eignung landschaftsökologischer Bewertungskriterien für die raumbezogene Umweltplanung. Books on Demand, Norderstedt

Yachi S, Loreau M (1999) Biodiversity and ecosystem productivity in a fluctuating environment: the insurance hypothesis. Proc Natl Acad Sci USA 96:1463–1468

Yaffee SL (1999) Three faces of ecosystem management. Conserv Biol 13:713–725

Zacharias D (1990) Flora und Vegetation von Waldrändern in Abhängigkeit von der angrenzenden Nutzung – unter Berücksichtigung auch der floristisch schwer charakterisierbaren Bestände. Verh Ges Oekol 19:336–345

Zalewski M (2000) Ecohydrology – the scientific background to use ecosystem properties as management tools toward sustainability of water resources. Ecol Eng 16:1–8

Zehler E (1981) Die Natrium-Versorgung von Mensch, Tier und Pflanze. Kali Briefe 15:773–792

Zeien H, Brümmer GW (1989) Chemische Extinktion zur Bestimmung von Schwermetallbindungsformen in Böden. Mitt Dtsch Bodenkdl Ges 59:505–510

Zeiler M (1996) Nähr- und Spurenelementkreislauf in einem eutrophen Hartwassersee mit saisonal anoxischem Hypolimnion (Belauer See, Schleswig–Holstein). Ecosys Suppl 11

Zeitz J (1991) Untersuchungen über Filtrationseigenschaften von Niedermoorböden mit Hilfe verschiedener Methoden unter Berücksichtigung der Bodenentwicklung. Z Kulturtechn Landes 32:227–234

Zeleny M, Pierre NA (1976) Simulation of self-renewing systems. In: Jantsch E, Waddington CH (eds) Evolution and consciousness. Addison–Wesley, Reading, Mass., pp 150–165

Zhou X, Slauenwhite DE, Pett RJ, Wangersky PJ (1989) Production of copper-complexing organic ligands during a diatom bloom: Tower tank and batch-culture experiments. Mar Chem 27:19–30

Ziegler R, Egle K (1965) Zur quantitativen Analyse der Chloroplastenpigmente. I. Kritische Überprüfung der spektralphotometrischen Chlorophyll-Bestimmung. Beitr Biol Pflanzen 41:11–37

Zimmermann H (1994) Untersuchungen zur Bedeutung der Ciliaten im mikrobiellen Nahrungsnetz des Belauer Sees. PhD thesis, University of Hamburg, Hamburg

Index

Printing: Krips bv, Meppel, The Netherlands
Binding: Stürtz, Würzburg, Germany